**Landscape and the
Nature in Exurbia**

Routledge Studies in Human Geography

This series provides a forum for innovative, vibrant, and critical debate within Human Geography. Titles will reflect the wealth of research which is taking place in this diverse and ever-expanding field.

Contributions will be drawn from the main sub-disciplines and from innovative areas of work which have no particular sub-disciplinary allegiances.

For a full list of titles in this series, please visit www.routledge.com.

9 **Ageing and Place**
   Edited by Gavin J. Andrews and David R. Phillips

10 **Geographies of Commodity Chains**
   Edited by Alex Hughes and Suzanne Reimer

11. **Queering Tourism**
    Paradoxical Performances at Gay Pride Parades
    Lynda T. Johnston

12 **Cross-Continental Food Chains**
   Edited by Niels Fold and Bill Pritchard

13 **Private Cities**
   Edited by Georg Glasze, Chris Webster and Klaus Frantz

14 **Global Geographies of Post Socialist Transition**
   Tassilo Herrschel

15 **Urban Development in Post-Reform China**
   Fulong Wu, Jiang Xu and Anthony Gar-On Yeh

16. **Rural Governance**
    International Perspectives
    Edited by Lynda Cheshire, Vaughan Higgins and Geoffrey Lawrence

17 **Global Perspectives on Rural Childhood and Youth**
   Young Rural Lives
   Edited by Ruth Panelli, Samantha Punch, and Elsbeth Robson

18 **World City Syndrome**
   Neoliberalism and Inequality in Cape Town
   David A. McDonald

19 **Exploring Post Development**
   Aram Ziai

20 **Family Farms**
   Harold Brookfield and Helen Parsons

21 **China on the Move**
   Migration, the State, and the Household
   C. Cindy Fan

22 **Participatory Action Research Approaches and Methods**
   Connecting People, Participation and Place
   Edited by Sara Kindon, Rachel Pain and Mike Kesby

23 **Time-Space Compression**
   Historical Geographies
   Barney Warf

24 **Sensing Cities**
   Monica Degen

25 **International Migration and Knowledge**
Allan Williams and Vladimir Baláž

26 **The Spatial Turn**
Interdisciplinary Perspectives
Edited by Barney Warf and Santa Arias

27 **Whose Urban Renaissance?**
An International Comparison of Urban Regeneration Policies
Edited by Libby Porter and Katie Shaw

28 **Rethinking Maps**
Edited by Martin Dodge, Rob Kitchin and Chris Perkins

29 **Rural–Urban Dynamics**
Livelihoods, Mobility and Markets in African and Asian Frontiers
Edited by Jytte Agergaard, Niels Fold and Katherine V. Gough

30 **Spaces of Vernacular Creativity**
Rethinking the Cultural Economy
Edited by Tim Edensor, Deborah Leslie, Steve Millington and Norma Rantisi

31 **Critical Reflections on Regional Competitiveness**
Gillian Bristow

32 **Governance and Planning of Mega-City Regions**
An International Comparative Perspective
Edited by Jiang Xu and Anthony G.O. Yeh

33 **Design Economies and the Changing World Economy**
Innovation, Production and Competitiveness
John Bryson and Grete Rustin

34 **Globalization of Advertising**
Agencies, Cities and Spaces of Creativity
James Faulconbridge, Peter J. Taylor, J.V. Beaverstock and C Nativel

35 **Cities and Low Carbon Transitions**
Edited by Harriet Bulkeley, Vanesa Castán Broto, Mike Hodson and Simon Marvin

36 **Globalization, Modernity and The City**
John Rennie Short

37 **New Economic Spaces in Asian Cities**
From Industrial Restructuring to the Cultural Turn
Edited by Peter W. Daniels, Kong Chong Ho and Thomas A. Hutton

38 **Climate Change and the Crisis of Capitalism**
A Chance to Reclaim, Self, Society and Nature
Edited by Mark Pelling, David Manual Navarette and Michael Redclift

39 **Landscape and the Ideology of Nature in Exurbia**
Green Sprawl
Edited by Kirsten Valentine Cadieux and Laura Taylor

Landscape and the Ideology of Nature in Exurbia
Green Sprawl

Edited by Kirsten Valentine Cadieux and Laura Taylor

NEW YORK   LONDON

First published 2013
by Routledge
711 Third Avenue, New York, NY 10017

Simultaneously published in the UK
by Routledge
2 Park Square, Milton Park, Abingdon, Oxon OX14 4RN

*Routledge is an imprint of the Taylor & Francis Group,
an informa business*

© 2013 Taylor & Francis
First issued in paperback in 2013

The right of Kirsten Valentine Cadieux and Laura Taylor to be identified as the authors of the editorial material, and of the authors for their individual chapters, has been asserted in accordance with sections 77 and 78 of the Copyright, Designs and Patents Act 1988.

All rights reserved. No part of this book may be reprinted or reproduced or utilised in any form or by any electronic, mechanical, or other means, now known or hereafter invented, including photocopying and recording, or in any information storage or retrieval system, without permission in writing from the publishers.

Trademark Notice: Product or corporate names may be trademarks or registered trademarks, and are used only for identification and explanation without intent to infringe.

Library of Congress Cataloging-in-Publication Data
Landscape and the ideology of nature in exurbia : green sprawl / edited by Kirsten Valentine Cadieux and Laura Taylor.
    p. cm. — (Routledge studies in human geography ; 39)
  Includes bibliographical references and index.
  1. Landscapes. 2. Nature. 3. Suburbs. I. Cadieux, Kirsten Valentine. II. Taylor, Laura Elizabeth.
  QH75.L277 2012
  508—dc23
  2012018114

ISBN13: 978-0-415-74761-5 (pbk)
ISBN13: 978-0-415-63715-2 (hbk)
ISBN13: 978-0-203-08477-9 (ebk)

Typeset in Sabon
by IBT Global.

Contents

List of Figures	xi
Preface	xv
Acknowledgments	xxi

1 Introduction: Sprawl and the Ideology of Nature   1
LAURA TAYLOR AND KIRSTEN VALENTINE CADIEUX

Relating Sprawl and Nature in Exurbia	2
Exurbia as a Cultural Landscape	8
Exurbia, Exclusivity, and Political Engagement as Influenced by the Ideology of Nature	11
Reading the Cultural Landscapes of Exurbia and Sprawl	15
Ecological Exurbia: Ecosystem Impacts at the Human-Wildlands Interface	17
Rural Studies: The Impact of Exurbia on Agriculture	18
Political Ecology and Political Economy: Nature from Critical Perspectives	19
Discomfort with Urban Modernity and the Ideology of Nature	21
Structure of the Book	22

2 Bridges in the Cultural Landscape: Crossing Nature in Exurbia   31
LAURA TAYLOR

Editor's Introduction	31
Bridges in the Cultural Landscape	35
A History of the Bridges of Churchville	37
Bridges Are an Intersection of Nature and Culture	51
Concluding Thoughts: Landscape Values in Planning	55

3	**Exurbia Meets Nature: Environmental Ideals for a Rootless Society** RICHARD JUDD	59
	Editors' Introduction	59
4	**Airworld, the *Genius Loci* of Exurbia** ANDREW BLUM	78
	Editors' Introduction	78
5	**Rewilding Walden Woods and Reworking Exurban Woodlands: Higher Uses in Thoreau Country** BRIAN DONAHUE	94
	Editors' Introduction	94
	The History of Walden Woods	100
	The Suasco Woodlands	110
	Ecological Stewardship	111
	Ecological Research	114
	Environmental Education	114
	Sustainable Forestry	115
6	**Sojourning in Nature: The Second-Home Exurban Landscapes of Ontario's Near North** NIK LUKA	121
	Editors' Introduction	121
	Exurbia and Cottage Country as Imagined Landscapes	129
	The Imagined Landscapes of Cottaging	130
	Cottaging in Ontario: Origins and Early Growth	135
	Postwar Growth and Contemporary Changes	141
	Getting Cottagers (and Others) to Take a Critical Perspective	145
	Conclusion: From Sojourning To . . .	150
7	**Design and Conservation in Québec's Rural-Urban Fringe: The Case of Lac-Beauport** GENEVIÈVE VACHON AND DAVID PARADIS	159
	Editors' Introduction	159
	Planning Context and Challenges	163
	Form, Population, and Spatial Representations	166
	History and Morphogenesis	171
	Mission, Approach and Design Orientations	173

	Design Proposals	177
	The Old Waterloo Settlement and Village Center	178
	Lac Neigette: New Residential Development	179
	The Chapel Sites	179
	Conclusion	181
8	**Time, Place, and Structure: Typo-Morphological Analysis of Three Calgary Neighborhoods**	**185**
	BEVERLY A. SANDALACK AND ANDREI NICOLAI	
	Editors' Introduction	185
	Urban Form Analysis—An Approach	189
	The Relationship Between Form and Nature Is Expressed in the Environmental Context, Conditions and Features of a Place	189
	The Spatial Relationships of Production, Maintenance, Transformation and Use of the Urban Forms are Expressed through the Land Uses and Functional Relationships	191
	The Relationships Between Built Forms Include Morphology, Typology and Visual Relationships	191
	Morphology and the Spatial Relationships between Built Forms	191
	Typology	192
	Visual Relationships	192
	The Evolving Urban Edge of Calgary	193
	Roxboro	195
	Glamorgan	200
	Lake Chaparral	207
	Conclusions	212
9	**The Imagined Landscape: Language, Metaphor, and the Environmental Movement**	**218**
	THOMAS LOOKER	
	Editors' Introduction	218
	The Crisis of Words	220
	Certainty and Uncertainty	224
	Balance vs. Chaos	229
	Thoreau's Art of Nature	234
	Wildness in Words	235
	"Walking"	235
	Taking the Measure of Nature	238

Walden, "The Pond In Winter"	238
Laboratories for Artists	239
Walden, "Spring"	239
The Book of the Earth	243
Walden, "Spring"	243
Conclusion: Re-Imagining Our Environment	245

10 The Mortality of Trees in Exurbia's Pastoral Modernity: Challenging Conservation Practices to Move beyond Deferring Dialogue about the Meanings and Values of Environments 252
KIRSTEN VALENTINE CADIEUX

Editors' Introduction	252
How Viewing Landscapes as "Pastoral" Makes Working Land Aesthetic	258
Reading Trees to Interpret Landscapes	258
The Mythology of Nature Symbolism: Imagining Wilderness and Immortality	263
The Amenitization of Nature	267
Forests as Desirable Habitat	269
Modern Pastoral	269
The Problems of a Pastoral Modernity	273
Where the Wild Trees Are: Real and Imagined Landscapes	276
The Vanity of Distant Wildness	276
The Illusion of Preservation	278
Commodifying Nature	279
From Useful to Beautiful: Aesthetic Paralysis in the Naturalized Rural Landscape	281
Living with the Pastoral: Death in the Very Midst of Delight	282
If Trees Die, What Happens to Idealized Landscapes?	283
Negotiating Pastoral Landscapes: Making Exurbia Inhabitable	286

Editors' Epilogue: An Agenda for Addressing Green Sprawl 295
LAURA TAYLOR AND KIRSTEN VALENTINE CADIEUX

Contributors	299
Index	301

Figures

2.1	Location map.	39
2.2	Settlement in Churchville, 1827.	40
2.3	Settlement in Churchville, 1848–1856.	42
2.4	Settlement in Churchville, 1877.	43
2.5	Inaugural trip on the new steel bridge, 1907.	44
2.6	Crossroads at the Churchville Bridge looking east.	45
2.7	North Churchville bowstring bridge.	45
2.8	Looking north at the Steeles Avenue bridge.	46
2.9	South side of Steeles Bridge, looking west, showing the river flowing in its new channel.	47
2.10	Heritage Conservation District.	48
2.11	Illustration from brochure for new subdivision to the west of Churchville.	49
5.1	Walden Woods protected open space.	98
5.2	Walden Woods surface geology.	101
6.1	Central Ontario, showing the built-up parts of the major urban centers, major roads, expressways, and "cottage country" (shaded) to the north and east of the city.	123
6.2	A modest cottage probably built in the 1930s (Crystal Lake).	124
6.3	An elaborate year-round "cottage" compound built in the late 1980s (Lake Muskoka).	125
6.4	Typical configuration of modest early twentieth-century waterfront cottages (Severn River).	127
6.5	Lakeside subdivision in the Kawartha district (Pigeon Lake).	127
6.6	The typical pattern of cottage buildings and roads.	131
6.7	Aerial view of a typical central Ontario cottage setting (Parry Sound).	131
6.8	A late 19th-century postcard showing a steamer picking its way through a log boom in the Muskoka Lakes.	136

6.9 A proudly posed late nineteenth-century postcard image of the kill at a Muskoka hunt camp. 137
6.10 1908 advertisement for the new Canadian Pacific Railway express service to the Muskoka Lakes. 138
6.11 An early advertisement for the Grand Trunk Railway service to Muskoka, boasting of the "splendid service" afforded by hotels "set in fragrant pines." 139
6.12 A "classic" nineteenth-century cottage compound nestled amongst the trees on an island in Lake Rosseau. 140
6.13 & 6.14 The "Muskoka Room." 141
6.15 An overstated example of the standard spatial arrangement in cottage country (Lake Joseph). 143
6.16 View in the "Port 32" lakeside subdivision in the village of Bobcaygeon (Pigeon Lake). 144
6.17 View of the main commercial strip in the east-end Toronto district known as The Beach. 149
6.18 A scene in the Beach district of Toronto showing in its built form how the metropolitan landscapes of cottage country can evolve over time. 149
7.1 Québec's metropolitan region locating the municipality of Lac-Beauport, just outside Québec's city limits. 164
7.2 Aerial photograph showing Lac-Beauport's different areas and their construction periods, as well as major landmarks, circa 2002. 167
7.3 Aerial photograph of Boulevard du Lac, leading to the ring road and the lake, circa 2004. 168
7.4 Typical residential developments built in the 1980s and 1990s, creeping up the mountain slopes, circa 2004. 169
7.5 Mental map drawn by a local primary school student. 170
7.6 Typical Lac-Beauport summer cottage. The Lavigueur house, circa 1908. 171
7.7 Montage of postcards from the 1940s showing different "wintering" sites in Lac-Beauport, including Le Relais. 172
7.8 Schematic sketch indicating the three sites of the design proposals. 177
7.9 Project 1—Civic node with new City Hall: existing conditions and proposed consolidation scheme. 178
7.10 Project 2—Lac-Neigette: existing conditions and proposed development. 180
8.1 At the scale of the city, the evolution of neighborhoods was mapped. 194
8.2 Roxboro Edges, functional relationships and street pattern. 196

8.3	Street Section: street trees (elm, interplanted with lilac) on boulevards separate the sidewalks from the streets.	197
8.4	Typical Roxboro Block—property line, figure ground, and model.	198
8.5	Roxboro Streetscape.	198
8.6	Roxboro House.	199
8.7	Glamorgan 1924.	202
8.8	Glamorgan 1999.	203
8.9	Glamorgan edges, functional relationships, and street pattern.	204
8.10	Typical Glamorgan Street Section.	205
8.11	Typical Glamorgan Block—property line, figure ground, and model.	205
8.12	Typical Glamorgan House.	206
8.13	Typical Glamorgan Streetscape.	206
8.14	Chaparral 1924.	208
8.15	Chaparral 1999.	208
8.16	Chaparral edges, functional relationships, and street pattern.	209
8.17	Typical Chaparral Street Section.	210
8.18	Typical Chaparral Block—property line, figure ground, and model.	210
8.19	Typical Chaparral House.	211
8.20	Typical Chaparral Streetscape.	211
10.1	Jefferson Forest's imagined aesthetic and entrance, with houses effaced by forest.	260
10.2	Jefferson Forest entrance, contrasting the advertised residential forest with the actual fringe of forest remnant and closely spaced houses.	261
10.3	Developer's map of the Jefferson Forest site alongside aerial image of the Jefferson Forest Site (after completion of Phase 1).	262
10.4	Autumn Grove, across Yonge Street from Jefferson Forest, represents itself using a similarly imaginary forest.	263

Preface

A well-known public intellectual once tried to convince us that his move out of the city did not count as sprawl because he had moved *beyond* sprawl—to exurbia, and to nature. As the keynote speaker at an international landscape architecture conference, in front of hundreds of people, he defended a claim we have heard countless times in both our professional and personal lives, a claim that made us ask ourselves what it is about the exurban landscape that has the power to exempt people from feeling like they are part of sprawl. In the tradeoff we think of as "green sprawl," people appear to want to trade in urban landscapes for exurban ones in an attempt to get closer to nature. This book traces our attempt to figure out what exurban landscapes might tell us about nature and the city, the way we construct home landscapes, and the ways our homes and nature relate to each other.

Our view of exurbia—and the premise of this book—has been shaped by our work with various aspects of urban sprawl and exurbanization. As academics and planners, we have followed the forms and ideas shaping urbanization in natural and rural areas over a set of careers spanning a wide swath of Canada and the U.S., and have spent the past decade comparing notes on what is happening in the landscapes of sprawl. What our colleagues say about sprawl and urban development often echoes the sensational treatments of sprawl in the newspapers: sprawl is either a scourge—taking over and fragmenting valuable natural habitats and countryside, stressing municipal services, and confirming a space- and energy- intensive automobile culture—or sprawl is the logical outgrowth of residential preferences for single family houses with yards, and, as such, is to be planned for and supported. Increasingly, we have noticed that, at some level, many of the people talking about these topics think *both* things are true—high-profile anti-sprawl activists live in large houses outside the city and government officials promoting "smart growth" insult dense planned developments as ugly—and they often do not seem to notice or care about the tensions or contradictions involved.

What we have come to recognize is that there is something about the exurbs that naturalizes the desire to live there. This natural quality of the desirability of green sprawl seems to make it difficult to talk about why and how people *decide* to create sprawl, and even enjoy the sprawl landscape they create. This is true even though many exurbanites contradict their demonstrated landscape preferences by denouncing sprawl as bad! Insisting that sprawl is a problem that *other* people have makes critical problems of urbanization very difficult to discuss—and discussion is a core feature of the kind of collaborative environmental management that holds some promise for negotiating ways to address the problems inherent in the landscape forms created by sprawl.

As we became fascinated with the story lines, paradoxes, and contradictions in the meanings we were reading and hearing discussed in the landscapes of exurbia, we became aware that our way of looking at and talking about exurbia and sprawl was not well represented in popular or academic discussions of those topics. A series of breakfast conversations with our colleagues about the cultural landscape of dispersing urban edges started spilling over not only into lunches but also into field trips that went increasingly farther afield. From the breakfast table in the (suitably named) Arbor Room at the University of Toronto, we diagrammed what we saw happening in the exurban landscape, and convened a diverse group of scholars whose work would help us get a handle on exploring the changing cultural landscapes of what we came to think of as "green sprawl." The following chapters were presented and discussed together in the resulting series of symposia in the context of bringing together a range of different approaches to exurbia to try to understand more broadly the processes, experiences, and landscapes involved in the relationship between exurbia, sprawl, and nature.

The main questions we probed are about the influence on the exurban landscape of the cultural desire for a relationship with "nature," and the way that the form and practice of the exurbs, and of contemporary sprawl more broadly, make it harder for society to understand or interact with the natural environment. Development beyond the city's boundary is seen, on one hand, to allow access to nature and to particular versions of the single-family home landscape but, on the other hand, to undermine urban ideals of dense, efficiently-serviced residential living and vibrant, friendly neighborhoods. Exurban sprawl allows the affluent access to nature, but denies and "ruins" the perception of nature's uninhabitedness for everyone else. Exurban attempts to transform unique ecosystems and cultural landscapes into idealized, naturalized residential settings end up, over time, blanketing the spreading metropolis with generic sprawl. The central paradox of green sprawl is the way these good intentions lead to devastating effects.

Considering the degree to which the debate on sprawl has come to concern the general population to such a great degree, a significant part of our motivation in preparing this volume has been not only to provide a new perspective on the sprawl debate for environmental planners and academics—to honestly examine people's desire for a residential experience of nature in exurbia in a non-polemical way—but also to attempt to engage a wider audience in this inquiry by providing a sympathetic if critical look at the phenomenon of nature-seeking. As green landscapes are increasingly categorized as a version of nature that should be beyond human intervention, we hope our conversation engages people who are interested in issues of life beyond the edge of the city and its promise of a relationship with nature. We have especially in mind those who participate in the production of exurban landscapes and those who are interested in what these landscapes mean, and how they might consequently be planned and managed.

While examining the residential landscapes of exurbia we kept hearing the same refrain: *we want nature where we live*. Many of the problems and also the solutions lumped together under the labels "anti-sprawl," "smart growth," and "urban sustainability" show up in the appeal of the landscape of exurbia. Looking at what exurbia represents to people—both the good and the bad—allows us to explore both the opportunities and shortcomings of green sprawl. In spite of current uncertainty in housing markets, the way that anti-sprawl ideals have been constructed appears to interfere with owning up to or exploring the importance of exurban ideology to the production and reproduction of North American residential landscapes. In bringing together a range of different approaches to exurbia and nature, our hope has been that a close look at exurbia could help reconceptualize the ideology of contemporary residential relationships with the natural environment. More explicit understanding of the way that at least parts of the exurban impulse represent efforts to create ideal environments could perhaps help people learn to understand, talk about, and practice their environmental goals more explicitly—and also to talk about more intentionally planning and managing their interactions with their environments. If we can cultivate such a conversation about exurbia and sprawl without necessarily lauding or condemning either, we might stop the finger pointing over who bears the responsibility for "sprawl." We could then start to consider how the kinds of intentions driving exurbia could be explored and mobilized in support of progressive social and environmental goals—the kind of goals that the ideal of exurbia as a place where people have equitable access to a high quality living environment might optimistically be said to represent.

The conversation in this book about the central paradox of the ideology of nature in exurbia started out with a set of nagging questions about the potentials and liabilities of the exurban construction of nature, about how ideas of nature shape the landscape of exurbia, and about how to think through the difficult question of how such ideas of nature drive the urbanization of the landscape beyond the city. As we have edited the chapters that follow, while also conversing with many other people thinking about the relationship between sprawl, exurbanization, and nature, we have realized that in order to explore such a complex set of questions, our approaches need to reach beyond existing disciplinary and common interdisciplinary approaches. We also came to see our task in this book as organized around modeling such work and such extended, politically engaged conversations needed to address the compelling challenges of green sprawl. Keeping this in mind, it becomes easier to see how specific differences between different disciplinary approaches and between the experience of different places need to be negotiated in the context of local socio-environmental knowledge cultures.

We had been thinking of the role of nature in exurbia as an irony: our casual name for the sessions was *"deeply connected" to the "natural landscape."* The connection offered seemed neither deep, nor natural, not really a connection to a landscape at all, but rather a gesture toward a symbolic version of such a connection. At first, this gesture seemed to us to contain too little invitation to think about what the desire for a connection to nature might mean, or too little substance for people to act on the gesture in a thoughtful way.[1] Early on in the series of symposia, we decided that this irony of nature in exurbia wasn't so much ironic as paradoxical. Exurbia itself was a manifestation of a larger cultural phenomenon of a paradoxical ideology of nature—an analytical (and geographical) entry point to a much larger set of processes. We also realized that identifying this paradox allowed us to talk about it and to work through the paradox in a much more productive way. Understanding the role of nature in sprawl became our goal. The quest for the residential version of nature that green sprawl promises distracts people from planning landscapes that provide access to nature for everyone. What processes, such as commodification, reproduce the power of this ideology of nature? How can such representations of nature be engaged and contested?

While editing our conversation into chapters, we encouraged each other to reach across disciplinary boundaries and to recognize each other's assumptions and to question them. Coming together from the disparate fields of urban design and landscape planning, geography, history, literature, and journalism, we may originally have harbored some notion that together we could assemble something like a comprehensive overview. Or at the very least, that we might assemble a list of important questions—and possibly even propose one or two solutions to the problems we were identifying. As each person presented his or her paper, we were increasingly struck by the way that sharing our backgrounds and perspectives allowed us to work deeper into our ideas and also brought together from seemingly disparate perspectives a remarkably coherent set of critiques of our central paradox: exurbia's promise of becoming deeply connected to the natural landscape. From a position of not really knowing how people in other disciplines were approaching sprawl landscapes, or what we had to offer each other, we discovered that talking about representations of nature in exurbia from a variety of perspectives helped us all articulate our questions that much better, and place them in a broader context, one that felt as if we might be able to move this conversation into domains where other people concerned about issues of exurbia, especially people who lived there, might join in. However, with the gaps that still exist between our different approaches, and with the constant change and uncertainty in the residential landscape more salient than ever, it is difficult to say what *solutions* to the problems posed by exurbia would even look like. We have become even more convinced, though—especially in the context of Canadian and U.S. political discourses that are divided, in significant part, by where people live—that it is crucial to get people thinking about why they sprawl, why they are bothered by sprawl, and how they act on desires to bring together ideals for home life and natural settings.

We do not assume that our analysis of the importance of the ideology of nature is the only interpretation of how sprawl works, yet we do believe that it is important to inhabit the paradoxes of the modern urban landscape, and to work through them, collectively, out loud, including many perspectives, especially those most likely to be marginalized (a challenging task we cannot emphasize enough, given its difficulty and necessity, and one we hope to approach differently in the future, using tactics from environmental justice work).

NOTES

1. Andrew Blum, Kirsten Valentine Cadieux, Nik Luka and Laura Taylor, "'Deeply Connected' to the 'Natural Landscape': Exploring the Places and Cultural Landscapes of Exurbia," in *The Structure and Dynamics of Rural Territories: Geographical Perspectives*, eds. Doug Ramsey and Christopher Bryant (Brandon, MB: Brandon University, 2004): 104–112.

Acknowledgments

We would like to acknowledge, first, Nik Luka and Andrew Blum, who talked through those early Arbor Room breakfasts with us, and whose imaginative understanding of urban edges catalyzed much of the conversation this book reports. Jenny Hall and Dick Judd shared their ideas (and labor) organizing special sessions at the meetings of the Canadian Association of Geographers and the American Society of Environmental Historians, and instigated our inviting people we thought would be interested in exploring exurbia's cultural landscapes. Those people, both those with chapters in this book and also Susan Preston, Johanna Wandel, Peter Nelson, Greg Halseth, Ken Beesley, Michael Troughton (now sadly deceased) and especially Michael Bunce, not only extended the conversation into a remarkably coherent constellation of additional questions, but many of them had already exerted formative influences on us, shaping the questions that got us started, and for both of these things, we thank them. Thanks also to the many colleagues in the Department of Geography and Planning at the University of Toronto who goaded and supported us through this book, as well as to the exurbia, amenity migration, and rural landscapes discussion participants over the last decade of meetings of the Association of American Geographers, Canadian Association of Geographers, and American Society for Environmental History. Special thanks are due to those who read and responded to drafts, including Maria Frank, Madison van West, Renata Blumberg, Zane Miller, Lloyd Irland, and especially Sandy Crooms, who envisioned this collection as a book and saw it through a remarkable number of iterations. Finally, thanks go to the chapter authors and to our families, for inhabiting this volume through all the enjoyable effort it took.

1 Introduction
Sprawl and the Ideology of Nature

Laura Taylor and Kirsten Valentine Cadieux

This book is about the role of nature in the production of sprawl. Nature is everywhere in cities, but often does not feel as present in the city as it does in "natural" landscapes of woods, hills, beaches, or grasslands outside of the city. A home in natural settings like these, somewhere beyond the city, has many quality of life advantages. But a collection of *many* such homes—sprawl—is highly problematic for reasons of environmental and social justice. This book explores ways that sprawl is a product (i.e. literally produced by) the ideology of nature. The ideology of nature, the taken-for-granted notion that escaping to nature is a good thing, is one of the reasons why sprawl is so typical of contemporary urbanization, particularly in the United States and Canada. The chapters in this book decode the landscape of sprawl, drawing on several different perspectives—including urban design and landscape planning, geography, history, literature, and journalism—to explore what an examination of the effects of ideologies of nature can tell us about sprawl.

The desire to live in daily contact with nature, far away from the cares of the modern city, is a familiar trope in Western culture. Just look to the glossy pages of lifestyle magazines promoting home renovation and travel to see how powerful the promise of a home in nature is. In a recent televised lottery ad, a man and woman are shown happily walking through a forest path, following the flight of a small bird. At the end of the path, they arrive home to their house surrounded by palm trees. The camera pans out and shows that the house is on a secluded island, intimating that this is a private retreat available to those who buy a ticket and win big. Similarly, newspaper ads show "artists' concepts" of new homes surrounded by trees, illustrating what is essentially a lie about the actual cheek-by-jowl subdivision in which the house will sit. These images provide examples of the premise of this book, calling into question why such fantastical natural settings are the symbolic markers of the best lifestyles imaginable, and exploring the implications of this problematic fantasy, which we see as a challenge impeding engagement with *urban* ecologies.

"Green sprawl" is our label for exurban residential development. We explore how important the ideology of nature is (and has been) in producing sprawl—where the ideology of nature is a recognized, yet understudied part of the culture and production of cities. Although the idea that many people make their residential choices based on proximity to natural areas is widely, if implicitly, accepted, why this is the case has not been critically explored. Unmasking the cultural preference for a home life in nature seems to us a necessary step in addressing the contemporary crisis of sprawl.

Each of the following chapters uses different case studies or objects of analysis to explore how landscapes like "exurbia," where homes are set in nature beyond the edge of the city, have become a social norm signifying the achievement of high quality of life. The social segregation and environmental impact that are masked by these landscapes seem particularly important at a time when more people than ever before live in cities—and while new attention is being paid to cultivating urban ecologies to better meet residents' goals.

RELATING SPRAWL AND NATURE IN EXURBIA

Technically, "sprawl" is defined as unplanned, unorganized development leap-frogging into the countryside. However, in popular use the term has come to refer to the dispersed metropolitan structure that North Americans love to hate, but where most of them live. In spite of valiant efforts over the last century, urban planning policy has not ended sprawl, despite fostering a widespread discomfort with sprawl, and perhaps with urban expansion in general—a problematic stance as the proportion of people living in cities continues to grow.[1] This failure to successfully address sprawl—the urban form that planners have branded as unequivocally "bad"—suggests that we may not be asking the right questions about how sprawl is a problem. Popular uses of the term "sprawl" have come to point to everything negative about urban expansion, a whole litany of problems that urban expansion creates for individuals, municipalities, agriculture, and the environment. At the same time, the term "exurbia" evokes the *good* qualities of residential living at the expanding metropolitan edge. In the large homes out in the peace of the countryside, the problems of the world, and particularly of urban modernity, are imagined to take place at a comfortable remove. This book delves into the relationship between nature and residential land use that helps explain this paradoxical contrast between sprawl as bad and exurbia as good by exploring the way that the ideology of nature produces the cultural landscape of exurbia as the attractive face of sprawl.

Exurbia illustrates the role that ideals of nature play in the process of sprawl. We call exurbia "green sprawl" because it is so recognizable as a landscape of housing set in a matrix of vegetation—"nature"—that is different from urban greenspace, which is usually vegetation and other "nature" set in a matrix of the built environment.

Built in opposition to planning policy and philosophy in most places it exists, exurbia is a particularly sprawling and resource-intensive residential settlement pattern, predicated partly on a desire for contact with nature and all that nature represents. Attempting to reconcile this desire for authentic experience of the natural environment with the modern, fundamentally urban living habits of North Americans, exurban residential environments, particularly in the housing boom of the 1990s and 2000s, became North America's most rapidly growing land use.

In this expanded exurbia, more people than ever before are able to live surrounded by fields and forest, birds and wildlife, seemingly far from the stresses and cares of work and society, but with all the modern conveniences, access to services, and connections to the city that modern life seems to require. These promises of simultaneous connection with nature and with the city are ones that the landscapes of exurbia seem to make. On one hand, North American culture sustains the dream of owning a home outside the city; on the other, sprawl symbolizes urbanization gone awry on a number of social, cultural, philosophical, environmental, and even ethical levels. Despite the centrality of this paradoxical promised experience of exurban landscape to debates around planning and the future of contemporary city-regions, the *experience* of landscape and exurban landscape expectations are seriously understudied and underacknowledged in planning discourse.

As the most sprawling form of settlement in North America, the exurbs are logically the target of urban planning policies against sprawl. Yet sprawl is the specter *against which the exurbs themselves are conceived*. The idea of sprawl as a scourge on the landscape that needs to be stopped is at odds with the ideals of a nature that is imagined to be found only outside of the city—ideals that motivate choices about where to live. The incursion of homes and infrastructure into farmland, rangeland, forests, or mountains can be seen in many places, often prompting the perception that the scenic countryside is no longer as boundless as it once was, and uneasiness about the finite nature of areas not developed for urban uses. Planners and decision-makers have responded by devising and implementing "smart growth" policies, and by promoting "new urbanism"—encouraging compact, mixed-use urban areas, intensification of existing urban areas, and the development of walkable and transit-oriented communities to reduce reliance on the automobile—all of which are supposed to preserve the environment at the city's edge by limiting the acres of countryside needed for urbanization.

In spite of efforts in communities across North America, urban growth has not been contained, and recent studies have concluded that growth outside of existing metropolitan areas is increasing, not decreasing as planners would like.[2] In this book, we explore the effect of the ideology of nature on residential settlement patterns as an understudied cause of sprawl. Normative definitions of sprawl, smart growth, and "good" planning policy tend not to consider the important role that landscape ideology plays in the kind of decision making that leads to exurbanization. Instead, they are based on theories and models of settlement that do not engage with the complexities of motivation and behavior. In spite of the attractiveness of nature imagined in the idealized countryside, urban planning definitions and spatial policies often do not capture the public imagination when it comes to making choices about where and how to live. This disconnect between public policy and household-level action leads to significant problems in negotiating in broader society what should be done in response to the challenges of urban growth. Landscape approaches may help address these land-use management problems: people seem to know about and have a lot more to say about their experience of landscape and of what they expect landscapes to include and to provide than they do about "environmental process" or planning requirements. This makes landscape a useful and promising bridge between people's experience and the kind of participation in discourse expected in collaborative environmental management processes.

Understanding how influential the lure of nature is in the reproduction of exurban landscapes enables a new view of the problem of sprawl. Sprawl is not merely about poor land-use planning or making the wrong infrastructure decisions, but is a major symptom of a societal understanding of modern urban landscapes focused on perceptions of lightning speed and immediacy, crowding and uncertainty, and loss of connection to nature. A central theme of society-environment scholarship over the past three decades is the idealization of nature as an effect of modernity on contemporary urban society. Much of this scholarship suggests convincingly that the distancing from nature within everyday life in the city results in the desire for the protection of a pristine nature, that allows one to escape the city, but that has the effect of *further* distancing nature instead of bringing it into everyday life.[3]

For one thing, the idea of nature, at least in modern Western culture, usually means an idea of an authentic physical world without human intervention.[4] The desirability of such nature is often not questioned. Critics of such a "naturalized" nature point out, however, the problematic ways that nature is used to motivate consumption, entrench social injustice, and falsely construct a world view where nature is at its best only where humans are excluded, or perhaps more accurately, only where *other* humans are excluded. Within the discipline of human geography, particularly, the concept of nature has been the subject of much analysis and discussion. The main thrust of the discussion considers the way that nature is socially constructed and the importance of understanding how society imagines and, as a result, behaves toward (or, rather, *within*) the natural world. Much discussion has taken place about how the social construction of nature produces particular kinds of space that are organized in large part by particular ideologies, spaces such as national wilderness parks—and the city itself.[5] Our interest is in extending this discussion about the ideology of nature to thinking about sprawl. Our focus is on the way that the ideology of nature motivates the people and institutional actors who produce and maintain the landscapes that constitute sprawl, exemplified by exurbia. The chapters concentrate on actions and work motivated by natural ideals, and on the embodiments of those ideals in material landscapes. Nature, as a theme, is itself naturalized in exurbia, as part of the tendency to respond to aversive aspects of the modern humanized city via recourse to an appealing, if unconsidered, nature. Many North Americans sympathetic to the environment are familiar with the sentiment, at least, of Henry David Thoreau's commonly repeated phrase "In Wildness is the preservation of the world."[6] Usually used to emphasize the need for preservation of literal wilderness, rather than cultural wildness, this phrase's play with literal and literary nature helps make more apparent the experiences and politics bound up in the multiple and competing expectations associated with nature.

The material form that the idealization of nature takes can be understood by examining the way that people escape from the city to nature as a major component of sprawl. This focus also highlights the kinds of decisions that people make in their everyday lives in relation to that idealization of nature—decisions that have very material effects on the landscape. Seen through the lens of a green and peaceful landscape, exurbia has a lot of popular appeal; understanding exurbia as a cultural landscape might help people concerned with sprawl figure out how to popularize their analyses of the problems of exurbia and sprawl, particularly those related to social and environmental themes. Understanding the influence of nature in the cultural landscape of exurbia may help to link public concerns with professional and academic planners' thinking about sprawl in a way that helpfully addresses the feeling that much North American urban containment is, at best, a parody. Even when urban boundaries are considered successful within their jurisdictional boundaries, sprawl often continues in the less regulated county next door. As many North American communities struggle to figure out how to reconcile urban dispersion with environmental and social (especially servicing) impacts, a productive space for discussion of these issues may be opened by considering the influence of the ideology of nature.

Beyond addressing the obvious paradox of continued conventional patterns of urban dispersion despite conventional concerns about sprawl, understanding the ideology of nature may also help address the obstacles faced by land-use planners and decision-makers, obstacles presented by fear and avoidance of significant contemporary social problems related to urbanization. Urban growth is complex and answers to problems of climate change, income disparity, and racism are not going to be definitively solved in the land-use planning process; after all, sprawl is a "wicked" problem.[7] Although natural landscapes may provide aesthetic buffers, the social and spatial challenges and difficulties of living a modern life (as discussed in Chapters 4 [Blum], 5 [Donahue], and 9 [Looker]) are as present today as they were in the early nineteenth century when Thoreau and his contemporaries were dealing with the specter of the early trains' effects on their rural residential environments. Popular discomfort with modern urban life is arguably worse, not better, today. Despite improvements in the material environments of the urban global North, the cumulative effects of traffic, air and water pollution, noise, and financialization have fanned alarmist strands of anti-urbanism. Anti-urban sentiments may be intensified by concern about serious (and often sensationalized) problems of urbanization in the global South.[8] And (as Chapters 3 [Judd], 4 [Blum], and 5 [Donahue] demonstrate) such anti-urbanism has been further influenced by what we are calling the ideology of nature.

We focus so intently on this ideology of nature because it provides a way to examine the cultural aspirations represented by exurbia. It also helps us explore what green sprawl reveals about the influence of popular struggles with the modern urban landscape on contemporary metropolitan and land-use governance. On the face of it, exurbia deserves many of the critiques it receives. Many of the central features of exurbia—very low density residential land use, extra-large energy-intensive homes, conversion of forested or rural land for housing, and long commutes—contribute considerably to the very problems that the social and environmental aesthetic of exurbia attempts to avoid. Exurbanization contributes to the continued dispossession of both rural and urban people by converting working landscapes to amenity landscapes (raising property values and taxes, and undermining social support and infrastructure for farming, forestry, and natural resource extraction at the same time that new exurban development requires massive natural resource inputs) and by relocating urban tax bases (while continuing to rely on at least some urban services) while exacerbating "uneven development," as exurbs are constructed as exclusive "landscapes of privilege."[9] Therefore, exurbia pushes out those involved in rural production at the same time that it abandons support for the city. In the contemporary world of gridlock and climate change, while state response to the problems of sprawl has been to contain urban areas and promote higher-density living and environmentally friendly infrastructure, the individual response for people with means has often been to flee to greener pastures.

Perhaps even more insidiously, exurbia separates people not only spatially but also conceptually from the impacts of their decisions about where and how to live. The naturalized cultural landscape of exurbia hides both urban and rural evidence of the processes of dispossession.[10] On an individual basis, exurbanites often say that their residential environments enable them to balance their lifestyles successfully with the material effects of their everyday habits. However, the beauty of and positive normative associations with green sprawl landscapes inhibit efforts to thoughtfully engage with the socio-environmental impacts of those lifestyle choices.[11] The actions and challenges associated with the ideology of nature are clearly not the only explanation for the processes and form that constitute exurbia. However, we believe the degree to which the ideology of nature influences attitudes toward and goals related to a high quality of life is often underestimated—particularly when investigating the details of justifications for green sprawl that may not obviously rely on natural ideals, per se, such as personal health, safety, privacy, secure financial investment over the long term, more homogeneity among neighbors, and also perhaps including lower taxes and less government oversight in property matters.

The challenges of reconciling homeowners' needs and desires through the union of city and country have been described in terms of the quest for the "middle landscape."[12] The middle landscape motivated Ebenezer Howard's Garden City ideal in the 1890s and has inspired generations of planners designing residential landscapes around the world from the mid-1900s to the present.[13] In theory, the middle landscape was supposed to achieve the best of both worlds, where inhabitants live outside of the city with access to the modern conveniences that the city provides; however, in practice, this idea has since been roundly criticized for producing sprawling suburbs. Despite this obvious dilemma, proponents of green sprawl often do not see exurban landscapes as part of the sprawl debate, but rather imagine exurbia as a middle landscape—imagined as *beyond* the suburbs that failed to develop in keeping with the ideals of the middle landscape. To the degree that participants in green sprawl (residents, planners, developers, land managers) do not acknowledge its problems, even amidst a robust conversation about conservation and sustainable communities, the insights of decades of discussion of this quest for the middle landscape are lost.

The ten essays in this collection explore the ideology of nature to gain insight into the aspiration for an everyday relationship with nature that seems to be so widely shared. The purpose of these essays is to address the possibilities opened up by self-conscious and collective consideration of the implications of the ideology of nature for managing sprawl. They range from historical narratives of the beginnings of exurbia and everyday relationships with natural environments to accounts of planners' interventions in urbanization and conservation to literary approaches to everyday experiences of nature and residential environments. These explorations of exurbia, nature, and sprawl model ways to engage in conversation about the landscapes that have been created as desirable and compelling recourses—and often escapes. We hope this conversation spreads widely—in community discussion, planning, and public negotiation of landscape decisions, personal reflection and observation, historical analysis of the shaping of the built environment, and negotiation of social and environmental values embedded in those landscapes.

Exploratory conversation about green sprawl seems particularly important where highly valued everyday environmental experiences are challenged by complexities of the modern urban landscape. People complain about sprawl because it impoverishes environmental experience, but negotiation over the management of urbanization regularly privileges expert prescriptions for arranging urban form, often to further market and state goals rather than to incorporate recognizable and valued aspects of people's everyday environments.[14] Several of the chapters describe exurban themes and places that the authors consider important because of the powerful shaping effect of people's relationship with nature on the management of residential landscapes. Each chapter author approaches the promises and problems of exurbia, nature, and sprawl from a different perspective, but all understand the charged tension between modern landscapes we wish we could change and the landscape ideals into which we'd like to transform them.

In the rest of this introductory chapter, we discuss the relationship between exurbia, sprawl, and the ideology of nature in the context of other work on exurbia. Looking at the ways that the ideology of nature makes exurbia different from suburbia, we consider how critical approaches to thinking about the ideology of nature help us understand the ways we reside in our environments—and the ways we wish we could. Reflecting on the widely felt angst over sprawl, and exploring the tension between natural environmental ideals and contemporary trends of urban dispersion, we consider pitfalls involved in the exurban attempt to apply an abstract ideology of nature to existing landscapes without acknowledgement of the desires embedded in those idealized images of nature or the clashes between the ideal and the real everyday environment. We make an argument about the importance of paying attention to what can be read in the cultural landscape of exurbia. We then extend our reading to proposals for intervention through spatial planning and landscape design. We build this argument in the context of considerable public concern about human impacts on local and global environments and particularly about impacts of residential consumption habits in industrialized nations.

EXURBIA AS A CULTURAL LANDSCAPE

The concept of "landscape" helps counter the tendency in everyday language to separate "natural" environments from more "social" environments. When we say *landscape*, we more readily convey the idea of the entire environment around us as something humans have helped to shape, and that includes and shapes us. The concept of landscape, often referred to as "cultural landscape" in the literature of academics and practitioners, further helps push back against the powerful tendency to think of the environment as primarily natural, except for human "impacts."[15] Landscape, as a concept, provides a lens for seeing the materialization of cultural processes in the environment, and keeps the dynamic relationship between people and their environments more evident in language, thought, study, and policy. Those authors who emphasize the necessity of understanding natural environments as inherently linked with culture remind us that nature is a concept inextricably tied to individual and societal experience and performance and language, not something separate from us.

Looking at a landscape as cultural means understanding that what is seen is mediated by imagination, experience, identity, and ideology.

Particular landscapes and views of the landscape are produced through the interaction of people, environments, and the specific conditions of a particular place at a particular time. These interactions, conditions, and influences—from zoning requirements to cultural norms and preferences to economic necessity—are often not visible in the most available views of a given landscape. Looking at exurbia through the interdisciplinary lens of cultural landscape studies helps us to see how land, buildings, roads, and other structures and elements are created through cultural processes responding to needs for shelter, for resources, to move across the land, to communicate, or to represent oneself and one's community to others. These material processes contribute to experiences of the landscape; they shape movements, activities, and understandings that occur there. All landscapes are shaped by cultures while they also help to shape culture; individuals and groups are influenced by the material world in which they find themselves and they modify places to better suit their needs. At the same time, they construct and negotiate meaning in that landscape.

This book contributes to this tradition of understanding society-environment relations by using landscape studies to decode sprawl. Our approach in developing this volume has been to use landscape studies as understood in cultural geography and related disciplines to look at the exurbs as produced by deeply embedded societal preferences for a home life outside of the city. The landscape approach deliberately looks at the material landscape as an artifact of culture worthy of study. The processes through which exurban landscapes are created can be understood in terms of the demographic, economic, and regulatory explanations that may be familiar to those who study dispersion, but these processes also include the motivations and choice-making of exurbanites and exurban land managers and developers. As demonstrated by the chapters in the book, the approach of landscape studies combines multiple theories and methods from several disciplines to study complex places like exurbia;[16] as a whole, the book embodies the landscape approach even if individual chapters are written from within specific intellectual traditions.

Working with the landscape approach, we argue that the ideology of nature is a major force in the production of landscapes of exurbia because ideology influences land management decisions. To analyze this influence of nature ideologies, we look to political ecology, an approach that considers environmental issues within their larger political and economic contexts. With its roots in thinking about environmental issues in the context of political economy and unequal environmental and resource access regimes, political ecology augments landscape studies to analyze how nature ideologies work—especially those that privilege particular ideals favoring certain groups over others. For instance, seeing a wooded area as a good residential building site rather than a habitat or a timber resource is an ideological evaluation of a landscape that favors exurbanites. Political ecology "explor[es] the symbolic meanings ascribed to lands and environments and how these imbricate with struggles over control and access to material resources . . . [by] interrogating the role of ideology and the importance of the cultural context in shaping the perceptions of scientists, policy-makers and bureaucrats."[17] As political ecologists have expanded their interest from more traditional case studies in "developing" areas of the global South to more recent studies investigating parallels between global South and North political ecologies, geographers and others using political ecology approaches have renewed interest in landscape studies.[18]

The perceptions of exurbanites, residential developers, and other land managers in choosing and shaping residential sites are critical to understanding the proliferation of sprawl in forms such as exurbia—and to the role of ideology in mediating environmental sustainability questions more broadly. Environmental perceptions are based on social constructions of nature, as what is seen as natural (or not) in the landscape is mediated by ideology. Exurbanites see an aestheticized nature, one that is associated with a good quality of life. But this is a privileged view enjoyed by those who tend not to be involved in the day-to-day production or extraction of natural resources necessary to build and maintain their homes. It is also thus an urban conception of nature that reaches beyond the city.

This ideal privileges particular types of knowledge and relationships with nature and discounts (and thus hides) others. The concept of the ideology of nature is useful for discussing the motivations behind the production of exurbia specifically, and green sprawl more generally, because the exurban perspective constitutes a significant "discursive field" within which cultural preferences for residential lifestyle are negotiated.[19] A particular discursive field—the discourse of what is culturally acceptable, or potentially even possible, to say about a topic such as nature—becomes common sense and subverts other ways of looking at the landscape, often without obviously doing so. Describing the power of the discursive field associated with the ideology of nature, Noel Castree sums up the stakes in how seeing nature as an amenity might, for example, confer to exurbanites individual property rights and freedoms, especially access to natural amenities, while necessarily limiting access to human and non-human others through privatization: "The power to say what nature is, how it works, and what to do (or not to do) with it is enormously consequential for people and the non-human world."[20] In the unequal power struggle over what can be said and done in the landscapes where urban and rural land uses meet, exurbanites have gained disproportionate power to shape the discourse, limiting the acceptable vocabulary of landscape management to domains having to do with nature and the picturesque, and vilifying ways of talking about landscape that are more oriented toward livelihood—while at the same time engaging in significant economic accumulation.

The landscape produced in exurbia in this way is defined by what cultural geographers call the *geographical imagination* of people who create and inhabit a particular kind of cultural landscape they identify as natural.[21]

Exurbia is an imagined landscape as it combines the powerful associations of the symbolic landscapes of countryside and wilderness with the material landscape of the urban fringe, which is often recently converted farmland or forest.[22] In describing exurbia as an imagined landscape, we do not question the reality of the landscape itself—exurbia is, of course, an actually-existing landscape, not just the figment of someone's imagination, geographical or not—but rather we emphasize the importance of the ideology of nature in the way exurbia is produced. The rich cultural content of the ideology of nature in the exurban landscape is often as important as the easier-to-see (and to legislate) *material* landscape. Describing exurbia as a cultural landscape does not provide a specific material or spatial definition of exurbia, but instead invokes recognizable exurban landscape characteristics and processes based largely on attachment to aesthetic ideals of natural living.

When we talk about "reading the landscape" as exurban, we acknowledge different approaches to understanding the ideology of nature and we recognize that different frames of reference for studying the phenomenon of exurbia have often been at odds. Tensions are particularly notable around seemingly incommensurate claims about the definitions, drivers, and impacts of exurbanization made by demographers, ecologists, and social scientists concerned with social constructions of natural environments. Although this book singles out the role of the ideology of nature in producing exurbia, we see this construct in conversation with work studying dispersion in other ways, particularly including studies of exurban demographic and ecological change.[23] And although the idea of reading the landscape is sometimes considered a mode of analysis critically opposed to more materialist readings, as editors, we use this idea to bring together and engage with the chapter authors and you, the reader—as well as to help frame a dialogue between competing investigations of green sprawl that form a richer dialogue than any one set of methods or approaches could provide. So what do people read from or read into the landscape of exurbia?

EXURBIA, EXCLUSIVITY, AND POLITICAL ENGAGEMENT AS INFLUENCED BY THE IDEOLOGY OF NATURE

The landscape of exurbia can be interpreted both in terms of the landscape qualities exurbanites are seeking and those they are seeking to avoid. We can see the influence of the ideology of nature in the kinds of landscapes that are selected as pro-nature and anti-urban by exurbanites, in the kinds of landscapes exurbanites produce, in policies for and activism over exurban landscapes, in representations of exurbia in the popular press, and particularly in advertisements—not only advertising exurbia itself, but also using exurbia to sell everything from financial services and the peace of mind associated with natural ideals to the accoutrements of lifestyle associated with natural exurban living, such as SUVs, patio furniture, and recreational equipment.[24] Exurban nature is often represented as spectacular scenery, or as a bundle of other attractive amenities of the rural setting. Studying exurbia as a cultural landscape helps add a critical perspective to the exurban reading of nature as a setting for desirable lifestyles, making the impacts of those lifestyles more visible, too.

When we use the phrase "ideology of nature" to describe the complex set of desires, ideals, and motivations that produce and reproduce exurbia, we treat these explicitly as ideology because many exurbanites see their own individual reasons for wanting to live in nature as obvious and uncomplicated—and often as personal and exceptional. Calling into question this unqualified acceptance of "natural" exurban landscapes helps to structure thinking about the outer reaches of metropolitan areas by highlighting the ideological components underpinning exurban processes of rural change, particularly in terms of the relationships between ideals and material landscape processes. In increasing use by academics, journalists, and laypeople, the term "exurbia" encompasses the settlement taking place "out there" (or "out here," depending upon where you stand). For many, exurbia is the "outer suburbs," the recent residential and commercial development beyond the suburbs proper. This is often a landscape complicated by pre-existing land users who resist further development. For others, the exurbs are those settlement areas way out in the hinterlands, seemingly completely separate from any central city. Then there are the individual exurban homes created one at a time, sometimes through severances of farm or rangeland, within commuting distance to a city, or at least within *tele*commuting distance, and perhaps within driving distance of an airport, taking advantage of sparsely populated amenity landscapes—until other amenity migrants follow.[25]

The utility of a named concept, such as "exurbia," for a specific category of settlement type or zone has been the subject of much debate, especially in academic literature. This debate has been spurred, in large part, by planners' use of the term exurbia to describe exurban settlement as different from suburban. Planners use specific categories to describe settlement zones in order to plan and provide services to the homes and businesses there, and to plan for mitigating the ecological, economic, and social impacts of such development. However, given the wide range of patterns of exurban settlement—and particularly the difference between urban-adjacent and far-flung exurbs—some people ask whether the category is really a useful tool at all. We argue that the idea of exurbia is indeed useful, not only because of its potential for physical description or as a land-use zone, but also as a social and cultural category.

Including much more than the physical settlement form, this broader understanding of exurbia as a cultural landscape can be useful for opening up opportunities for people to recognize, acknowledge, and explore the role of landscape and cultural ideology in the settlement choices that they and others make. Acknowledgement of social and cultural values in the landscapes of nature and of exurbia could focus planning and land-use negotiations, by illustrating how definitions of what constitutes nature are narrow and often very ethnically white.[26] Explicitly recognizing and countering this ideology in exurbia and natural land-use management discourses could help address some of the environmental justice and equity problems inherent in exurban escape landscapes portrayed as alternative modes of dwelling. Although not focused on the *disamenities* that exurbanization exacerbates by furthering racialized segregation, this book takes a step in the direction of environmental justice by exploring the way that the ideology of nature appears to legitimate the exclusivity—and high impact—of exurbia.[27]

The way that alternatives to sprawl are often couched in terms of sustainable urbanism provides an illustration of where a broader cultural landscape understanding of what exurbia represents may be useful. For example, in cases where growth management calls for intensification within built-up areas of the city, and also in cases where greenfield development is unavoidable, sustainable urbanism suggests that new communities should be built with homes close together, at high densities, with their form supporting walking and public transit.[28] In these proposed sustainable alternatives to sprawl, whose purpose is represented in terms of protecting nature from being paved over by city development, access to natural environments for residents—a key promise of green sprawl landscapes—often falls down the list of priorities. Where Frederick Law Olmsted saw daily access to nature in city parks, or in new subdivisions, as a necessity of city life, planners today see the need to accommodate for *ecological functions* of nature more than those aesthetic and cultural functions people often value in nature.[29] The implication is that in the desire to stop sprawl, the focus of practical research and policy has been on normative theories of the city, and has not been about the pressure that concern about sustainability might exert on exurbia—or on nature-valuing households who might choose exurban migration for exactly the same reasons they might support policy measures to stop sprawl.

People are concerned about the sustainability issues involved in exurban settlement patterns, but broad consensus in the field of sustainability activism and study is emerging that acknowledging, understanding, and working with the culture of how people inhabit their environments is crucial to any changes we may wish to see widely adopted.[30] Therefore, addressing exurbia as a landscape of interest to sustainable growth management advocates might be more successful if they were to consider the landscape perspective in the production of green sprawl as a way to bridge experiential, ideological, and material understandings of environment.

How landscapes are constructed to be read by other people in the landscape—and how different people actually read them—becomes especially important considering that exurbia is used (sometimes defiantly, sometimes unwittingly) by some people to entrench power relations (e.g. "keep out") and to perpetuate social difference (e.g. me here, you there). The reproduction of natural landscape benefits makes particular landscapes more accessible to some than others, both through more obvious mechanisms, such as property values, and also through less obvious mechanisms, such as subtle indicators of who—and what kinds of cultural activity and performance—belong in a place.[31] Similarly, the designation of some landscapes as "natural" and hence "sustainable" benefits some more than others, not only in terms of access to environments of widely varying amenity or disamenity, but also in terms of the social relations developing around sustainability initiatives, for example rewarding exurban amenity ranchette owners for carbon sequestration—as they commute by air across the country.[32]

These currents of power in place making—and the inequality of access that makes exurbia so often a landscape of the rich, characterized by the distinction that an exclusive landscape like exurbia offers—often make people uncomfortable talking about the ideology involved in exurbia, sprawl, and nature. The difficult ethical issues involved in the way that access to natural residential landscapes is negotiated are daunting, especially when juxtaposed with the apparently straightforward and *natural* desire to just have a nice place to live.[33] Further, the fact that exurbanites live beyond urban containment boundaries while everyone else is supposed to live within them for "smart growth" reasons seems unfair—especially when the rhetoric for protecting all that greenspace claims it for everyone's benefit. All of these tensions complicate and inhibit public discourse on sprawl, exurbia, and nature in residential environments. Adding to inhibition in talking about sprawl and nature is the ironic paradox underlying exurbia: many of the very people making urban growth boundaries and agitating against sprawl live *out there*, in the exurbs.[34] Understanding exurbia and sprawl more in terms of natural ideals could help clarify the terms and goals of research and policy related to exurbia and sprawl.

Protests against sprawl merit examination not only in their own right, but also because of how upset people get over questions of nature and residential land use. A combative emotional tone often accompanies conflicts over development and conservation, for instance in debates in public meetings over plans for urban expansion. The depth and prevalence of this emotional tone may rest on the paradox that is perhaps the most defining quality of exurbia. The ideology of nature represented by exurbia gets people riled up precisely because so many people recognize, share, and respond to the desire to live in natural settings, and yet also at the same time recognize the threat posed by residential development. The promises of the ideology of nature resonate with the reasons that people protest against sprawl—particularly as similarities become evident between the kind of nature people find desirable enough to want to protect (even if they want to "protect" it by possessing it themselves) and the kind of nature they think needs to be protected from *everyone else's* desires. The chapters that follow consider a range of issues and themes in the landscape of exurbia; many also suggest methods to read the landscape to better engage such issues and themes in discourse and practice. We put this book together because we felt that the cultural landscape approach has a lot to offer in thinking about how to engage with exurbia and would be a useful contribution to the efforts to deal with the negative effects of the downsides of exurbs—urbanization, urban deconcentration, and sprawl—on societies and environments. The chapter authors present varied perspectives on reading and engaging the issues of the exurbs when confronting the desire to escape from modern urbanity and retreat to nature.

Observing and participating in the discourses of sprawl as environmental planners, teachers, researchers, and community members, we have grappled with the ways that negotiations and conflicts over sprawl issues are tangled up with emotional and exaggerated representations of land use at the edges of metropolitan regions. The most vehement objections to the kind of land use and landscape ideology we see in exurbia come as the battle cries of urbanists against sprawl and as pleas by ecologists and green activists in favor of nature and against further residential and industrial development. By conflating the goals of controlling development and protecting the natural environment, the pro-nature, anti-urban development position tends to deepen the conceptual separation between the places where people live and the places where they don't. This nature-focused view of exurbia may also perpetuate the tendency to imagine unpeopled nature, and to ignore crucial urban environmental justice questions exurbia appears to sidestep, but with which exurban lifestyles are highly, if invisibly, entangled.[35] By taking these issues seriously, but also emphasizing the need to resist deepening and polemical rifts, the authors in this collection offer grounded ways to engage with the actual material environment while also examining its imaginative and ideological layers. Treating the layered nature of landscape seriously by understanding the relationship between these material, imaginative, and ideological layers may help counter the tendency to make nature abstract, out there, disconnected from the people whose lives and livelihoods are bound up in it, and a panacea for the problems of modernity.

READING THE CULTURAL LANDSCAPES OF EXURBIA AND SPRAWL

With our particular perspective on the role of the ideology of nature in sprawl, we collected essays to deliberately discuss the lack of focus on the ideology of nature in discussing sprawl, using exurbia as an illustrative example. Introducing this collection of essays, having broached the idea that an ideology of nature is implicated in urban growth and change, we address the most common preconception that we (the collective of authors of this book) have found in talking to people about the exurbs: that as planners and academics we must think that exurbs and sprawl are pernicious and that we are consequently studying exurbs in order to do away with them. In fact, in many ways the opposite is true. Although we agree with many critiques of sprawl, we see in the landscape of exurbia both the expression of important cultural themes and also a resistance to address problems associated with sprawl. As discussed in the previous section, our intention is to bring forward for discussion concerns about exurban nature and the power of the ideology in the acceptance of a particular lifestyle, as a step in imagining more sustainable and equitable futures, beyond—but also including—the city.

In the rest of this introductory chapter, we provide a brief introduction to some of the strands of thought that have shaped the way that landscapes of sprawl and exurbia are read. A strikingly high percentage of texts that strive to "read" exurbia situate themselves with reference to the introduction of the term by A.C. Spectorsky in 1955. Spectorsky's book *The Exurbanites* named the "exurban" phenomenon as something distinct from suburbs— and as profoundly bound up in the emergence of what would come to be described as a highly mediated, "Post-Fordist" urban modernity. Although Spectorsky described communities of large "country homes" within a sixty-mile commuting arc of Manhattan, exurbia has since spread far beyond the New York exurbs he described. The natural landscape ideals associated with exurbia in the post-WWII American northeast have become common currency in the housing industry, with exurban-style housing estates becoming increasingly common not only in metropolitan areas with obvious links to Western ideologies of nature, but also in places where the lineage of landscape influence on exurbs may be more unexpected, and harder to read, such as in China and Latin America.[36] Spectorsky's book has remained the starting point of so many discussions of exurbia for a number of reasons. Even if he did not invent the idea behind "exurb," in his insightful description of exurban residential choices, he created an evocative portrait of the motivations of people who were both living out their exurban dream and also representing this dream of getting out of the city to millions of others via the culture industry.[37]

Spectorsky's exurbanites were arguably some of the most influential people in America at the time. These were the people who owned and managed America's communications industry—advertising, book publishing, film, and television—in New York, the center, in Spectorsky's account, of production of the public's desire for goods and services. These were people, Spectorsky said, who moved from the city to the countryside because they were hooked on the lifestyle promises associated with its landscape. Fed up with urban landscape as a manifestation of all that was wrong with modernity, they sought respite (and social distinction) in the natural landscapes of exurbia. That so many exurban analysts return for their starting point for discussions of exurbia to a minor popular text, rather than to the considerable academic work conducted since, suggests that Spectorsky's way of capturing the ideology and aesthetic of the exurban landscape ethos speaks to people.[38] It also demonstrates how people understand exurbia as being as much a state of mind as it is a settlement pattern. In characterizing what he called the "limited dream" of this ethos, however, Spectorsky was not optimistic about the prospects of this search for quality of life in the exurban landscape. He tried to demonstrate, instead, that in buying into the belief that an idealized natural residential landscape would provide the transformation they sought, exurbanites sold themselves a "bill of dubious goods by means of [their] own advertisements."[39]

The exurbs today retain their powerful effect on the public imagination, simply by example, whether their residents mean to or not, whether they are trying to stop sprawl or simply escape it. As discussed above, new home ads use powerful images of secluded homes within leafy settings or commanding spectacular views. Exurbs mobilize a vocabulary of nature whose physical manifestation looks a lot like sprawl. The vocabulary of sprawl draws upon a nature-based mythology of what constitutes a better place to live. Given the predictable need that will continue to arise—to escape tomorrow the built environment created today—the conversation about green sprawl that we propose aims to address the contemporary desire for a haven amidst a sometimes forbidding built environment by considering how in reaching for nature, people reproduce the very thing they feel a need to escape.

Ecological Exurbia: Ecosystem Impacts at the Human-Wildlands Interface

For instance, the resettling of rural areas by amenity-seeking urbanites has brought with it ecological impacts that are of concern especially in the field of conservation biology.[40] As much as the productive ranching or farming landscape may have involved significant human transformation of landscapes, there is still habitat in the forests, wetlands, and on marginal land and steep slopes. Exurban incursions into these landscapes change ecological systems of the productive-natural setting into new amenity-natural configurations with interesting results in terms of habitat, competition, and conflict. Although our collection does not include essays from natural science perspectives, and although no comprehensive picture of exurbanization has emerged from this sector, the large amount of work using the concept of exurbia to express concern over exurban effects has played an important role in reifying exurbia in the public and academic imagination.[41] An exemplary minority of this work demonstrates the analytical power of considering the ideological role of nature more explicitly, considering in a more balanced manner both the negative *and* positive effects humans have on their habitat ecologies—along with the differences in material outcomes that follow from the way people think about the landscape.[42]

Despite hopes for its possibilities, it is clear that the continued effect on settlement patterns of the ideology of nature is problematic. Exurbia represents a promising bundle of benefits like fresh air, peace, and quiet. In contrast to the exurban landscape's promises are the issues of sprawl that are being extensively studied: loss of open space, productive agricultural land, and the scenic countryside; lack of housing diversity in a landscape of predominantly single family homes; unwise use of increasingly scarce energy resources; inability to support transit use; traffic congestion and air pollution; impediments to the revitalization of downtowns; lack of community centers; and social inequity caused by a focus of new building provision on upscale markets and by dispossession of people with prior claims to exurbanizing landscapes.While it is difficult to say how much the cultural landscapes of exurbia have been influenced by the genre of lifestyle advertisements Spectorsky spoofed back in the 1950s, his warnings about the failings of exurbia have become common tropes: dissonance between the expectations associated with the natural landscape of dispersed living and its everyday realities, in addition to pressing debt and the impacts of commuting. Writing at the very beginning of the post-war period, which saw the realization of sprawl, Spectorsky was not the first or only one to articulate concerns with the beginning of the repopulation of rural areas by urbanites.[43]

Rural Studies: The Impact of Exurbia on Agriculture

Using an impacts-based conceptual framework similar to many ecological studies, the broad field of rural studies (usually centered on rural sociology, agricultural economics, history, and geography) has perhaps done the most to further the understanding of exurbia as a rural problem. Although tending to be under-integrated with *urban* studies addressing sprawl, the field of rural studies has provided a more complex, social, and comprehensive view of exurbia than any other field, especially in terms of the impact of sprawl on the sustainability of rural economic systems, especially agriculture. As the strongest cell of exurban academic studies, studies of the urban-rural fringe, in their attempts to understand how to mitigate exurban impacts, have done the most to counter with specifics of landscape and land use the tendency to view the environment into which exurbia grows as an abstracted field of idealized nature.[44]

The recognition that the population has been dispersing because people wish to live closer to nature often appears to have been accepted in an uncomplicated way in the literature on exurbia and sprawl.[45] Taken as a whole, the literature on the analysis of exurbia and its impacts fails to ask "why" about the pull of nature; although often claimed as a motivating factor, this pull has rarely been investigated. Many writings on exurbia assert that planners should think about preserving the ruralness and naturalness of exurbs, but these pieces stop short of examining the motives and politics behind such a move.[46] Precedents do exist for designing with nature in the city, including the work of Olmsted as well as McHarg's and Arendt's "conservation" development strategies—forms that have been suggested for over fifty years (since "exurbia" became a word, in fact) that perhaps better reflect the ideal exurbia seems to attempt to achieve.[47] In building on these bodies of academic work, we seek to explore how academic work can address or reconcile the fact that exurbia is sprawl, on the one hand unpacking the way that the preservation of ruralness and naturalness encourages green sprawl, and on the other hand exploring whether the aspirations of exurban rhetoric, representations, and practices could be incorporated into less problematic residential settlement forms.

Political Ecology and Political Economy: Nature from Critical Perspectives

The compelling need to address the ideology of nature in exurbia arises as increasing numbers have followed the pacesetting exurbanites to take up some version of the exurban lifestyle. Some research shows that while many exurbanites express concern about the consequences of sprawl, they don't see the lifestyle associated with the exurban landscape as producing sprawl, instead often striving to protect from sprawl the promises of this natural-seeming landscape.[48] Given the ubiquity of calls for nature preservation in anti-sprawl campaigns, it may seem odd to equate the desire to connect with nature with sprawl. But in this central paradox of the exurban ideology of nature, sprawl landscapes and anti-sprawl rhetoric both privilege an abstracted nature. We are encouraged by recent strands of research examining the conviction that sprawl is bad for nature and should be avoided, and the way this leads people to sprawl even farther.

By approximating the ideal of natural landscape so well, the landscape of exurbia effectively becomes naturalized and shrinks from analytic exploration. The house in the woods, or the ranchette on the prairie, or the house on the beach are, for many, unquestionably desirable. Things that seem natural, as cultural critic Raymond Williams has pointed out in his etymology of "nature"[49] seem just "natural," as if they are supposed to be that way—something's nature is just the way it is. And this, we argue, is part of the potent lure of exurban nature, and also a good reason to look closely at exurbia. Natural things are appealing because they appear to be the way they should be, without people having messed them up, without requiring people to make choices about how they should be. Part of the slippery nature (so to speak) of the ideology of nature is its capacity to naturalize the very human choices and histories it necessarily involves. Many "natural" landscapes that are desirable for residential use have been modified by human activities; when these landscapes are naturalized (i.e. seen as natural), it becomes easier to leave unquestioned the political intents and choices involved in the production of those landscapes.

Cultural geographers and political ecologists have picked up on this disparity, since most environmental protection legislation is written in terms of value to science (and through science, indirectly to people), rather than in terms more directly related to meaningful people-environment relationships. Both approaches have drawn heavily on the critical social theory of Karl Marx, and particularly upon Neil Smith's *Uneven Development* and Raymond Williams' *The City and the Country*, to explore the ways in which ideals and material manifestations of nature are constructed and produced.[50]

In the past two decades, social scientists and historians have taken up with great vigor the intimate examination of the relationship between ideology and space whose foundations were laid in English-language literary criticism by figures such as Leo Marx and Raymond Williams, and the translation of the French social critic Henri Lefebvre has greatly popularized the consideration together of ideology, environment, and everyday life.[51] The basic story these critics have put forward—that the environments of everyday life are highly mediated by imagination and ideology—has contributed an important understanding of the way nature is abstracted as it is imagined and idealized. Consequently, the pastoral lens so often used to view landscapes such as exurbia must be examined more explicitly for the imaginative richness it makes visible, enabling us to understand landscapes not just as physical manifestations that can be fully apprehended or protected in terms of prescriptive policies based on natural science paradigms.

The centrality of political and cultural motivations to this contemporary work has translated into a rigorous application of theoretical and literary constructs to material situations. Even extremely abstract critical social theories, like those of the Frankfurt School, are being taken up by grounded practitioners such as architectural historian Hilde Heynen, landscape architect Simon Swaffield, and rural sociologist John Fairweather, to address the way that discomfort with urban modernity and idealization of the pastoral play out in everyday struggles over the way the environment is shaped (this is taken up in more detail in Chapter 10 [Cadieux]).[52] Part of the value of this work is its accessibility as an entry point into understanding what can be gained by discussing the ideology of nature, not just its ecological makeup—stories of how subdivisions are often shaped according to pastoral ideals provide a frame of reference most people can share.

A particularly important point of crossover between this theoretically informed exploration of how ideologies inform material decisions in the environment is represented by Peter Walker, Louise Fortmann, and Patrick Hurley's study of the political ecology of the exurban Sierra.[53] Walker, Fortmann, and Hurley help extend the understanding of what is exurban, or of what counts as urbanization in areas explicitly defined for their nonurbanness. By conscientiously bringing the analytical tools of political ecology to the questions of exurbia, they also make a convincing case for the need to understand the way that meaning is negotiated and contested in the landscape together with an understanding of the ecological changes brought about through these negotiations and contests.

Discomfort with Urban Modernity and the Ideology of Nature

If escape from the generic placelessness of the modern urban landscape is a problematic aspect of sprawl, several of the chapters in this book consider in detail what counterpoints to escape can be found in the ideology of connecting with nature. For example, in Chapter 10, Cadieux discusses active practices of inhabitation that make places suitable for dwelling, and that build environmental values into places. Judd, Luka, Vachon and Paradis, and Sandalack and Nicolai also take up in their chapters the specifics of how tensions between urban escape and attempts to inhabit nature play out materially in the exurbs. In seeking to understand the basis for the contemporary ideology of nature, several chapter authors turned to Thoreau for clarification on how to think about embracing the contradictions sprawl presents.

Although Thoreau did not address sprawl as such, his exhaustive examinations of what it takes to inhabit places and to aspire to build connections with nature provided a central gathering point for our thoughts and conversations while we wrote this book—and particularly while we grappled with important contradictions and tensions in the difference between the symbolic and material nature of exurbia. Chapter authors Blum, Donahue, and Looker discuss Thoreau's suitability as a model of the creative exploration required to inhabit or to dwell. Thoreau's historical context relative to sprawl itself makes him an even more illustrative model—he wrote during (and became part of) the great turn toward the aesthetic foundations of the ideology of nature we discuss, as part of the manifestation of urban-escape ideology that would be passed down from the Romantic era. Our readings of the divide Thoreau so eloquently bridges—between the approach of modernity with the train and the retreat of the exurbanites—owe a great debt to the argument championed in 1964 by Leo Marx in *The Machine in the Garden*, his analysis of technology and the American pastoral ideal. Marx is one of the central scholars to address explicitly the tensions and paradoxes of the natural response to modernity, and to apply that critique of what he calls "the American ideology of space" specifically to exurbia.[54]

The critiques of exurbia and sprawl we build on contain many questions that remain outside the scope of our analysis of the ideology of nature. The following chapters also do not take on many of the questions we were dismayed to find unanswered as we set about assembling this work, including questions of segregation, climate equity, and the role of amenity property acquisition in volatile housing markets and in the face of continuing agricultural restructuring. Nor do these chapters offer the comprehensive methodological advice needed to bridge the varied disciplines represented here to answer these questions along with those we have discussed above.

However, better understanding the function of the ideology of nature in all these domains may contribute to efforts to transform North American ideologies of space, and to reimagining green sprawl as a site of increased social and environmental equality, justice, and sustainability.

STRUCTURE OF THE BOOK

We would here like to present a brief overview of the key themes of the collection in terms of the way we have ordered the chapters' presentation. Each of the chapters begins with a brief editors' note that extends this introduction and highlights the value of that chapter's particular approach. In the chapter following this introduction, Laura Taylor (Chapter 2) looks at a case study of an exurban landscape at the edge of Toronto, and introduces ways of thinking about how the ideology of nature is negotiated in landscape transformation as countryside becomes exurbia, which is then engulfed by the city. Following her analysis of the legitimacy of landscape values in planning processes, Richard Judd (Chapter 3) discusses how nature and place have become commodified and generic, and traces what subscribing to nature has meant for patterns of urban dispersion. Asking what happens in the process of this commodification, Judd describes how privileging a nature that is not in accord with everyday life turns people away from the complexity, difficulty, and dissonance of modernity to escape into an imaginary nature. Building on Judd's analysis of the material processes and infrastructures that support everyday life, Andrew Blum (Chapter 4) considers the relationship between the local and the remote in the construction of the special "place" qualities of exurbia.

Continuing from Blum's consideration of the infrastructure supporting everyday life, Brian Donahue (Chapter 5) looks at how nature has become idealized, making the possibility of managing real landscapes just about impossible, using the Walden Woods of Henry Thoreau as an evocative example. Where Donahue uses an American example to explore how we might better reconcile our ideals of nature with the material processes of our everyday lives, Nik Luka (Chapter 6) uses a Canadian one: cottage country in Ontario, where the natural ideal is explicitly sought as a respite from everyday life in the city. In the next chapter, Geneviève Vachon and David Paradis (Chapter 7) contrast the failure to incorporate the desire for nature into design that Luka describes, and provide an example of a planning process in which the desire for many different kinds of nature is negotiated and accommodated—if not without difficulty.

The value of being able to discuss, question, and negotiate the motives and ideology underlying design decisions allows us to see in the next chapter by Bev Sandalack and Andrei Nicolai (Chapter 8) how natural landscape ideals are imposed on ideologically unappealing (or unmarketable) landscapes. Presenting a morphological history of three neighborhoods, each at the edge of an expanding city, Sandalack and Nicolai's account demonstrates the pitfalls involved in the attempt to apply abstract ideals of nature to existing landscapes without considering the desires embedded in the image of nature or the clashes between the ideal and the real everyday environment. Asking about the origins of the ideals that overlay the everyday environment, and about the way that the impulse toward nature is framed through language, Thomas Looker (Chapter 9) combines an exploration of dissatisfaction with everyday environments with an analysis of "sound-bite thinking." If we have difficulty finding the words to discuss environmental values, Looker suggests, we fall into a tendency toward the fundamentalist thinking that created the pastoral–modernity and city–country binaries in the first place. In the last chapter, Valentine Cadieux (Chapter 10) explores these binaries in the way that symbolic and imagined landscapes are mobilized in the experience of everyday material landscapes like exurban yards and gardens.

These chapters have grown out of a long and rewarding conversation between the authors. With these essays we hope to participate in (and to spur many who have hesitated to participate in) the larger conversation about what to do with exurban landscapes and our common environment more broadly: not only what *not* to do with it, but how to carry out the tasks that enable us to do anything with it, of figuring out how to translate our needs, desires, and environmental imagination into a landscape we cannot just live in, but can inhabit, and can do so equitably. Can the environmental impulse be mobilized in other ways than through the consumer ideology of nature within which exurbia is embedded? These essays explore the potentials as well as the liabilities of the exurban ideological construction of nature. What is accomplished by highlighting the ironies of exurbia? By understanding the processes and ironies of green sprawl, greater insight is gained into *where* and *how* and *why* people create meaning in sprawl landscapes, and in all of the landscapes we strive to inhabit.

NOTES

1. There is a large literature on the issues of sprawl. For a shorter overview, see Peter Gordon and Harry W. Richardson, "Are Compact Cities a Desirable Planning Goal?" *Journal of the American Planning Association* 63 (1997): 95–106, doi: 10.1080/01944369708975727. For a broad discussion see Oliver Gillham, *The Limitless City: A Primer on the Urban Sprawl Debate* (Washington, DC: Island Press, 2002) or Robert Bruegmann, *Sprawl: A Compact History* (Chicago: University of Chicago Press, 2005). For examples of the use of the rhetoric of sprawl to stress the need for alternatives, see Andre Duany, Elizabeth Plater-Zyberk, and Jeff Speck, *Suburban Nation* (New York: North Point Press, 2000) and Peter Calthorpe, *The Next American Metropolis* (New York: Princeton Architectural Press, 1993). For a discussion of urban form and design alternatives, see Mark Roseland, *Toward Sustainable Communities: A Resource Book for Municipal and Local*

Governments (Gabriola Island, BC: New Society Publishers, 2009) and Douglas Farr, *Sustainable Urbanism: Urban Design with Nature* (Hoboken, NJ: Wiley, 2007). For a recent examination of exurban boom and bust trends that helps contextualize contemporary exurban sprawl, see William Frey, *Population Growth in Metro America since 1980: Putting the Volatile 2000s in Perspective* (Washington, DC: The Brookings Institution, 2012).

2. For contemporary examples of communities held up as models of growth management, consider Portland, Oregon as discussed in Alex Marshall, *How Cities Work: Suburbs, Sprawl, and the Roads Not Taken* (Austin: University of Texas Press, 2001) and in Peter A. Walker and Patrick T. Hurley, *Planning Paradise: Politics and Visioning of Land Use in Oregon* (Tucson: University of Arizona Press, 2011) with discussion of planning in Oregon, including Portland. Also consider the model of Toronto as expressed in Ontario Ministry of Public Infrastructure Renewal, *Growth Plan for the Greater Golden Horseshoe* (Toronto: Queen's Printer, 2006), https://www.placestogrow.ca. Analyses by Peter Gordon and Harry W. Richardson, "Exit and Voice in U.S. Settlement Change," *The Review of Austrian Economics* 17 (2004): 187–201, and Alan Berube, Audrey Singer, Jill H. Wilson, and William H. Frey, *Finding Exurbia: America's Fast-Growing Communities at the Metropolitan Fringe* (Washington, DC: The Brookings Institution, 2006), conclude that growth outside of existing metropolitan areas was increasing significantly prior to the economic crisis, and, along with Frey, *Population Growth in Metro America since 1980*, suggest that sprawl will continue to be a pressing problem even if trends of considerable reinvestment continue in more central urban areas.

3. William Cronon, ed., *Uncommon Ground: Rethinking the Human Place in Nature* (New York: W.W. Norton, 1996), Leo Marx, *The Machine in the Garden: Technology and the Pastoral Ideal in America* (New York: Oxford University Press, 2000 [1964]).

4. For discussion and critiques of the Western ideology of nature, see: Simon Schama, *Landscape and Memory* (London: Harper Collins, 1995); Cronon, *Uncommon Ground* (especially chapters written by Cronon, Carolyn Merchant, and Ann Whiston Spirn); Denis Cosgrove, *Social Formation and Symbolic Landscape* (Madison: University of Wisconsin Press, 1998 [1984]); Alexander Wilson, *The Culture of Nature* (Oxford: Blackwell, 1992), John Rennie Short, *Imagined Country: Environment, Culture and Society* (Syracuse: Syracuse University Press, 2005 [1991]); and William M. Denevan, "The Pristine Myth: The Landscape of the Americas in 1492," *Annals of the Association of American Geographers* 82 (1992): 369–385.

5. Regarding the production of space, see Henri Lefebvre, *The Production of Space*, translated by Donald Nicholson-Smith (Oxford: Blackwell, 1992 [1974]). For discussion of the effect of the wilderness ideology in producing national parks, see William Cronon, "The Riddle of the Apostle Islands: How Do You Manage a Wilderness Full of Human Stories?" *Orion* (May–June 2003): 36–42, and Roderick P. Neumann, "The Production of Nature: Colonial Recasting of the African Landscape in Serengeti National Park," in *Political Ecology: An Integrative Approach to Geography and Environment-Development Studies*, eds. Karl S. Zimmerer and Thomas J. Bassett (New York: The Guilford Press, 2003): 240–255. Cronon and Neumann both discuss efforts by park managers to erase traces of human settlement from park spaces in an effort to inscribe the wilderness landscape ideology in the actual landscape. We argue that similarly, an exurban ideology of an aestheticized wild nature works for exurbanites in producing their home spaces. In geography, there has been both recent and longstanding interest in theorizing the

role of nature in the production of the city (see for instance Matthew Gandy, *Concrete and Clay: Reworking Nature in New York City* (Cambridge, MA: The MIT Press, 2003), Nik Heynen, Maria Kaika, and Erik Swyngedouw, eds., *In the Nature of Cities: Urban Political Ecology and the Politics of Urban Metabolism* (New York: Routledge, 2006), and Michael F. Bunce, "Rural Sentiment and the Ambiguity of the Urban Fringe," in *The Rural-Urban Fringe: Canadian Perspectives*, Ken B. Beesley and Lorne H. Russwurm, eds. (Downsview, ON: Department of Geography, Atkinson College, York University, 1981), but these have not interrogated the role of the ideology of nature in producing contemporary settlement patterns. In talking about cities and the "urban," we often use "urban" in its ecological sense to refer to environments that are "human-dominated ecosystems" (N.E. McIntyre, K. Knowles-Yánez, and D. Hope, "Urban Ecology as an Interdisciplinary Field: Differences in the Use of 'Urban' between the Social and Nature Sciences," *Urban Ecosystems* 4 (2000): 5–24); in this sense, although the exurbs are not *as* urban as cities might be, they exhibit signs and effects of urbanization such as impacts to the quality and quantity of surface water runoff with downstream and aquifer effects, as well as intentional and unintentional changes to habitats and thus species composition and interactions.

6. Henry David Thoreau, "Walking," *Atlantic Monthly* (1862). The same familiarity with Thoreau does not seem to underlie Canadian environmental aesthetics. In Ontario at least, this aesthetic of wildness may instead have been influenced by the forest paintings of the Group of Seven and a different literature as touched on in Nik Luka's Chapter 6. American (over)use of Thoreau quotations is discussed in Thomas Looker's Chapter 9.

7. "Wicked" problems are difficult not only to solve, but even to define, as they involve dynamism, uncertainty, and complex interdependencies, see Horst W.J. Rittel and Melvin M. Webber, "Dilemmas in a General Theory of Planning," *Policy Sciences* 4 (1973): 155–169.

8. Mike Davis, *Planet of Slums* (New York: Verso, 2006).

9. Neil Smith, *Uneven Development: Nature, Capital and the Production of Space* (New York: Verso, 2010 [1984]), James Duncan and Nancy Duncan, *Landscapes of Privilege: The Politics of the Aesthetic in an American Suburb* (New York: Routledge, 2004), Ralph E. Heimlich and William D. Anderson, *Development at the Urban Fringe and Beyond*, Agricultural Economic Report No. 803 (Washington, DC: U.S. Department of Agriculture, 2001), and Naomi Klein, *The Shock Doctrine: The Rise of Disaster Capitalism* (Toronto: Knopf, 2008).

10. Patrick C. Jobes, *Moving Nearer to Heaven: The Illusions and Disillusions of Migrants to Scenic Rural Places* (Westport, CT: Praeger, 2000), Pierce Lewis, "The Invasion of Rural America: The Emergence of the Galactic City," in *The Changing American Countryside*, ed. Emery N. Castle (Lawrence: University Press of Kansas, 1995): 39–61.

11. Robert L. Thayer, *Gray World, Green Heart: Technology, Nature, and the Sustainable Landscape* (Hoboken, NJ: Wiley, 1994).

12. Marx, *Machine in the Garden*, Peter G. Rowe, *Making a Middle Landscape* (Cambridge, MA: The MIT Press, 1991).

13. Ebenezer Howard, *To-morrow: A Peaceful Path to Real Reform* (London: Routledge, 2003 [1898]), Robin M. Leichenko and William D. Solecki, "Exporting the American Dream: The Globalization of Suburban Consumption Landscapes," *Regional Studies* 39.2 (2005): 241–253.

14. The constructs of modernity and particularly of post-Fordism have been used to help make sense of the way that urban form is influenced by the relationship between land-use planning and politics, city- and state-making, and market

actors; these constructs are particularly helpful for understanding what many people understand to differentiate exurbanization from more traditionally "Fordist" suburbanization, by focusing (often via economic geography) on the triad of economic actors, policy actors, and investment circuits bolstered by the commodification of landscape and nature. See Hilde Heynen, *Architecture and Modernity: A Critique* (Cambridge, MA: The MIT Press, 1999), Bent Flyvbjerg, *Rationality and Power: Democracy in Practice* (Chicago: University of Chicago Press, 1998), Edward C. Relph, *The Modern Urban Landscape* (Baltimore: Johns Hopkins University Press, 1987), Roger Keil and John Graham, "Reasserting Nature: Constructing Urban Environments after Fordism," in *Remaking Reality: Nature at the Millenium*, eds. Bruce Braun and Noel Castree (London: Routledge, 1998): 100–125, Ismael Vaccaro, "Theorizing Impending Peripheries: Postindustrial Landscapes at the Edge of Hyper-Modernity's Collapse," *Journal of International and Global Studies* 1 (2010): 22–44, L. Anders Sandberg and Gerda R. Wekerle, "Reaping Nature's Dividends: The Neoliberalization and Gentrification of Nature on the Oak Ridges Moraine," *Journal of Environmental Policy and Planning* 12 (2010): 41–57, Manfred Perlik, "The Specifics of Amenity Migration in the European Alps," in *The Amenity Migrants: Seeking and Sustaining Mountains and Their Cultures*, ed. Laurence A.G. Moss (Cambridge, MA: CABI, 2006), R. Alan Walks, "The Social Ecology of the Post-Fordist/Global City? Economic Restructuring and Socio-Spatial Polarisation in the Toronto Urban Region," *Urban Studies* 38 (2001): 407–447, Helga Leitner, Jamie Peck, and Eric S. Sheppard, eds., *Contesting Neoliberalism: Urban Frontiers* (New York: Guilford Press, 2007).

15. Trevor Barnes, "Cultures of Landscape: A Tribute to James Duncan" (panelist at the annual meeting of the Association of American Geographers, Washington DC, April 14–18, 2010); see also Trevor J. Barnes and James S. Duncan, *Writing Worlds: Discourse, Text and Metaphor in the Representation of Landscape* (New York: Routledge, 1992). For an introduction to cultural landscape studies, including a chapter-length bibliography of core and classic texts, see Paul Groth and Todd W. Bressi, eds., *Understanding Ordinary Landscapes* (New Haven: Yale University Press, 1997), and John Wylie, *Landscape* (London: Routledge, 2007). For more discussion of culture/nature, and literature on a range of ways people understand the environments around them through landscape approaches, see Braun and Castree, *Remaking Reality*, Heynen, Kaika, and Swyngedouw, *In the Nature of Cities*, George Robertson et al., eds., *FutureNatural: Nature/Science/Culture* (London: Routledge, 1996), W.J.T. Mitchell, ed., *Landscape and Power* (Chicago: University of Chicago Press, 1994).

16. Groth and Bressi, *Understanding Ordinary Landscapes*, is a particular exemplar of the landscape studies approach.

17. Roderick Neumann, *Making Political Ecology* (London: Hodder Education, 2005): 7.

18. For examples of the recognition of landscape studies within political ecological analysis, see Peter Walker and Louise Fortmann, "Whose Landscape? A Political Ecology of the 'Exurban' Sierra," *Cultural Geographies* 4.10 (2003): 469–491, Lynn Huntsinger and Nathan Sayre, "Introduction: The Working Landscapes Special Issue", *Rangelands* (June 2007): 3–4, Reginald Cline-Cole, "Knowledge Claims, Landscape, and the Fuelwood-Degradation Nexus in Dryland Nigeria," in *Producing Nature and Poverty in Africa*, eds. Vigdis Broch-Due and Richard A. Schroeder (Stockholm: Nordiska Afrikainstitutet, 2000), Paul Robbins, *Political Ecology: A Critical Introduction, 2nd edition*. (Hoboken, NJ: Wiley, 2012): see, e.g. 29–20, 98, 103.

19. Terry Eagleton, *Ideology: An Introduction* (London: Verso, 1991): 44, provides a good introduction to this concept of Foucault's, although see also Slavoj Žižek, "The Spectre of Ideology," in *Mapping Ideology*, ed. Slavoj Žižek (London: Verso, 1994): 1–33, for an argument pressing discourse analysis not to abandon "the problematic of ideology."
20. Noel Castree, *Nature* (New York: Routledge, 2005): 32.
21. Derek Gregory, *Geographical Imaginations* (Cambridge: Blackwell, 1984) and, more recently, the concept of "environmental imaginaries" captures the notion that these are contested normative ideals, see Richard Peet and Michael Watts, eds., *Liberation Ecologies: Environment, Development, Social Movements* (London: Routledge, 1996).
22. For literature on issues of amenitization of "working landscape," see James McCarthy, "Rural Geography: Globalizing the Countryside," *Progress in Human Geography* 32 (2008): 129–137, Steven Wolf and Jeffrey Klein, "Enter the Working Forest: Discourse Analysis in the Northern Forest," *Geoforum* 38 (2007): 985–998.
23. See Laura Taylor, "No Boundaries: Exurbia and the Study of Contemporary Urban Dispersion," *GeoJournal* 76 (2011): 323–339, doi: 10.1007/s10708-009-9300-y and Hannah Gosnell and Jesse Abrams, "Amenity Migration: Diverse Conceptualizations of Drivers, Socioeconomic Dimensions, and Emerging Challenges," *GeoJournal* 76 (2011): 303–322, doi: 10.1007/s10708-009-9295-4, for a review of these approaches. David L. Brown and John Cromartie, "The Nature of Rurality in Postindustrial Society," in *New Forms of Urbanization: Beyond the Urban-Rural Dichotomy,* eds. Tony Champion and Graeme Hugo (Aldershot: Ashgate, 2004): 269–283; Peter B. Nelson, James Nicholson, and E. Hope Stege, "The Baby Boom and Nonmetropolitan Population Change, 1970–1990," *Growth and Change*, 24 (2004): 526–544.
24. See also Chapter 3 by Richard Judd.
25. Daniel Brown, Kenneth M. Johnson, Thomas R. Loveland, and David M. Theobald, "Rural Land-Use Trends in the Coterminous United States, 1950–2000," *Ecological Applications* 15 (2005): 1851–1863, Arthur C. Nelson and Thomas W. Sanchez, "Debunking the Exurban Myth: A Comparison of Suburban Households," *Housing Policy Debate* 10.3 (1999): 689–709, William E. Riebsame, Hannah Gosnell, and David M. Theobald, "Land Use and Landscape Change in the Colorado Mountains I: Theory, Scale, and Pattern," *Mountain Research and Development*, 606 (1996): 395–405.
26. The ideology of nature in exurbia is influenced by a particular white Anglo-historical trajectory. See, e.g., Carolyn Merchant, "Reinventing Eden: Western Culture as a Recovery Narrative," in *Uncommon Ground: Toward Reinventing Nature,* ed. William Cronon (New York: W.W. Norton & Co., 1995): 132–159.
27. David Harvey, *Justice, Nature, and the Geography of Difference* (Malden, MA: Blackwell Publishers, 1996), Lisa Sun-Hee Park and David Naguib Pellow, *The Slums of Aspen: Immigrants vs. the Environment in America's Eden* (New York: NYU Press, 2011).
28. Farr, *Sustainable Urbanism*, and Roseland, *Toward Sustainable Communities*.
29. Witold Rybczynski, *A Clearing in the Distance: Frederick Law Olmsted and North America in the Nineteenth Century* (Toronto: Harper Collins, 1999), Paul H. Gobster, Joan I. Nassauer, Terry C. Daniel, and Gary Fry, "The Shared Landscape: What Does Aesthetics Have to Do with Ecology?" *Landscape Ecology* 22 (2007): 959–972.

30. Thomas Dietz and Paul C. Stern, eds., *Public Participation in Environmental Assessment and Decision Making* (Washington, DC: National Research Council, 2008).
31. A considerable body of work traces the tensions between egalitarian ideals and ethics associated with American-style middle-class suburbs and the exclusions involved in those landscapes, based on race, class, gender, sexuality, and physical ability, see, e.g., Dolores Hayden, "What Would a Non-Sexist City Be Like? Speculations on Housing, Urban Design, and Human Work" in *Women and the American City,* eds. C. Stimpson, et al. (Chicago: University of Chicago Press, 1980): 167–84; Iris Marion Young, "The Ideal of Community and the Politics of Difference," in *Feminism/Postmodernism,* ed. Linda J. Nicholson (New York: Routledge, 1990): 300–23; Harvey, *Justice, Nature, and the Geography of Difference.*
32. David D. Diaz, Susan Charnley, and Hannah Gosnell, *Engaging Western Landowners in Climate Change Mitigation: A Guide to Carbon-Oriented Forest and Range Management and Carbon Market Opportunities,* General Technical Report, PNW-GTR-801 (Portland, OR: U.S. Department of Agriculture, Forest Service, Pacific Northwest Research Station, 2009).
33. For more discussion on the power of ideology mobilized in this way, refer to Žižek, "The Spectre of Ideology," where he writes (p. 23), "it is not possible to isolate any 'objective' social process or mechanism whose innermost logic does not involve the 'subjective' dynamics of class struggle; or—to put it differently—*the very 'peace', the absence of struggle, is already a form of struggle,* the (temporal) victory of one of the sides in the struggle" [emphasis in original].
34. Kirsten Valentine Cadieux, "Competing Discourses of Nature in Exurbia," *GeoJournal* 76 (2011): 341–363, doi: 10.1007/s10708-009-9299-0.
35. Klein, *Shock Doctrine,* Thayer, *Gray World, Green Heart,* Harvey, *Justice, Nature, and the Geography of Difference,* Cronon, "Riddle of the Apostle Islands."
36. Leichenko and Solecki, "Exporting the American Dream," Martin Regg Cohn, "In Shanghai Suburbs, Bigger's Better; Canadian Architects Make Their Mark; Some Criticized for Forsaking Tradition; Sprawl Spreads East," *Toronto Star,* October 17, 2004.
37. See, e.g. Walter Firey, "Ecological Considerations in Planning for Rurban Fringes," *American Sociological Review* 11 (1946): 411–23.
38. The success of the television show Mad Men—set in this landscape in this era—may also demonstrate the important role of an ambivalently dramatized exurbia in the popular imagination of this era. And the way almost all citations discuss Spectorsky's "coining" of the term "exurbanite" illustrates some of the boundedness in how discourses function!
39. A. C. Spectorsky, *The Exurbanites* (New York: J.B. Lippincott Co., 1955): 261.
40. A collection that provides ecologists' views of the difficulty of incorporating humans in ecology: M. J. McDonnell and S.T.A. Pickett, eds., *Humans as Components of Ecosystems: Subtle Human Effects and the Ecology of Populated Areas* (New York: Springer Verlag, 1993). For a convincing account of why political and cultural contexts are necessary to ecology, see P. Robbins, "Research Is Theft: Rigorous Inquiry in a Postcolonial World," in *Philosophies, People, Places, and Practices,* ed. G. Valentine and S. Aitken (Thousand Oaks, CA: Sage, 2005): 311–324.
41. Exurbia is a zone of ecological impact often used by biologists; for a broad treatment from this perspective, see David M. Theobald, "Landscape Patterns of Exurban Growth in the USA from 1980 to 2020," *Ecology and Society* 10 (2005): 32.

42. See also Gordon A. Bradley, ed., *Land Use and Forest Resources in a Changing Environment: The Urban/Forest Interface* (Seattle: Washington University Press, 1984).
43. See Taylor, "No Boundaries" for a discussion of counterurbanization, and see also, for example, Firey, "Ecological Considerations in Planning for Rurban Fringes," David Riesman, "The Suburban Dislocation," *Annals of the American Academy of Political and Social Science* 314 (1957): 123–46. For reviews of these impacts of exurbanization, see Richard A. Miller, "Exurbia's Last Best Hope: With Acreage Zoning, Towns on the Urban Fringe Are Trying to Stem the Tide of Urban Sprawl; but Zoning Is Only Part of the Answer," *Architectural Forum* 108 (1958): 94–97; Jean Gottman, *Megalopolis: The Urbanized Northeastern Seaboard of the United States* (New York: Twentieth Century Fund, 1961).
44. E. Melanie DuPuis and Peter Vandergeest, eds., *Creating the Countryside: The Politics of Rural and Environmental Discourse, Conflicts in Urban and Regional Development* (Philadelphia: Temple University Press, 1996); Owen Furuseth and Mark B. Lapping, *Contested Countryside: The Rural Urban Fringe in North America, Perspectives on Rural Policy and Planning* (Brookfield, VT: Ashgate, 1999); Ivonne Audirac, *Rural Sustainable Development in America* (Indianapolis: John Wiley & Sons, 1997).
45. For a comprehensive review of the exurban literature, see Taylor, "No Boundaries" and Cadieux, "Competing Discourses of Nature in Exurbia."
46. In an influential planning article highlighting the problems exurban development may pose for planning, "the lure of the countryside" was identified but not questioned by Richard Lamb, in "The Extent and Form of Urban Sprawl," *Growth and Change* 14(1983): 40—nor was the attraction of the rural lifestyle interrogated by Judy Davis, Arthur C. Nelson, and Kenneth J. Duecker in "The New 'Burbs: The Exurbs and Their Implications for Planning Policy," *Journal of the American Planning Association* 60 (1994): 45–59.
47. Ian McHarg, *Design with Nature* (Garden City, New York: Doubleday/Natural History Press, 1969); Randall Arendt, *Rural by Design: Maintaining Small Town Character* (Chicago: Planners Press, 1994). Both have great standing in terms of sound planning, but neither has been implemented in any more than token ways; see Chapters 8 (Sandalack and Nicolai) and 7 (Vachon and Paradis) for accounts both of the usefulness of this way of planning and of the difficulty in getting communities (even those who like the idea) to carry it through. However, conservation subdivision design is not without its problems, see, e.g. Patrick T. Hurley and Angela Halfacre, "Dodging alligators, rattlesnakes, and backyard docks: a political ecology of sweetgrass basket-making and conservation in the South Carolina Lowcountry, USA," *GeoJournal* 76 (2011): 383–399.
48. Cadieux, "Competing Discourses of Nature in Exurbia," Jobes, *Moving Nearer to Heaven*, John Darwin Dorst, *The Written Suburb: An American Site, an Ethnographic Dilemma* (Philadelphia: University of Pennsylvania Press, 1989).
49. Williams calls nature "perhaps the most complex word in the language." Raymond Williams, *Keywords: A Vocabulary of Culture and Society* (Glasgow: Croom Helm, 1976): 184.
50. Williams, *Keywords*, Smith, *Uneven Development*. Bruce Braun, Noel Castree, David Demeritt, Bruno Latour, and Paul Robbins have been particularly prolific and central figures in debates around the construction and production of nature.
51. In addition to Marx and Williams, Roderick Nash's 1982 *Wilderness and the American Mind* (New Haven: Yale University Press) was an important,

although not particularly critical, step in this field, and William Cronon, Lawrence Buell, and Terry Eagleton have played central roles in bringing aspects of this conversation to diverse audiences. Lefebvre's most influential works in English have been his 1974 *The Production of Space* (trans. Donald Nicholson-Smith, 1991, Oxford: Basil Blackwell) and his 1970 *The Urban Revolution* (trans. Robert Bononno, 2003, Minneapolis: University of Minnesota Press).
52. Heynen, *Architecture and Modernity*; Simon R. Swaffield, and John R. Fairweather, "In Search of Arcadia: The Persistence of the Rural Idyll in New Zealand Rural Subdivisions," *Journal of Environmental Planning and Management* 41.1 (1998): 111–127.
53. Walker and Fortmann, "Whose Landscape?"
54. Marx, *The Machine in the Garden* and "The American Ideology of Space" in *Denatured Visions: Landscape and Culture in the Twentieth Century*, eds. Stuart Wrede and William Howard Adams (New York: Museum of Modern Art, 1991): 62–78.

2 Bridges in the Cultural Landscape
Crossing Nature in Exurbia

Laura Taylor

EDITORS' INTRODUCTION

The study of ordinary landscapes is a focus of cultural geography and at its best can reveal societal attitudes and political choices in ways otherwise hidden from view. Where historical accounts of interesting places tend to focus on social histories and downplay the role of space within which change takes place, landscape study situates that space as a central character. In this chapter, the landscape of Churchville, an Ontario riverside village, is the central character in a narrative of exurban change. Laura Taylor, a planner and geographer, draws out the felt experience of sprawl, which is often difficult to discuss as something beyond mere nostalgia for the landscape sprawl has replaced. Recounting the history of settlement as the intertwinement of nature and culture, Taylor explains the exurban impulse to escape urban areas as a reaction to contemporary urbanization, which creates landscapes devoid of everyday relationships with nature.

In the chapter, Laura Taylor describes her encounter with the reconstruction of a bridge within a larger modernization of transportation infrastructure, a moment where the possibility for experiential engagement with the river landscape is lost. Taylor is an assistant professor in the Faculty of Environmental Studies, York University. A background in heritage planning has informed her approach to a central issue of exurbanization: the co-production of an idealized countryside landscape beyond the city's edge with the urban infrastructure needed to service the growing demand for this landscape.

Responding to the tendency for people to move to exurbia in part because they value everyday encounters with landscapes that retain more layering of nature and culture than modern urban planning tends to allow, Taylor also calls into question the efficacy of the planning tools available to retain that layering and sense of place. The extension of urban infrastructure is based on rational planning for convenience and efficiency. The reaction to this planning mindset in this case was heritage designation of the old bridge and village setting that allowed people to express their values for the landscape in terms of their appreciation for its historical attributes, but which helped to freeze a place into a narrow idea of what that landscape could be.

Part of the way such designations operate is by *not* allowing expression of value for the ordinary landscape, the landscape of everyday experience that changes as people live their lives in it, the landscape that gathers the attachments that will become heritage—or will become hindrances to the smooth efficiency of modern planning. As an urban planner accustomed to witnessing change in landscapes such as the one discussed, Taylor challenges the reader to think about alternatives to both the rational approach and the local reaction to consider how negotiated compromise could help to break the stalemate between urban and rural. This case study is an example of the lack of compromise in Ontario's settlement history and in contemporary metropolitan planning. The new bridge (which could have instead been designed to express the landscape setting) unequivocally states that the need to move through the space is greater than the need to experience that space through an everyday engagement with nature and history.

The ideology of nature in exurbia that is the subject of this book is a version of the search for a way of living in the world that is "in-between" the urban and rural; a way of living that is not all one thing: city/modernity/culture on one hand or country/pastoral/nature on the other. But the story in this chapter suggests that perhaps there is no in-between landscape between the city and the wilderness, an irony that fuels the desperate search for just such a balance.

This chapter situates the book's conversation in the landscape approach to thinking about the ideology of nature. It provides an intentionally accessible introduction to the appeal of landscape ideology to think about exurban change and the difficulties facing contemporary planning at the exurban fringe.

—Kirsten Valentine Cadieux and Laura Taylor

■ ■ ■

"Every day trucks and cars take people and the things they need from place to place on roads. But who built the roads and how did they do it?"[1] So starts a picture book about huge earth-moving machines and giant trucks that is a favorite of my small boys. Even as an urban planner, I have always taken roads and highways pretty much for granted as a necessary, although often bleak and placeless, part of the city's system of movement of people and goods. It was not until I began to seek out road construction sites so that the boys could see the picture book machines at work that I began to truly appreciate their power to reshape the landscape and the frequency and scale of this reshaping, especially in newly urbanizing areas. With these awesome machines, earth is moved, massive structures are created from steel and concrete, and miles of sculptured asphalt surfaces are laid down.

Rarely are the roads that we come to watch being built completely new, rather they are two lanes being widened to four-plus-turn-lanes, or they are being straightened or leveled, or they are acquiring a new bridge. While the boys are riveted by the bite of the excavator and the ceaseless train of dump trucks, I find myself looking off in the other direction and thinking about the history of that particular site which is being ripped apart. What story is embedded in the landscape about the ecology of the place that has sustained it and about the people who have moved through, settled, and resettled the land? The lay of the land, the pattern of fields, the growth of native and exotic trees, the type of lumber used to build the fences, and the farmhouses and other structures and relics taken together tell a story. The landscape holds many clues about settlement histories and about changing natures and cultures and I cannot help but wonder what it means that we are willing to displace this heritage for more lanes of fast-moving traffic.

Not long ago I was driving (without the boys this time) through an area at the edge of metropolitan Toronto on the outskirts of Brampton, a large city of over 500,000 residents, once a pioneer village, then a suburban bedroom community, and now a full-fledged city in its own right.[2] Brampton is one of the fastest-growing places in North America, expected to double its population in the next twenty years, and is rapidly transforming its countryside to make room for new residents and businesses.[3] The area of Brampton's Credit River Valley through which I was driving is slated for development and is currently the subject of much heated urban planning debate because the planned development threatens the scenic rolling farmland of continuous historic settlement that is the home of many long-time residents and more recent exurbanites.

I was enjoying the winding drive down in the valley along the river through the town of Churchville, one of the original European settlements that grew up at the river's edge, and now a small, quiet area of houses nestled in the valley. Here the forest hides many of the houses and gives an intimate, close feeling in contrast to the open fields of corn up on the tablelands. The settlement area follows the river from north to south; the "out of town" houses further north are secluded, becoming more closely built in a ribbon along the road until the center of the village proper is reached to the south. I was coming around the bend of Eldorado Park, a spot for recreation since the railway in the mid-nineteenth century first brought tourists out from Toronto on the weekend for a swim in the river and dances at the fairground, when my drive abruptly ended at the noisy and dusty site of the reconstruction of the Steeles Avenue Bridge. The historic village road had been severed by the widening of the roadway and the construction of the massive new bridge over the river. I knew from previous visits that there was more to Churchville on the other, south, side of Steeles Avenue. There, the historic village center straddles the original river crossing made permanent almost 200 years ago. Sitting there, in my car, I could clearly see evidence of the trade-off that had been made: to improve the vehicular capacity of Steeles Avenue, to connect Brampton with its burgeoning suburbs, and to the global marketplace beyond, the landscape of Churchville had been severely degraded, losing much of its meaning by literally being split in two.

In the months following my trip that day, I was not able to shake this feeling of being profoundly moved by the transformation in the landscape. Witnessing the construction of the bridge opened up a new line of thinking for me. Here I could see stark trade-offs creating fixed results, imprinted on the landscape.[4] Clearly, with the upper hand is the choice that has been made to accommodate urban expansion—justified by the doctrine of economic growth, implemented by rational planning at a regional scale—a choice that has propelled the expansion of the road and construction of the bridge. The road and the bridge represent the expansion of transportation and communications networks and improved movement of people, goods, and services (see also Chapter 4, Blum); the roadside construction sign proclaims: "Building Together for a Stronger Ontario" with the implicit understanding that road improvements bring economic prosperity. The road and the bridge also satisfy our expectations of freedom of mobility when we're in the driver's seat: because we choose to drive as often and commute as far as we do, the infrastructure is provided to serve us.[5] On the losing side is what is being traded away: experience of nature, cultural heritage, and sense of place. These trade-offs are what I would like to pursue in this chapter.

In this book, *Landscape and the Ideology of Nature in Exurbia: Green Sprawl*, the essays address exurbia, the land of very low-density residential land use and monster homes on converted rural land at and beyond the edge of metropolitan areas. Of specific interest in this book are the motivations behind exurban settlement. Perhaps it is a reaction to the pace and chaos of modern life in the city that motivates a desire to move further and further away, to escape to a better life closer to nature. Perhaps it is just a desire for a bigger yard, scenic views, and fewer neighbors. If part of the motivation does have to do with being close to nature, then we should wonder why people feel they have to move out of the city to find it. Exurbia, I believe, is an example of a tendency of certain members of society to be attracted by the idea of living closer to a more natural nature than is thought to be found in the city, only to find that the impact of moving into it destroys it.[6]

With this complicated relationship with nature in mind, this chapter explores the current transformation of Churchville's landscape around the Steeles Avenue Bridge. Here is an exurban place at the edge of Toronto's metropolitan area that is undergoing urbanization. Neither urban nor rural, this visible, physical landscape is a space in between, a jumble of urban-serving and rural-seeking uses that can only exist in proximity to the city. Yet, here, as elsewhere, one finds strongly held perceptual differences between country and city that bubble to the surface when visible changes to the landscape are made. Hundreds of new homes and businesses are planned to be built in this next and final urban expansion area within the municipal boundary of the City of Brampton, along with the stores, restaurants and infrastructure (of which the Steeles Avenue Bridge is one piece) to serve new residents and workers. Ideas of landscape are sharply contested here, as a largely rural landscape of farms, orchards, and scattered homes is transformed. Where, until recently, the scenic river valley and farmland attracted people who sought to move out of the city, the rebuilding of the bridge signals a major change in how this landscape will be experienced. Here, with the bridge in sight, is evidence of a place where everyday encounters with nature will be replaced with a quick drive back and forth to Brampton.

The city's edge is a landscape of intense change between city and country, two very different ways of inhabiting the landscape. Unfortunately, change is often seen as a trading down of rural productive and amenity values for crass, modern urbanization (in the form of cookie cutter houses and big box stores). I do not think it has to be this way (in attitude or physical design) but it has proven difficult to achieve some kind of balance between the two.[7] This study of the Steeles Avenue Bridge in Churchville is not merely to point out what I perceive to be its failure in fitting into the landscape, but to explore what could have been. There are two main points that I want to make. The first is that the Steeles Avenue Bridge by its insensitive transformation of Churchville's landscape has upset pre-existing relationships between the river and community and thoughtlessly precludes the possibility of future ones. Secondly, because everyday transformations such as the Steeles Avenue Bridge restrict connections to the landscape and nature, they trigger the desire for people to move further out. These failures in planning and engineering such typical everyday landscapes are what compel the exurban impulse—the desire to escape, which this book is about.

BRIDGES IN THE CULTURAL LANDSCAPE

The study of landscape is a conceptual approach to make sense of the natural and cultural processes that converge at a point in time in a particular place (such as Churchville) to produce the material environment that we see and touch and experience. I find it useful to think of landscape "as enmeshed within the processes which shape how the world is organized, experienced and understood."[8] In this way, the landscape can provide powerful evidence of cultural influences and political decisions. Geographers Trevor Barnes and James Duncan discuss landscape as a "medium of particular discourses." They contend that landscape can be read in many of the same ways as literary text where many different lines of thinking and concepts come together as an artifact (a book or a landscape) that is then interpreted (read) by the people who use it. So landscape is not only evidence, but also part of an ongoing dialogue about living in and making sense of the world. The way people read a landscape is based not only on clues in the material environment about cultural expectations for behavior, but by the culturally conditioned imagined landscapes in their heads. How we experience landscape, how we perform in that landscape, and how we talk about our experiences together inform understanding of the relationship between nature and culture—landscape is "one moment framed by and constitutive of larger discourses."[9] In other words, the landscape provides clues to how we think about things, including cultural attitudes and beliefs if we look at how landscape is created over time and then how people use it and make sense of it. Churchville's story is written in its landscape, with the Steeles Avenue Bridge being one of the most recent chapters.

Some readers may think that experience of landscape is not important. I recently had a scuffle with a planner friend that left both of us confused. I was discussing the importance of urban design guidelines for a new suburban retail strip. He said, why bother? Through his experience developing sites for a national chain of big box hardware stores, he found that the landscaping just died soon after it was put in, and anyways, who cares about making the parking lot for a hardware store inviting? My defense was that every transformation in the landscape should be designed with its impact in mind: aesthetic impact (beauty, but negotiated), connection to nature and the landscape, and an attempt to anchor ourselves in the world as a counterpoint to the frustration and the chaos that is at times the experience of modern life. The wasteland of Home Depot parking lots is what triggers the escape to somewhere else (to punctuate this, imagine the impact of a billboard ad for a beach holiday or cottage resort in a parking lot such as this). For me, landscape is the everyday environment, our surroundings, which we see and experience, *including* Home Depot parking lots—and arterial roads and bridges. Landscape is not just a pleasant picture of somewhere else painted in a frame. It is a real place that is created by many minds and hands, a place that is full of symbols of their intentions for the people who will use that place, such as a paved walkway or a stop sign, and other symbols, such as blank concrete walls, that people may not consciously "see," but which will nevertheless make their impacts felt on feelings and moods. Kevin Lynch, a famous professor of urban studies, tried to get at this aspect of environment by calling it "sensuous," as in people's "sensing" of not only when and where to walk and drive, but what time of the day and year it is, and the feeling of being connected to places in space and time. A blank concrete wall provides few such clues (which is why it invites graffiti). How a particular transformation fits and relates to an existing landscape can support or frustrate sense of place. "Landscape," J. B. Jackson, who pioneered the study of everyday landscapes through his teaching of the study of landscape at Harvard and Berkeley for many years, said, is "something more than beautiful scenery," it is something that we are a part of, something from which we derive our identity.[10]

What is experienced as perhaps superficial the first time you visit a place (a bridge perhaps), tends to have deeper meanings with each subsequent visit. Each person will make his or her own meaning in a place, which may be a stable meaning if that person re-enacts similar behaviors in that space over time, or s/he may have a more fluid relationship with that space as the body and mind change and experience new things over time. This relationship becomes very interesting when many people have the same reaction to a place, when they find similar meanings over time, and then begin to modify the environment through their behaviors in and conversations about that place so that a landscape that began by informing *them* becomes changed through their interpretations. Studying the history of a particular landscape allows us to see how these various interpretations are created, and how they change and morph over time.[11]

I took the opportunity of being so moved by the reconstruction of the bridge to research the history of Churchville in the context of the local area's historical development along the Credit River. The historical texts, maps, and pictures found in the local archives and locally produced histories provide some perspectives on Churchville's historical growth. A heritage conservation district study looked at changes in the community based on the architectural record. In addition, I have begun conversations with area residents about their experience of the landscape and their memories of changes that have occurred. And my own recent history with the landscape, punctuated by the drive into the construction of the bridge that day, but also subsequently layered by many additional drives and walks in the intervening months, has informed me as well.

Churchville's history is unique in the eyes of its residents, but it has a very typical southern Ontario settlement history. Like many relatively recent European settlements in North America, this area today is the setting of some very rich and still highly visible layers of evidence of previous occupation related to the Credit River. Focusing on the crossings of the Credit River in the village of Churchville helps to explore the layering of historic time in the landscape, revealing the traces of how the landscape has influenced settlement and how individuals have shaped and inhabited the landscape over time.

A HISTORY OF THE BRIDGES OF CHURCHVILLE

The recently completed Steeles Avenue Bridge is one of four bridges in Churchville that span the Credit River. It is generously proportioned with six lanes and a median. It serves as a major connection between the newly urbanizing areas of Brampton and its employment centers as well as with the rest of the Toronto metropolitan area. It is a piece of the city, a symbol of the future invasion, constructed quickly and surreptitiously as if during the night by scouts and engineers in advance of the infantry of single detached homes and phalanxes of big box stores.

From this bridge, you can see the past: a short distance to the south, in the center of this very modest little town, the single-lane steel pony-truss bridge built in 1907 crosses the river. This is the focus of the oldest part of Churchville and is now incorporated into a heritage conservation district.

Looking the other way, to the north, but far enough away to be out of sight of the Steeles Avenue Bridge, there is a single-lane concrete bowstring arch bridge built in the 1930s, now starting to crumble. Further north still, the Eldorado Park Bridge is made of wood, more recently reconstructed and used mostly by picnickers. Two other known historic crossings are now faint ruins and no longer visible. The first is at the site of the original pioneer sawmill, and the second used to connect from the center of town up to the old Steeles Avenue schoolhouse.

The Credit River is nearly 100 kilometers long from its headwaters near Orangeville in the north to where it empties out into Lake Ontario at Port Credit in the city of Mississauga. The river valley is a significant feature in the otherwise gently rolling landscape of southern Ontario, the deep valley created by meltwater some 14,000 years ago when the glaciers retreated.[12] In the Churchville area, the river channel now flows through a broad flat floodplain. It has been greatly altered over the years by the creation of the settlers' mill dams in Churchville and other communities up and downstream and by frequent flooding largely curtailed by the construction of flood prevention measures instituted after the creation of the Credit Valley Conservation Authority in the 1950s.[13] The river channel was recently redirected again as part of the reconstruction of the Steeles Avenue Bridge.

Archeological evidence shows that people have been living in the Churchville area for about 7,000 years. The remains of an Iroquois settlement on the river date back to between 900 and 1,500 CE. The present landscape holds no clues for the casual observer of how the landscape was changed by this inhabitation over time. Archeological records of past finds or potential areas of aboriginal settlements are not available to the public, so as to prevent looting and disturbance. Certainly the river would have been a major resource for the Iroquois and others, as well as a thoroughfare when frozen in the winter.[14]

The village of Churchville was built around a sawmill that was constructed by Amaziah Church in about 1815 before the area was purchased from the Mississauga Anishinaabe tribe (who had conquered the Iroquois in southern Ontario around 1700) and subsequently surveyed. He was a "late" United Empire Loyalist from Virginia fleeing the "rough justice" meted out by victorious American patriots to loyalists of Britain and the King. His "new" land was part of the "raw, backwoods colony" that was the British-held province of Upper Canada, a colony struggling to rebuild after successfully repelling the American invasion of the War of 1812. The mill was built on a branch of the Credit River in a heavily forested area where the valley floor was relatively wide and flat. When the surveyors made it through the forest a few years later from York (the original name of present-day Toronto) they indicated that the Third Line road would have to curve around to bypass the mill—according to the grid of the survey, the road right-of-way would have run straight up the river through Church's mill. The survey was completed in 1818, Church added a saw and grist mill in 1819, and the lands in the immediate area of the mill were purchased by the first white settlers in 1821 and 1822, beginning in earnest the major reshaping of the landscape from forests of old-growth trees to open farmland. Church's mills were among several that were constructed up and down the river. The survival of early settlers depended on the river to provide power to their mills and for food (fishing and game hunting).[15]

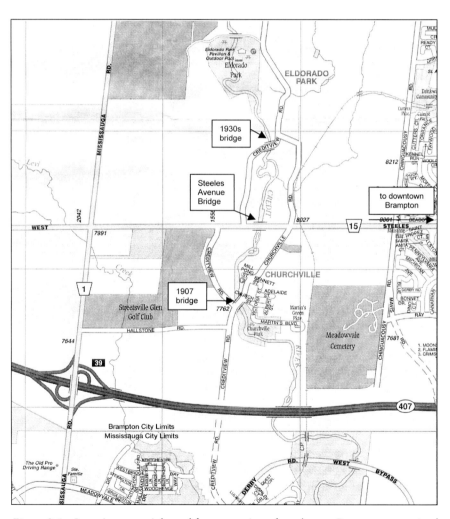

Figure 2.1 Location map. Adapted from a section of south-west Brampton contained within the Canadian Cartographics Corporation copyrighted Brampton Street Map

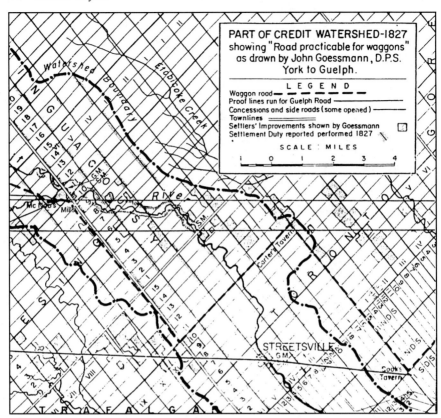

Figure 2.2 Settlement in Churchville, 1827. Source: Province of Ontario, Department of Planning and Development, Credit Valley Conservation Report 1956. Parts I–VI. (Toronto: Queen's Printer, 1956).

The town grew quickly and was at its peak in the late 1840s and 1850s, reaching a population of 400, never to be as vibrant again. There appear to have been five crossings of the river in Churchville's heyday, all originally made out of wood: one major crossing where the 1907 bridge is today, one at the site of the mills, one further north leading to the schoolhouse, then north of Steeles Avenue a crossing at a narrow meander, and one at Eldorado Mills. With the river at its heart, the community of Churchville grew with a road and settlement pattern that for the most part did not follow the survey grid. If a road had followed the grid along the Third Line north-south public right-of-way, it would have been underwater for a distance; instead, the public road split east and west of the river, joining up to follow the surveyed right-of-way heading to the north and to the south, where the river meandered away, allowing the grid to be followed. The east-west public right-of-way that would much later become Steeles Avenue was not opened either; again, the grid was abandoned in favor of the river roads.[16]

The demise of the village began with the decline of the lumber industry in the mid-1800s as the forest here and in Ontario generally was depleted. Then came the railway era, the line bypassing Churchville on its way out from Toronto, with a stop instead in the village of Brampton. Soon after, Brampton was chosen as the seat of Peel County government, another blow. Where once the two prosperous villages (no different from others in Toronto's hinterland) were isolated from each other (a day's journey in good, dry weather on the muddy, stump-strewn pioneer roads), the advent of the railway brought industry and more settlers to Brampton and the county government brought the courthouse and jail, and with them more jobs. In the mid-1870s, a major fire destroyed much of Churchville, and the damaged buildings were not rebuilt. In 1877, the Credit Valley Railway was constructed through Peel County with a stop just at the edge of town, but it arrived too late to have much of an impact (see Figure 2.3 showing three bridges in south Churchville; see Figure 2.4 showing the heavy hand of the grid on the landscape even where the road allowances were unopened). The Canadian Electric Line streetcar was opened in 1917 running between Guelph and the Junction in Toronto with a stop near the center of Churchville and another a little further north to Eldorado Park. The line was popular in the summer and many people built cottages along the river and visited the park in Eldorado to go swimming and enjoy the merry-go-round and Ferris wheel. Unfortunately, the line closed in 1931, leading to further decline. The last exodus from Churchville occurred in 1959 with the cancellation of the Avro Arrow contract, Canada's first (and last) supersonic fighter jet, which was built close by at the edge of Pearson International Airport to the southeast. According to the heritage conservation district study, about half of Churchville's population were among the 14,000 people laid off. It was not until the 1980s, when the urban areas of the City of Brampton to the east and the City of Mississauga to the south had expanded to the extent that Churchville and the surrounding countryside were within commuting distance, when a resettlement of the village began.[17]

The present-day landscape of Churchville bears strong traces of the pioneer settlement and successive change. The landscape also hides some of its history, allowing the observer to forget the experiences of previous occupants. River crossings are like this. Bridges seem enduring, seem to fix time if they are memorable, but "lost" bridges, now artifacts submerged or replaced, take their stories with them.

In the central part of Churchville today is the bridge that was built in 1907 (see Figure 2.5). It is a single-lane steel pony-truss bridge that still looks pretty much like it did when it was built (although its abutments were rebuilt in the late 1970s after the bridge had been closed for some time due to a particularly

42 *Laura Taylor*

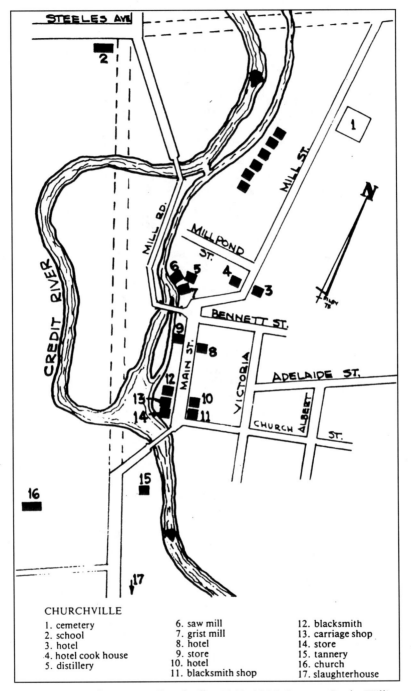

Figure 2.3 Settlement in Churchville, 1848–1856. Source: Cook, William E. Meadowvale & Churchville. (Cheltenham: Boston Mills Press, 1975).

Bridges in the Cultural Landscape   43

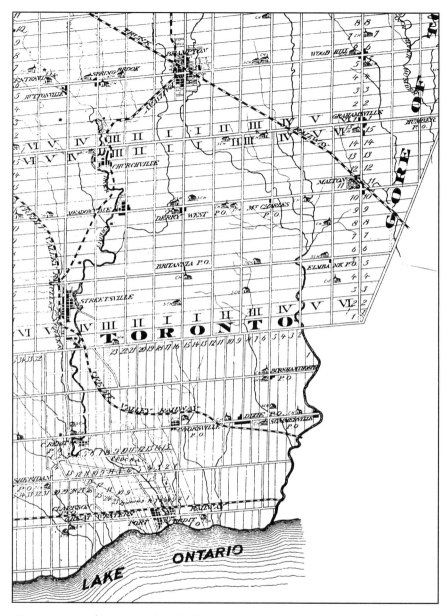

Figure 2.4 Settlement in Churchville, 1877. Source: Illustrated Historical Atlas of the County of Peel, Ont. Original 1877. Reprint edited by Ross Cumming. (Port Elgin, Ont.: Walker & Miles, 1971).

crippling flood). This crossing of the Credit River has existed since the first survey of the area in 1819 and the steel bridge replaced a succession of wooden bridges spanning the river in this location.[18] Recently refurbished with a new coat of paint (a startling shade of non-receding green), it has been upgraded with roadside curb markings reading 1998 (see Figure 2.6). It is a single lane bridge, not too long, and, as I discovered, it is comfortable to walk across, although I felt it necessary to keep an eye out for the fairly frequent automobile traffic. The sound of rushing water fills the air. The bridge spans the river from bank to bank; nestled down in the floodplain of the river valley I felt secluded (had I escaped?) from the outside world.

Steeles Avenue finally crossed the river from east to west in the 1960s with a two-lane three-span steel girder bridge,[19] the grid of the survey *still* colonizing a century and a half later. Although shown as a road allowance since the first survey in the early nineteenth century, Steeles Avenue, heading west from the suburbs of Brampton, had always ended at the river, requiring travelers to head south to cross at the 1907 pony-truss bridge or, later, to the north to cross at what is now the 1930s cement bowstring bridge (see Figure 2.7). Churchville's settlement and its movement patterns had been shaped by the peculiarities of the river and the landscape. With the river at the center—the focus of the mill and related general store and hotel—access was by means of north-south roads, one to the west of the river and one to the east. The 1907 pony-truss bridge was a crossroads to the south and the bowstring bridge was at a crossroads to the north. The river was the organizing element of the settlement, so much so that in the wintertime the roads were less favored than the frozen river itself for getting around.

Figure 2.5 Inaugural trip on the new steel bridge, 1907. Source: Peel Archives.

Bridges in the Cultural Landscape 45

Figure 2.6 Crossroads at the Churchville Bridge looking east. Source: L. Taylor.

Figure 2.7 North Churchville bowstring bridge. Source: L. Taylor.

The 1960s Steeles Avenue crossing disrupted the original patterns of movement as they related to the landscape. The original two-lane 1960s bridge was reportedly still in good shape when the new present-day high-capacity bridge was built. The new bridge is a 6-lane-plus-center-island, $2.5 million work of engineering, which lifted the bridge up out of the valley to improve sight lines (leveling out the grade change dipping down into and up out of the valley), and realigned the river to flow in the middle of the channel instead of meandering off to the west side (see Figures 2.8 and 2.9). In reconstructing the bridge, a schoolhouse that had marked the southwest corner of Steeles and Creditview Road since the 1870s was demolished (despite its heritage designation) taking with it the story of the historical Third Line route south to the village.

Churchville today is split into two communities: one north of Steeles Avenue and one to the south that is designated as a heritage conservation district. North of Steeles Avenue, the area has a typical rural Ontario atmosphere that one might expect to find: a scattering of contemporary or updated homes of exurbanites interspersed with less grand and less manicured properties. There are no commercial uses here, just residential properties and Eldorado Park (still a large regional park centered on the river). South of Steeles Avenue is the village of Churchville, which has a vibrant feeling with some new homes on existing lots, some lovingly restored older homes dating back to the original settlement, and some eclectic homes locked in the never-ending process of renovation. It is quiet here, with the sound of the river ever-present, under a canopy of large trees. However, as always, changes are occurring. Based on conversations with local residents, I suspect the village is enjoyed with a fishbowl awareness of the changes occurring around it. The entire area including Churchville is part of the large, rapidly urbanizing municipality of the City of Brampton. Highway 407, a major interregional expressway, was recently constructed south of the village, severing Churchville's historic ties

Figure 2.8 Looking north at the Steeles Avenue Bridge. Source: L. Taylor.

Figure 2.9 South side of Steeles Bridge, looking west, showing the river flowing in its new channel. Source: L. Taylor.

to Meadowvale village, another old neighbor. A large subdivision of single detached homes has been constructed to the west, and another to the northeast. One local resident referred to the community as the "hole in the doughnut" of suburbanizing Brampton.

This "hole in the doughnut" is presently protected by a heritage conservation district designation that limits new development—I am told there are only 5 or 6 lots left—and limits changes in public spaces including the 1907 pony-truss bridge and the roadways. Anticipating the oncoming suburban wave, the people of Churchville sought to protect the village from redevelopment pressures by adopting the designation of "heritage conservation district," which controls the changes that can occur in the landscape and buildings in the area. The reasons given for the designation were: historical significance as an important nineteenth-century milling center; scenic merit as a particularly attractive rural landscape setting in the Credit River Valley; architectural interest in the surviving rural, vernacular, frame, and masonry buildings scattered throughout Churchville; and, archaeological significance derived from a number of pre-historic archaeological sites. The idea behind heritage preservation is that heritage is everything that has come before us, everything that we as a society have inherited—buildings, literary works, artifacts, and oral traditions. In the physical environment, that which we inherit includes the landscape: the remnant natural environment (that which we did not or could not change) as well as everything that has been produced as a result of human occupation of the land.

The "landscape" is the ordinary environment that we move through in our everyday life. It is comprised of all of the material things that we see and experience whether in an urban or rural setting. This includes topography, trees and other plants, and the structures that people have added, including all types of buildings, settlement patterns, pathways and roadways, and all other material bits of human settlement. The landscape in its entirety is a living, evolving artifact created by cultural processes (i.e. people and their ideologies) over time. It is tangible and visible and provides us with a text from which to better understand the culture that created it. But in Churchville's case, in the newly designated "Hamlet of Churchville," the notion of landscape as heritage ends at the borders of the district designation. There was no attempt made in the design of the abutting subdivisions to mesh with the village through lotting arrangements or movement patterns (see Figure 2.10). The subdivision to the northeast, at the boundary it shares with the heritage conservation district, has attempted to be sympathetic in its design where it abuts Churchville, but it does not try to be part of it. To the west, the new development has not been sensitive to its setting, although it markets itself as doing so. Figure 2.11 is an advertising flyer for the new development that flaunts its proximity to natural and cultural heritage. The new subdivisions are separate from, rather than integrated with, the river valley and the village. The backyards of the

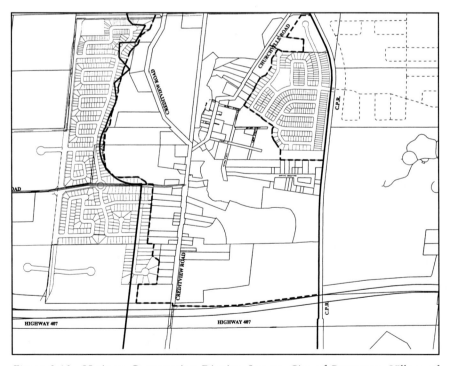

Figure 2.10 Heritage Conservation District. Source: City of Brampton, Village of Churchville Heritage Conservation District Study: The Proposed District Plan (1990).

Bridges in the Cultural Landscape 49

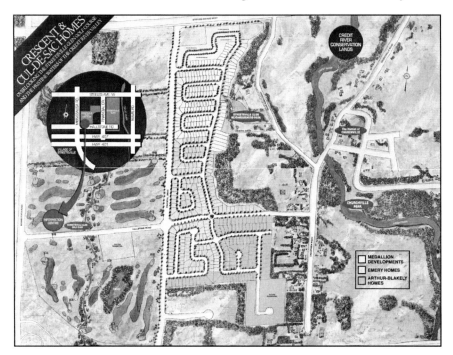

Figure 2.11 Illustration from brochure for new subdivision to the west of Churchville. Source: Guidelines Advertising Ltd.

homes that immediately abut the lip of the valley (a land use planning line determined by the "top of bank") are fenced off to limit environmental damage by new homeowners. The residents of these new homes, so separated, will not easily make attachments to the river and the landscape.[20]

Heritage conservation districts attempt to freeze a particular shared meaning. The landscape setting is "read" as desirable by the people who frame the legislation (the community, the area planner, the heritage consultants, the local architectural conservation society, and municipal council). Transformations that occur as a landscape suburbanizes are perceived to be such a trading-down that it is better to freeze a place in time rather than allow it to continue to modernize and evolve. Cultural geographers James and Nancy Duncan studied the landscape conservation practices in the affluent suburb of Bedford, New York and concluded that this "aestheticization" of landscape is exclusionary and homogenizing. To agree that a landscape is picturesque and that its scenic value is worthy of protection belies the "necessary interdependence of its very existence with other processes (economic, political, or social)."[21] Understood as a text, the landscape can be seen as a way of producing knowledge and as an "archive of official memory." In the context of heritage conservation, that which is chosen to be conserved becomes the community's memory of its past. The sites and buildings and artifacts that are chosen for retention, and the way in which they are retained, have an enormous impact on the way that the community collectively remembers its past and it is often idealized—preserving heroic stories of the elite members of the community and skipping stories of other people and times. It also artificially (and wrongly) limits the space and the landscape that is privileged as heritage—the landscape of the village is heritage while the landscape of the new homes is not.

Nevertheless, the natural setting of the river and the heritage character of Churchville are the stuff of marketing dreams for the developers of the new adjacent subdivisions. The promotional literature is photoshopped in hues of green and promotes the river and hamlet setting. A brochure reads: "Secluded and pristine. Overlooking the flowing waters and the beautiful lush landscape of the Credit River valley, Streetsville Glen is an exclusive upscale world in perfect harmony with the natural environment." The desire of people to connect with nature is captured neatly here. For those who can afford it, "harmony" can be bought.

But even if this promise is what brings people here in the first place, the experience of the material landscape will break it down over time. As is the case with most new home developments, the pre-existing agricultural landscape was removed and re-graded before house construction started. Except at the back fences of the most expensive ravine lots, the new landscape of the subdivision will not relate to any natural or cultural landscape heritage that came before. Due to poor design, it will not be easy to walk into the valley or the village from the new homes. The spaces of Churchville will co-exist, but they will not relate. People will drive over the new Steeles Avenue Bridge on their way to work and shopping, unaware of the promise of nature flowing by in the river valley below. If there is no daily engagement with the landscape of the river and the valley, if people do not walk over or under the bridge, or into the valley, the promise as advertised is broken.

The effect of societal ideologies and imagination on the landscape of Churchville and its surrounding subdivisions is profound. These are real and imagined landscapes at the same time. And they both respond to the promise of the middle landscape, that middle ground between city and country. Churchville as a heritage conservation district every day retreats further from the present and represents an idealized past whose historical story I have briefly recounted in this chapter. Enveloped by Brampton's suburbs it is, and will remain, the rural past. It is defiantly anti-city in its legislated rejection of further modernization, change, or diversity. The new suburban homes came packaged with the ideal of a retreat awash in history, connected to nature in the countryside. Both require a sustained imagination that suppresses the reality of the urban support networks, displaced wildlife, raw labor, and exclusion that makes them possible.

I began this chapter contemplating the Steeles Avenue Bridge, a piece of one of those urban support networks, part of the interregional transportation network northwest of Toronto. But the bridge is also intimately local.

As part of the local landscape, it influences the landscape of Churchville's heritage district, the exurban residential areas to the north and the new suburban areas all around. Having provided the landscape context of the bridge in these previous few pages, I would like to focus on the bridge as an artifact of the cultural landscape again, more closely this time, to see what questions arise.

BRIDGES ARE AN INTERSECTION OF NATURE AND CULTURE

Bridges tell a story about contemporary culture, where personal mobility is valued over local landscape heritage and sense of place. A bridge is not merely a structure made of wood or stone or steel, a bridge is an artifact in and of the cultural landscape: material, tangible, and above all, publicly accessible for reflection and contemplation. The bridge is a symbol representing the imagination and technological triumph and sheer determination of the culture that built it: the Golden Gate Bridge in San Francisco, the Brooklyn Bridge in New York City, and the Millau Bridge in southern France are such symbols on a grand and famous scale. But the act of bridging has an even deeper meaning. A bridge is a point in space, a moment in time, where the character of cultural processes is played out. The bridge represents the intersection of the land and people, where the natural processes of landscape formation and the cultural processes of settlement meet. The bridge is a snapshot that is laden with meaning, if we choose to stop and look. It is a glimpse of society at a given moment in time: its values, its priorities, and its failures. The bridge itself is a work of engineering representing the civilization that constructed it and the society that uses and maintains it. As a work of architecture, it expresses the imagination and sensibilities of that same society.

And for what is the bridge being used? The movement of goods and people across the bridge (the vehicles they use, the speed at which they travel, and the direction in which they are going, depending on the time of day, the time of week, and seasonal variations) gives evidence about the differences of people who occupy and move through the space. From a vantage point on a bridge, a bridge that is built to allow travel over a river or the sea, the movement of the water can be seen and heard, perhaps at high water, perhaps at low, marking the influence of the broader hydrogeological processes at work. The layers of geologic time are sometimes apparent in the banks where the water meets the land and this illustrates the previous ferocity or calm of the water as it has eroded the banks or has filled in with new land over time. (Bridges cross other flows than water, most often other roads or pathways, and allow similar contemplation of different kinds of processes).

Where once the temperament of the landscape and natural forces—the topography, the soil, the seasonal textures, the weather, the strength of the river—decided when, where, how, and at what speed people could cross, now such strength has been subdued. Impermanent bridges were first built by those seeking to cross by felling tree trunks and using rafts, then timber bridges and finally permanent stone bridges. Steel and concrete were introduced scarcely more than a century ago. Today the new six-lane Steeles Avenue Bridge in Churchville dominates the landscape and is a technological triumph over the landscape. No matter what the seasonal or daily conditions, the bridge is there to use. But the alteration in sense of place has been enormous. The simplest bridges made of rope and rough-cut timber can carry people and goods over the river; those travelers might be well aware of the elevation, their own weight and that of others on the bridge, the wind and the weather, and of the view from their "unnatural" vantage point. Technologically complex bridges now can erase much of the feeling of traveling through thin air. They do not allow views of the landscape from car windows because of concrete "safety" structures, and have a very heavy footprint on the land. Speed is now limited only by a person's ability to control a vehicle (and fear of getting caught). The changed relationship between landscape and technology is made apparent through the medium of the bridge.

As with other structures of modern life—buildings, factories, roads, houses—the bridge is a product of larger processes of making products, the flow of money, and the movement of goods and people that make everything else possible in contemporary Canadian society.[22] The organization of space around this river crossing has changed then, over time. The geographical space of that crossing defines "the relationships between people, activities, things, and concepts."[23] The spaces that are created by these larger processes define to a great extent what can take place in those spaces, making it difficult to do anything else, thus reproducing the dominant system of modernist planning against which so much of exurbia is defined. The bridge, a product of the processes that require the movement of people and goods over the river in that place, once built, will then frame the experience here. The size and investment in the bridge assume that the space is fixed this way for some time to come. The architecture of the bridge is coded to ensure that the high-speed movement of cars and trucks is all it will be used for. The Steeles Avenue Bridge is an artifact of the mathematics of population and employment forecasts, of spatial calculations for the future location of jobs and housing, of the volume of vehicle trips at peak periods that needs to be accommodated during the life of the bridge. If not pushed to do otherwise, science and engineering will not only determine the number of lanes and the posted speed, but will design a structure based on the safest vehicle sight lines and turning radii, as well as controlling water flow and flooding below. I'm not nostalgically suggesting that we return to old ways, that a wood bridge is preferable, but that we could have (in this and other instances) put our technology *also* toward the creation of a delightful bridgescape which would heighten the experience of the river and the landscape for the local community and for passers-by.

In the absence of a push for aesthetic values, the spaces of modern urban areas are often criticized as being inhospitable.[24] This bridge, as reconstructed, is certainly part of the problem. As a standardized urban element in a rural area, the bridge does not try to respond to its setting, but drops itself in. As a structure that will be used every day, it has utility and its function can be read clearly: the smoothness of the asphalt, the generous width of the lanes, the concrete barriers all say "drive fast this way." There is no opening into the space of the river where an experience of nature, even a fleeting one out the window, or a spontaneous stop, might be possible. This is not a structure meant to elevate human experience. It is not a structure to which memories will easily adhere.

Churchville's story, like that of many other settlements in southern Ontario, is written in the landscape by the intersections where the grid of the early nineteenth-century lot survey met the curve of the natural feature of the river. The grid roads, defined as ruler-straight by the survey, had to bend to cross the river at a point of least resistance, first at a fordable point and later where a bridge could easily and safely be constructed. The first bridges were made of wood, swept away every spring with the swelling of the river. Bridges of steel and concrete are much more permanent, safer for those using them, but also defiant of natural forces: the river still rises and ebbs with the seasons, but goes largely unnoticed.

No longer is it the river that frames the experience here in Churchville, it is the bridge. No one who cherishes safety would experience the new Steeles Avenue Bridge on foot, given the speed of passing vehicles and the lack of any pedestrian-scale street fixtures that would encourage strolling and viewing. In a car on this bridge, one has no sense of moving over the river, of being suspended in mid-air as is the case with many delightful bridges—there is no grade change from solid land to the bridge structure; in fact the bridge from above feels and looks like land.

The view of the bridge from afar is interesting as it is an unadorned concrete modernist structure that has an aesthetic based on its unapologetic functional construction as an artifact of a major regional road. The bridge is the product of regional planning and regional planning is something of which Ontarians are very proud. Steeles Avenue is a regional road, an arterial owned and maintained by the Region of Peel (the former county). The Region would have known twenty years ago (or more) through regional population and employment forecasting that this countryside would be urbanized. The provision of a transportation system to accommodate long-term growth is clearly one of the *raisons d'être* of regional municipal government, virtually precluding any real debate over the relative merits of different alternatives to crossing that do not have this imperative in mind. Steeles Avenue, the concession road right-of-way identified in the 1818 survey, is not recognized as anything but a functional piece of this abstract system.

In Raymond Williams' words, "there is no case in which the priorities of a capitalist system have not, from the beginning, been built in," and

Churchville's case is no different. The original settlement of Churchville is a colonial product of a capitalist system based in Britain that exploited and remade the countryside in its own service. The building of the Steeles Avenue Bridge signifies that the capitalist mode of production is alive and well. The building of the bridge represents to regional government, exactly in line with Williams' critique, "so clear a profit that the quite different needs of local settlement and community are overridden."[25] And certainly, the bridge shows no compromise in this local landscape. The rise and fall of Churchville, from the industry of the original mill turning raw timber into lumber, is an example of the unevenhandedness of capitalism[26] over the years. The site of the mill, once humming with prosperity serving the British lumber market, is now derelict. The railroad passed Churchville by and the jet-assembly labor market came and went. Now a global market fuels the expansion of Brampton's suburbs and Churchville's landscape has been made over to house in-migrants from around the world to support Toronto's strong manufacturing and knowledge-based economies. The Steeles Avenue Bridge is a product of our society and stands as a stark reminder of the choices we have made, or let be made without us.

Is this where we want to be? The beneficence provided by technology could be put toward improving the experience of this landscape. Instead, the mechanical provision of a piece of infrastructure that is bleakly "city," contemplated and constructed in the same cultural moment as the "anti-city" heritage conservation district, encapsulates an inability to find a middle-ground in remaking our contemporary landscapes. If landscape is a text to be read, then the bridge, the village, and the subdivisions together tell a story about place and placelessness.[27] The juxtaposition of the bridge, the heritage district, and the new subdivisions exemplify ambivalent societal values based on our continuing struggle to make sense of the war between ideologies of city and country. Deliberate compromise between the modern and the pastoral in any one of these landscapes would have helped to break the tension between the two.

This chapter has so far presented the story of how a bridge came to be, in the context of the settlement of a particular place, focusing on the diminishing role of the natural landscape in shaping that history. My interest in the bridge stems from a broader desire on my part to understand the longing to escape to nature that permeates North American culture. No wonder people want to leave if the landscape of their community is littered with structures like this. No wonder people are against urban expansion if they are trading in a rural aesthetic of the green and pleasant countryside for this. The bridge is an agent of the desire to escape: finding no connection to nature here, no wonder people dream of a house in the country and a cottage in the woods.

The exurban choice is not often imagined as an in-between, but as a country choice in opposition to the city. Making an individualized decision to fulfill a desire to live in nature, to engage with it through the planting of trees and management of succession (as discussed by Cadieux in Chapter 10) is a choice, a "self division" Williams calls it, to divide time and energy between the country and the city. It may be a compromise, but without a full accounting of the costs. Hidden from the spectacular view are the processes that have made it all possible.[28] The Steeles Avenue Bridge draws me into this discussion because it is an example of an external impact of exurban growth (high-speed travel from the country through Churchville to the city). At the same time, it has a negative localized effect that severs the tenuous connection to nature that may have existed in Churchville. Scholars of environmental philosopies Bruce Braun and Noel Castree see a need to be "attentive simply to the social and ecological consequences that everywhere are intertwined in everyday practices." The Steeles Avenue Bridge is an otherwise unremarkable structure with remarkable effects.[29]

CONCLUDING THOUGHTS: LANDSCAPE VALUES IN PLANNING

This essay is about looking at a landscape as a means of exploring how individuals and society negotiate the modern world. Landscape is a powerful subject for study because it is visual and visceral; it is made up of tangible material artifacts that confront the senses. This landscape will have many different readings as it is experienced by the people who live in the historic village center, those who have recently moved into the newly built homes, those who seek out the river and tarry there, and those who do not dwell even for a moment on their way somewhere else. The example of the Steeles Avenue Bridge reconstruction, for me, is a chance to explore cultural attitudes and choices. Here in this landscape I am able to contemplate at once the simultaneity of the global and the local, the preciousness of built heritage, and the desire to connect with nature all being played out in one place. As an urban planner used to guiding change in landscapes such as this, it challenges me to think about missed chances in this current moment in the evolution of this landscape.

This is not a static place; the landscape is dynamic and has constantly changed and shifted in the months since I first drove through. The evidence of the bridge construction has itself faded away: the carpet of sod has taken hold and is competing with the weeds, the new black asphalt has faded to grey (thanks to the salt and sand of winter), and the river flows underneath disguising past change by seeming so natural. The encroachment on the village by the adjacent subdivisions is nearly complete. This is now the taken-for-granted landscape. What story does it tell?

> *If the end of the world befall—*
> *and chaos smash our planet to bits,*
> *and what remains will be this*

> bridge, rearing about the dust of destruction;
> then, as huge ancient lizards are rebuilt
> from bones finer than needles, to tower in museums,
> so, from this bridge, a geologist of centuries
> will succeed in recreating our contemporary world.
> —Vladimir Mayakovsky "Brooklyn Bridge" 1925

NOTES

1. Georgie Adams and Peter Gregory, *Highway Builders* (Buffalo: Annick Press, 1996).
2. City of Brampton Economic Development Office. "Census Bulletin #1" (www.city.brampton.on.ca [Accessed July 26, 2012]).
3. Russell B. Mathew, Hemson Consulting Ltd. Personal communication. May 29, 2002.
4. See David Harvey's *The Condition of Postmodernity* (New York: Wiley-Blackwell, 1990).
5. I'm aware that what I have characterized here as freedom of mobility may be countered by others who believe that we are prisoners of our cars because we cannot travel without driving due to the urban structure of roads and expressways. For more discussion, see Geoff Vigar, *The Politics of Mobility: Transport, the Environment and Public Policy* (London: Spon Press, 2002).
6. See Laura Taylor, "No Boundaries: Exurbia and the Study of Contemporary Urban Dispersion," *GeoJournal* 76 (2011): 323–339, DOI 10.1007/s10708-009-9300-y for a review of scholarship on exurban processes. See William Cronon's edited volume *Uncommon Ground: Rethinking the Human Place in Nature* (New York: W.W. Norton, 1996) for a discussion of how we connect with the material reality of nature—trees, water, animals, and scenery and the values and ideas we have attached to these. See also Adam Rome, *The Bulldozer in the Countryside: Suburban Sprawl and the Rise of American Environmentalism* (Cambridge: Cambridge University Press, 2001) for a discussion of the phenomenon of out-migration.
7. In the decade or so that I have been a consulting urban planner, I have found few examples of places where the city is extended into the countryside in a manner that truly preserves the cultural landscape. In Ontario, we do a very good job at protecting heritage buildings as well as areas identified as environmentally significant, including landscapes and habitats associated with wetlands, watercourses, and woodlots but then treat the remaining "developable" lands as blank slates without regard for scenic values, landform conservation, or landscape heritage. This disregard for landscape is in spite of alternatives described by Ian McHarg in his classic *Design with Nature* (Garden City, New York: Doubleday/Natural History Press, 1969); and Randall Arendt's *Rural by Design: Maintaining Small Town Character* (Chicago: Planners Press, 1994), *Growing Greener: Putting Conservation into Local Plans and Ordinances* (Washington, DC: Island Press, 1999), and more technical texts such as William Marsh's *Landscape Planning: Environmental Applications* (Hoboken, NJ: Wiley, 2010). Locally, see also Ontario Ministry of Municipal Affairs and Housing, *Oak Ridges Moraine Conservation Plan* (Toronto: MMAH, 2002), which describes planning to retain existing landform.
8. Susanne Seymour, "Historical Geographies of Landscape" in *Modern Historical Geographies*, eds. Brian Graham and Catherine Nash (Harlow, England: Pearson Educated Limited, 2000), 214.

9. Trevor Barnes and James Duncan, eds., *Writing Worlds: Discourse, Text & Metaphor in the Representation of Landscape* (London: Routledge, 1992), 8; James Duncan and Nancy Duncan, "(Re)reading the landscape," *Environment and Planning D: Society and Space* 6 (1988): 117–126; Richard Schein, "The Place of Landscape: A Conceptual Framework for Interpreting the American Scene," *Annals of the Association of American Geographers* 87 (1997): 676. For a synthesis of landscape scholarship, see John Wylie, *Landscape* (London: Routledge, 2007). For more discussion of landscape as text, see Donald Meinig, ed., *The Interpretation of Ordinary Landscapes: Geographical Essays* (Oxford: Oxford University Press, 1979); Denis Cosgrove, "Prospect, Perspective and the Evolution of the Landscape Idea," *Transactions of the Institute of British Geographers* 10 (1985): 45–62; and James Duncan, *The City as Text: The Politics of Landscape Interpretation in the Kandyan Kingdom* (Cambridge: Cambridge University Press, 1990). A great next step for this research would be to explore non-representational theory and the emotional geographies of those inhabiting and moving through this space to delve into the experience of the contradictions in this particular landscape, see Wylie, *Landscape*.
10. Kevin Lynch, *Good City Form* (Cambridge, MA: The MIT Press, 1981); John Brinckerhoff Jackson, *Discovering the Vernacular Landscape* (New Haven: Yale University Press, 1984), 42.
11. Edward Relph, *Place and Placelessness* (London: Pion, 1976); Edward Casey, "Between Geography and Philosophy: What Does it Mean to Be in the Place-World?" *Annals of the Association of American Geographers* 91 (2001): 683–693; Jeff Malpas, *Heidegger's Topology: Being, Place, World* (Cambridge, MA: The MIT Press, 2006); and Kevin Lynch, *What Time Is This Place?* (Cambridge, MA: The MIT Press, 1976).
12. Credit Valley Conservation. *Integrated Watershed Monitoring Program, 2001 Report* (Mississauga, 2002).
13. Ontario Department of Planning and Development, *Credit Valley Conservation Report 1956. Parts I–VI.* (Toronto: Queen's Printer, 1956), also available online at http://creditvalleyca.ca.
14. Ibid.
15. William E. Cook, *Meadowvale & Churchville* (Cheltenham: Boston Mills Press, 1975); Huttonville Book Committee, *From "Wolf's Den" To Huttonville and the Pioneers Who Made It Possible: Circa 1800 and Beyond* (Guelph: Ampersand, 1996); Desmond Morton, *A Short History of Canada*, 5th ed. (Toronto: McClelland & Stewart, 2001), 36. The Town of York was renamed the City of Toronto in 1834.
16. Andrew Tham, Judy Anderson, and Jill Chen, *Churchville: Reminders of the Past* (Brampton Heritage Board, 1984); Cook, *Meadowvale & Churchville*.
17. Cook, *Meadowvale & Churchville*; Tham et al. *Churchville*; Canadian Broadcasting Corporation, *The Avro Aircraft: Canada's Broken Dream* (http://archives.cbc.ca).
18. The heritage conservation district proposed plan agrees that the present steel bridge was built circa 1907, but does not indicate a primary source. The photo of an "inaugural'" crossing of the bridge by a carriage is identified by hand as 1907 in the Golden family scrapbook in the Region of Peel archives and in conversation with a local resident.
19. Personal interview with Project Manager, Engineering and Construction, Region of Peel, April 3, 2002.
20. Personal interview with Policy Planner, Planning and Building Department, City of Brampton, March 13, 2002.
21. James Duncan and Nancy Duncan, *Landscapes of Privilege: The Politics of the Aesthetic in an American Suburb* (London: Routledge, 2004).

22. Harvey, *The Condition of Postmodernity*.
23. Harvey, *The Condition of Postmodernity*, 216.
24. Edward Relph, *The Modern Urban Landscape* (Baltimore: The Johns Hopkins University Press, 1987).
25. Raymond Williams, *The Country and the City* (London: Chatto and Windus, 1973), 293–294.
26. See David Harvey's work on the "uneven geographic development" of capitalism, *Spaces of Hope* (Berkeley: University of California Press, 2000).
27. Relph, *Place and Placelessness*.
28. It is outside the scope of this chapter to look at how the social construction of wilderness complicates our ability to engage with nature at home. Environmental historian William Cronon says that if we cannot face the issues symbolized with country and city on their own terms, there can be no in-between, no "middle ground in which responsible use and non-use might attain some kind of balanced, sustainable relationship." In *The Trouble with Wilderness* his argument focuses on the social construction of "wilderness" that is a romantic ideology where the most natural places are not where humans live. The central paradox for Cronon is that "wilderness embodies a dualistic vision in which the human is entirely outside the natural. If we allow ourselves to believe that nature, to be true, must also be wild, then our very presence in nature represents its fall. The place where we are is the place where nature is not" (pp. 40, 45–46). To move into what is perceived to be a natural environment makes it seem less natural by our very presence, fuelling the fetish for a yet more natural existence (a cottage perhaps?) a little further out.
29. Bruce Braun and Noel Castree, "The Construction of Nature and the Nature of Construction" in *Remaking Reality: Nature at the Millennium* (New York: Routledge, 1998) 7, 33.

3 Exurbia Meets Nature
Environmental Ideals for a Rootless Society[1]

Richard Judd

EDITORS' INTRODUCTION

Why are idealized natural places so attractive? In this chapter, Richard Judd, a professor of history at the University of Maine in Orono, describes the historic setting and conditions for the ideology of nature in exurbia. He follows the rise of American environmentalism from its roots in early post-war suburbs through the emergence of nature resorts, sketching a compelling picture of a consumer-oriented relationship with a generic idea of nature.

Judd's research on the environmental and social history of conservation movements provides for this book a broad perspective on the pastoral view of green sprawl. He has written several books, including *Common Lands, Common People: The Origins of Conservation in Northern New England*. This chapter emerged from his work with Christopher Beach on *Natural States: The Environmental Imagination in Maine, Oregon, and the Nation*. In this essay he traces the emergence of the exurban ideal—"a home detached from the metropolis and blended into a natural environment"—through three historical strands. First, the larger political stage is set by the shift of environmentalism away from grassroots political participation to a more commodified investment in nature. Second, the transformation of the pastoral ideal from a complex and dynamic working landscape associated with rural folk to a timeless and pristine natural setting helps explain how an abstract nature distant from everyday life becomes imagined as a haven—and a place to escape to "out there." Third, this narrative shows how a commodified natural ideal is produced through a combination of hands-off environmentalism, commercialized and abstracted regional appeal, and the association of the consumption of natural amenity with elite status—in residences, resorts, and even clothes. Understanding the coalescence of specific ideologies of nature in their historical context underscores the value of the methodological approach of environmental history.

This chapter explores the apologia behind exurban expansion in the 1970s and places it in the context of an evolving pastoral ideal. An ideology of nature began to assert political cogency in the post-war years as

Americans constructed a vision of liveability founded on romanticized notions of a "middle ground" linking city and country. This vision inspired concerns about commercial blight and water pollution. As urban problems were exacerbated in the 1960s, traditional pastoral constructs—the humanized forest-and-field landscape—gave way to visions of nature divorced from human activity, as a timeless and untouched wilderness. An ideology of the pastoral re-entered the environmental dialogue in the 1970s with the rise of exurbia and the construct of an occupied, post-industrial landscape, this time exemplified by spa, resort, and residential developments offering a highly abstracted and commodified sense of place and nature. Environmentalists found it difficult to achieve a balanced perspective on this new pastoral ideal. As a form of low-impact economic activity, exurbia was less disruptive than traditional resource-extractive industries, and many hoped that exurbanites would abandon the self-indulgence of their suburban counterparts to become "caring conservers," working to preserve quality of life in simple natural surroundings. Others found this post-industrial nature ideal ironically artificial, destructive, and elitist.

Judd raises an important set of questions for this volume. What aspects of the experience and production of exurban places have encouraged and discouraged people from actively taking part as "caring conservers" in the shaping of their residential and larger environments? This chapter provides a backdrop to understanding the strength of landscape ideology in dealing with change at the exurban fringe, for example in the contentious political fight between exurbanites and ranchers detailed by Peter Walker and Louise Fortmann in "Whose Landscape?" where new exurbanites seek to mold the Sierra ranchlands of California to fit their ideals of nature, reimaging the working ranch landscape as natural environment.[2] The clashes between the ideal and the real everyday environment exemplify the effect of the ideology of nature in exurbia.

—Kirsten Valentine Cadieux and Laura Taylor

■ ■ ■

The post-World War II years brought to maturity America's first urban majority: women and men who grew up in a residential environment far more artificial than it was natural. Yet it was this generation that touched off the most comprehensive and sustained grass-roots defense of nature in American history. Even more remarkable, these urbanites and suburbanites forged their defense of nature around places that most would never visit: Echo Park in Dinosaur National Monument, the Everglades in Florida, and Maine's remote Allagash River, for instance. The nationwide battles to save these and other natural treasures point to an environmental consciousness that clearly transcends the local. Understanding why Americans fought so hard for natural places so remote from their residential experience helps us understand how the environmental movement has inspired so many for so long, and across such wide social and political divisions.

In *Bulldozer in the Countryside*, historian Adam Rome advances this inquiry by tracing the origins of environmental thinking to the suburb. Rapid post-war development, he shows, triggered a series of environmental crises, and coming to grips with these problems laid the foundations for a critique of unbridled urban-industrial growth. While this approach offers a compelling new interpretation of the origins of environmental consciousness, it overlooks an important step in the development of the grass-roots movement: the shift from resolving these "backyard" suburban issues to protecting rivers, wilderness areas, or old-growth forests located far from the suburbs. How, in short, did the experience of place in one locale lead to the idealization of place in another—perhaps on the far side of the country? The answer to this question lies in the dynamics of suburban culture, and particularly in the way suburbanites and later exurbanites fashioned their attachments to nature so as to sustain their environmental commitments in an increasingly mobile and rootless society.

As Rome suggests, the burgeoning post-war suburb was wrapped in powerful cultural symbols relating to the idea of universal middle-class homeownership. But the cultural roots of suburbia are much deeper than this dramatic post-war expansion might suggest. As Kenneth Jackson pointed out some time ago, the suburb was predicated on a mid-Victorian vision of isolated cottages, winding tree-shaded lanes, and semi-rural settings reminiscent of an older agrarian nation. This pastoral ideal, a blend of Romanticism and Jeffersonian agrarianism, took shape in the early nineteenth century and changed as the suburb itself was transformed by new values and new transportation technologies.[1]

The years after World War II saw a flowering of pastoral literature that incorporated these romantic and Jeffersonian themes against a backdrop of anti-urbanism. The Great Depression, which coincided roughly with America's emergence as a predominantly urban society, suggested to some that cities were economically unsustainable, and the rise of totalitarianism in the 1930s drove home the terrifying political implications of an alienated urban mass constituency. These dark musings played into an increasingly negative assessment of American cities, a mood epitomized by a generation of urban-based *film noir* theater productions emphasizing paranoia and human powerlessness and vulnerability. Social critics like C. Wright Mills, Eric Fromm, and Lewis Mumford implicitly or explicitly contrasted the organic culture of small-town life with the artificial and alienating elements of the urban world, and regionalist writers across the country accentuated this anti-urban mood by mythologizing the rural landscape as a liberating and authenticating alternative to urban life. Rejecting the sprawling metropolis as a model for authentic living, Americans adopted the suburb as a pastoral median between city and nature.[2]

In basic terms, the pastoral ideal was a nostalgic wish to return to simpler times, but in the regionalist literature of post-war America, it became much more nuanced. The post-war pastoral included folk images that had been crystalized in the 1930s, when, according to Warren Susman, the American intelligentsia "set out to become 'an unlearned class,' to assimilate the culture of the 'people.'" The plays of Clifford Odets, the novels of John Steinbeck, the WPA murals, the folk celebrations of artist Thomas Hart Benton and songwriter Woody Guthrie, the photographic and literary representations by James Agee and Walker Evans, and numerous other 1930s cultural expressions captured "an old and enduring vision within American political culture: the idea of a commonwealth at work." This democratic humanist tradition provided release from the anxieties of global instability and urban transformation.[3]

In the romanticized rural landscape of post-war America, nature and folk were inseparable: two creative forces interacting in intimate and harmonious balance. Nature's seasonal or diurnal rhythms become the rhythms of work and play, and landscapes shifted seamlessly from garden and field to meadow and forest. Unlike William Allen Whyte's perpetually ineffectual "organization man," the rural folk inscribed their way of life on the land in patterns of work handed down from generation to generation, and these deeply textured landscapes epitomized America's quest for authenticity. The promise of redemption in this pastoral image helped define the quality of life in post-war America.[4]

This pastoral idea played an important role in the post-war suburban migration, but it also helped shape post-war environmental politics. As nature philosopher Rene Dubos once said, there are two landscapes capable of eliciting the spirited political and cultural defense we associate with environmentalism: "one is primeval nature undisturbed by man. . .The other is one in which man has toiled and created, through trial and error, a kind of harmony between himself and the physical environment."[5] Although wilderness campaigns took precedence in later politics, the pastoral landscape played an important part in earlier phases of the environmental movement, and both images—pure nature and pastoral constructions—command allegiance in environmental politics and literature today. Tracing this pastoral construct as a source of nature appreciation in a changing metropolitan world is an important part of the story of modern environmentalism, since it links the consciousness-raising effects of backyard suburban problems to a broader vision of Americans at one with nature.

This creative interface between the natural and the artificial was the promise Americans carried with them to the suburbs in the years after World War II. Aware of nature's wholesome influence on the folk, Americans anticipated a better life in their suburban settings. Rightly planned, suburbia's compact residential designs would not transcend or efface nature; rather they would absorb nature's authenticating qualities. In its ideal form, the suburb would realize a harmonious blend of artifice and nature by bringing the city and the country together, incorporating nature into suburbia's quality of life. Towns would be rooted in the land, as they were in the planning utopias of Ebenezer Howard and Frank Lloyd Wright.[6]

Given these anticipations, the post-war pioneers on the "crabgrass frontier" were bound to be disappointed. Explosive development along the urban fringe resulted in what one planner called "the virtual elimination of city limits for practical...purposes." Urban problems quickly found their way into the large housing developments that characterized suburban expansion in these years, fusing the suburb to the city rather than to nature. These all-too urban problems were further compounded by the untried premises of this new residential design. Unregulated growth, untested forms of construction, and inadequate infrastructures, as Rome describes them, undermined the American Dream.[7] Developers obliterated open space, overextended sanitary facilities, built septic systems in unsuitable soils, and overtaxed municipal wells. Some developments simply doused sewage with chlorine and pumped it into ditches along the county roads. Subdivisions spread rapidly into jurisdictions having no building or zoning codes, resulting in a "hodge-podge of shacks, junk yards, and beer taverns," over which municipal or county governments exerted no effective control. Nor was life in the country as idyllic as appearances suggested. In the mixed rural neighborhoods on the suburban frontier, new residents came in contact with real-life country land-use practices that violated the romantic impressions of nature they brought with them to the suburb, and they deluged health officers with complaints about vacant fields used as "dumping grounds," odors from hog farms, mosquitoes breeding in abandoned gravel pits, smoke from burning trash or sawmill waste, and pollution in nearby streams.[8]

As Adam Rome shows, these disappointments set the scene for the environmental movement. Environmentalism entered suburban consciousness as a nostalgic challenge to the politics of growth, lodged among a population frustrated by the defaced landscapes around them and anxious about the personal issues raised in books like Sloan Wilson's best-selling *The Man in the Gray Flannel Suit* (1955) and Paul Goodman's *Growing Up Absurd* (1960). Harboring an equally dark assessment of the "doomed cities" and the disorganized rural landscapes looming on their horizon, besieged suburbanites began questioning their quality of life.[9]

The water pollution campaigns of the 1950s and 1960s helped transform this pastoral vision into environmentalist rhetoric. Polluted waters posed serious health threats to urban and suburban residents, but pollution also violated the expectation that cities, suburbs, and nature were essentially compatible. Suburbanites joined with fish and wildlife organizations, tourist promoters, and women's organizations to restore a "harmonious visual and physical relationship between nature and the urban environment," as one local environmental commission put it, and as these rivers were restored, they brought the blessings of nature back into the built environment, bridging the distance between nature and residential space. Once hidden behind freeways, scrap yards, mills and factories, and floodplain jungles, cleaner rivers beckoned once again as "natural beauty spots," around which the landscapes of city and suburb could be reanimated. Inspired by the pastoral imagery of the naturalizing river, planners designed riverside parks, greenways, waterfront gentrification projects, and recreational harbor or docking facilities that would serve as green hearths for the urban and suburban household.[10]

In the mid-1960s, environmental consciousness extended to remote natural features as well. Frustrated in their search for livability in the suburb, Americans learned to incorporate the more distant natural world into their quest for quality of life, using the automobile and the interstate highway to bridge the distance between residential place and the natural landscape. Flooded with nature images in the mass media, Americans took up outdoor activities like hunting, fishing, camping, boating, and hiking in pure natural settings. Since the suburbs had failed to provide the sense of authenticity and freedom Americans associated with the working rural landscape, they turned to more distant rural landscapes and adopted these far-off forests, lakes, rivers, mountains, or seacoasts as important elements in their sense of place. In contrast to suburbia, the experience of an old-growth forest, a national park, or an idyllic lakeside resort seemed immutable, lucid, and authenticating, providing the essential ingredients in the American quality of life.

The rise of popular ecology and wilderness ideology changed the perception of these distant landscapes. The success of Rachel Carson's early works—*Under the Sea-Wind*, *The Edge of the Sea*, and *The Sea Around Us*—reflected America's growing interest in nature, and Carson's books, like other popular natural histories written in the 1950s and 1960s, encouraged a view of humans as foreign to the natural landscape. In addition, the neo-Malthusian "limits to growth" discussion, centered around Paul Ehrlich and later Donella H. Meadows, added to the popular understanding that human activity was separate from, and even antagonistic to nature.[11] At the same time, the real hinterland economy was changing as well. New forms of logging, ranching, and farming, with heavier machinery, more petrochemical applications, and larger landholdings, transformed rural and small-town America.[12] Gone was the agreeable mix of garden, orchard, and field crops, barnyard fowl, and the family cow, and this loss of self-sufficiency brought a loss of rural innocence: in the urban mind, the rural inhabitant no longer sustained the idea of the folk in the pastoral landscape. The pastoral was redefined, not as a working rural landscape, with all its complexities and disappointments, but as a timeless natural setting.

These social and intellectual trends brought to fruition a new vision of the rural world, still valued in terms of authenticity and permanence, but no longer defined as a harmonious blend of human tradition and natural process. This in effect put more distance between the suburbanites and the natural forms that defined their quality of life. Early post-war Americans had defined rural America as a union of folk and nature, a kind of human-ordered climax community of simple country homes and neatly tilled fields set in a context of forest, mountain, and meadow. In the new pastoral perspective, nature was abstracted from rural activity as a timeless and pristine world, best experienced through solitary recreational activity.[13] Over time, the connection to the rural landscape became more remote and abstract, but no less compelling as an escape from the psychic confines of the city.

Although geographically remote from nature, Americans fought to insulate this world from the disorders that threatened their immediate residential world, since identification with a remote and somewhat abstracted recreational icon had become essential to the livability of the suburb. If these "God-given retreats" were degraded, they understood—if the forest was cut, the habitat lost to development, the river degraded by dams or pollution—the city and suburb would also lose their "desirability as a place to live."[14] The expansion of the recreational hinterland made the suburb livable, and at the same time made the rural world beyond these profaned places inviolate.

The environmental movement benefited from these cultural changes, but during these years, it was undergoing a difficult transition. After the rapid gains of the late 1960s and early 1970s, the political focus of the movement was shifting from grass-roots mobilization to interest-group lobbying, largely at the federal level, and this "Washingtonization" discouraged many local advocates. Daniel D. Chiras, a sympathetic critic, noted the symptoms of declining grass-roots vitality: a growing emphasis on mass-mail recruiting; a political base composed of "soft members" willing to give money in lieu of personal effort; and a narrow concentration on "imageable" issues likely to appeal to ephemeral public interest.[15] Chiras faulted the big environmental organizations for these changes, but in fact, the movement was caught up in broader trends in American politics. The Vietnam War fractured the American liberal consensus, as did the centrifugal forces in the anti-war, civil rights, and feminist movements, and the burdens of post-Vietnam and post-Watergate cynicism made Americans less receptive to grass-roots organizing. The inflationary spiral brought on by the war and the Arab oil embargo complicated political proposals that affected taxes, budgets, or consumer prices, and as Stephanie S. Pincetl points out, the decade saw a "deliberate depoliticization of politics" in major party platforms and ideologies. In addition, women were moving into the workforce in greater numbers, and had less time to devote to civic associations, and people in general were becoming more mobile and less attached to the community organizations that had launched the environmental movement two decades earlier. Finally, the waning of the counterculture deprived grass-roots environmentalism of a rich source of visionary thinking.[16] The mid-1970s saw the end of a unique era of popular political participation in America.

Grass-roots activists also confronted more perplexing issues, many of them lacking the intense anti-corporate focus that animated classic environmental campaigns in the previous decade. Many activists accepted the idea that problems could be solved through the new institutional arrangements created by the first wave of state and federal environmental laws, but as they moved from creating to enforcing regulations, they found the tried and tested political formulas of the earlier years less effective.[17] Persistent problems like non-point source pollution, soil erosion, habitat loss, and acid rain had much higher populist flash points, and here arcane technical expertise counted more than grass-roots mobilization.[18] Mainstream organizations moved in the direction of greater ecological sensitivity and wilderness purity in the later 1970s, and while these issues drew a passionate response from a core group of supporters, they were less likely to garner the broad public support generated during the 1960s and early 1970s.

In addition to these political changes, Americans seemed to be less rooted in place. Suburban mass construction discouraged a strong identification with residential space, and regional connections were becoming tenuous as well. With American cities in crisis, suburbanites distanced themselves from the adjacent metropolis, becoming even more insular. This geographical alienation, and the easy interchangeability of the mass-produced suburban world, eroded consciousness of community and place.[19] In his best-selling *Greening of America*, Charles Reich described post-war Americans as obsessed with mobility and change; they demolished and rebuilt neighborhoods before they could infuse these places with meaning and tradition; they allowed generic commercial considerations to dominate regional architecture, ate food that bore no relation to the land, and lived and worked in "small, identical rectangles" with artificial air, weather, and sound. Having lost their connection to the land, they filled the void, according to Reich, through obsessive consumerism.[20]

Despite the ebbing sense of place and the decline in civic consciousness, the indices of public support for environmental issues remained steady. The explanation for this persistent popular commitment seems to rest in the ways suburbanites adjusted their attachment to nature to their new residential realities, and particularly to the link between nature appreciation and consumer culture. "Sentimentalizing" natural landscapes—creating symbolic places to answer for empirical reality, as Lawrence Buell puts it-linked older pastoral values to new forms of nature appreciation emerging out of the material culture of the 1970s.[21]

The commodification of nature had its roots in a post-war development historian James T. Farrell called the "Great Compromise," a concession in which American labor accepted degraded work in exchange for high wages and expanded consumer choice. The compromise, according to Farrell, was an essential ingredient in the economic prosperity of the 1950s and the emphasis on materialism that defined the decade. In the 1960s, many American youths rejected this materialistic and commercial impulse by identifying nature as a cultural counterpoint to the synthetic world embraced by their parents. Late in the 1970s, suburban culture helped to unite this taste for the organic with a burgeoning consumer culture in ways that reintegrated nature and residential place.[22]

The rise of the L.L. Bean Company of Freeport, Maine, highlights these trends in nature appreciation and consumer culture. The company's outdoor image—a crucial element in its commercial success—was grounded in the founder's engaging zest for the Maine woods, which appealed to a metropolitan sporting set seeking closer connections to nature through recreational purchases. Leon L. Bean, born in 1872 in Maine's western hill country, grew up with a passion for hunting and fishing, and with a partner in 1912 he opened a small outdoor clothing store in Freeport. By the 1920s Bean's iron-clad guarantee of satisfaction, his common-sense approach to outdoor wear, and his enthusiasm for Maine had gained him a steadfast clientele for both in-store and catalog sales. In the post-war years, Bean's store in Freeport emerged as an icon for the eastern outdoors sporting set, personifying a masculine engagement with the Maine woods through hunting, fishing, and camping. When Bean's grandson Leon Gorman became president in 1967, the company accentuated the nature-as-fashion concept among a new generation of young, well-off, style-conscious customers looking for elegant yet sturdy outdoor gear and clothing with a palpable connection to the real outdoors. A livelier catalog, a younger sales force, and a renovated retail establishment in Freeport catered to the new nature shopper. By 1975, with sales topping $29 million, L.L. Bean was doubling in size every three years. Gorman fully recognized the commercial value of abstracted Maine woods images among suburbanites seeking quality of life: "the whole idea of the state pervades our catalog, and I think when people buy the Maine hunting boot, they are buying a little of Maine too."[23]

Like the traditional city "sports" who stopped in Freeport en route to Maine's northern woods or western lakes, Bean's new catalog customers were metropolitan, well-off, and well-educated, and even though their understanding of nature was less place-specific, they could appreciate the nature symbolism interleaved among the catalog's boots, jackets, and canoes. Since catalog purchasing transcended geographical boundaries, consumers around the world could actualize their love of nature by adopting "Maine" outdoor apparel. Retaining the founder's emphasis on assured quality and his passion for the outdoors, Gorman generalized the Maine image to appeal to a national—indeed international—clientele. L.L. Bean, like the many outdoor catalog companies that followed in its wake, reinforced the idea of commodified nature as an alternative to natural place.

This abstracted and commodified nature appealed particularly to a new suburban elite that adopted the environmental and counterculture apparatus of the late 1960s as objects of consumption. Journalist Joel Garreau found the ultimate expression of this trend in suburban Marin County, California:

> The unspeakably hip/chic suburban community north of the Golden Gate...thrives on tall redwoods, octagonal barns, alfalfa sprouts, walls made of planks nailed on the diagonal, hanging plants, and the highest achievement of the modern economic version of people taking

in each other's laundry: crafts. In places like Marin. . .there is no end to the cozy shops featuring locally made pottery, woodworking, and leather-tooling. Much of it is marvelously innovative and of great technical quality, but that sometimes gets lost in the overpowering grooviness of it all. Like too much health food, it can make you sick.[24]

In this class-conscious consumption of nature and place in suburban settings like Marin County, we can see in bold relief the dynamics of nature appreciation in an age when fewer people had direct connection to nature and place.[25] The fusion of nature and suburban consumer culture helped set the new elite apart from the ordinary suburbanite, but it also helped transform the suburban pastoral for all suburbanites: Adaptable to any location, nature consumption became an expression of suburban placelessness. Sensitive to this change in public consciousness, *Audubon* editors began including lavish pictorial spreads in their magazine, depicting the "artistry of nature" to highlight endangered species and threatened places. Buying Bean boots or perusing *Audubon* magazine did not translate directly into environmental activism, but the link to nature appreciation was nonetheless important. Publications like *Audubon* blended with a broader array of commercial images that helped sustain environmental commitments through the 1970s.[26]

These trends were closely linked to another feature of mid-1970s suburban change: young, upper-income families moving out beyond the limits of the classic suburb to new exurban homes. Beginning in the later 1970s, small towns outside the metropolitan umbra became new growth centers feeding on the decentralization of work, the commuting two-worker household, the limited-access commuter highway, and a growing preference for larger house-lot size. This exurban shift was facilitated by the emergence of new knowledge-based industries like pharmaceutical testing, computer software design, engineering, marketing and financial consulting, and information-processing services, along with the out-migration of light industry and craft production, all of which required less work time in a traditional downtown setting.[27] Popular ecology—separating nature from the humanized landscape—had distanced the suburban residence from the landscape that defined its quality of life; exurbia once again brought nature and residential place together.

Exurbia was part of the ongoing search for a way to bridge the distance between nature and residential space, and this, too, had implications for environmental consciousness. As historian Samuel Hays points out, modern environmentalism was an extension of the post-war consumer revolution. Since the beginning of the twentieth century the American economy shifted from the production of necessities to the production of conveniences, and finally to the production of amenities. Post-war Americans viewed nature as an amenity, but since clean air, clean water, and unspoiled wilderness could not be purchased in the marketplace like other amenities, they pursued these values through collective or political action.[28] In the 1970s, some Americans chose to purchase these natural amenities privately. Like boots from L.L. Bean, a home in the country was an expression of commodified nature. And like suburbia, the exurban ideal—a home detached from the metropolis and blended into a natural environment—was built on an image of romanticized nature. This exurban world, however, differed from the post-war pastoral in de-emphasizing the working folk in the natural landscape. As historian Lloyd Irland explains, the "ability of the urban [world]...to import its corn from Iowa, its lumber from Canada, its oil from Iraq, and its steel from Japan led its citizens to believe that they had been liberated from dependence on natural resources."[29] Disassociating the hinterland and the metropole, exurbanites idealized the countryside on different terms: as a pastoral landscape without a working relation to the city, premised on a belief that traditional activities like farming, mining, lumbering, ranching, or fishing were no longer the core of the rural economy. The countryside was re-colonized as residential space, free of any functional connection to the land.[30]

This postindustrial premise was captured in a work titled *Regions of Opportunity* by writer and real-estate speculator Jack Lessinger, who labeled the new rural landscape "penturbia," the fifth great demographic shift in American history, following after the transatlantic, westward, urban, and suburban migrations. An unabashed apologist for exurban development, Lessinger thought that reintegrating nature and residential space in this fashion would transform middle-class values. Suburbanites were self-indulgent and insular; penturbanites would relearn community bonding by seeking quality of life in open spaces and simple natural surroundings, substituting "gifts bestowed by nature" for the mindless material consumption encouraged by suburban living. Exurban citizens would build their own homes, sew their own clothes, raise their own organic foods, and "volunteer their loving labor as a substitute for expensive services in schools and hospitals and in ministering to the young, the elderly, and the handicapped." In Lessinger's wildly optimistic vision, exurban living would spawn new forms of environmental activism. Land-use regulations would prevent runaway growth; "intangible notions" like peace of mind and natural beauty would supersede economic exploitation of the land; and wildland amenities would inspire a shift from saving endangered species to saving entire ecosystems as habitat for plants, animals, and penturbanites.[31]

Lessinger saw his exurbanites as a force working to preserve the pastoral character of rural America, but in reality, exurbia, like suburbia, fell short of this ideal. In the eyes of most environmentalists and planners, exurbia was, contrary to Lessinger's appraisal, a planning disaster, bringing traffic congestion, higher property taxes, lake and pond eutrophication, visual blight, and a "feeding frenzy" in rural real estate that forced many locals out of the market. According to planner Evan Richert, Lessinger's penturbia came with a "built-in contradiction": it dismissed traditional suburbanites as "mass consumers," but glorified their upscale counterparts in sprawling exurbia, "who want two-to-five-acre lots out near farms and along rivers and lakes." Although premised on a quest for rootedness, exurbia was in fact the most rootless of communities, according to environmentalist Murray Bookchin. It was an entirely new social geography "based on the automobile, the suburban shopping center, and a high-income population that depends upon the city economically but is completely severed from it culturally."[32]

The environmental implications of exurbia are indeed contradictory. Exurbanites chose to purchase nature as a private enclave, surrounded by the small farms, pastures, woodlots, orchards, rivers, and lakes Americans have always included in their a quality-of-life ideals. But while Lessinger overlooked the glaring environmental and cultural drawbacks of this choice, his hopes for a renewed appreciation of nature bear some scrutiny. While exurbanites purchased a nature most environmentalists would scorn, these conflicting values did not preclude collective action on other environmental fronts—land-use planning, community recycling programs, annual contributions to the Sierra Club, and perhaps more.

The trend in selling nature to mobile suburbanites was epitomized in a new type of western destination resort village that emerged in the 1970s, mainly oriented around recreational pursuits like golfing or skiing. Pioneered by developers in Sun Valley in the 1940s, the recreational-residential resort village was typically sited near a river, lake, coast, or mid-elevation valley that offered isolation, regional-theme scenery, high-speed auto access, and open land suitable for subdivision. Around a series of ski lifts, golf courses, or other recreational facilities, developers assembled a mix of high-density condominium units, town houses, patio houses, and mini-estates. A wildlife area, working ranch, or "live stream" routed through the community represented the regional experience.[33] Here again nature was highly abstracted and universalized. River, mountain, or seacoast backdrops were carefully cultivated to accent the integration of nature and residential space, and on site, nature was abstracted as a generic ecosystem of artificial ponds, transplanted shrubs and trees, and winding paths through manicured lawns—nature reinvented as decor to provide a total recreational experience virtually independent of surrounding communities and landscapes.[34]

The developments were, as Robert Fishman wrote about the older country club ideal, "institutionalized means to define the social boundaries of suburbia"—expressions of suburban upper-class identity. And like the country club, the western ski resort made "nature...the instrument of social snobbery," attracting a painfully self-conscious middle- and upper-middle class clientele seeking temporary refuge from the identity-eroding city or suburb.[35] Footloose, monied, and linked to metropolitan centers by air and express-highway travel, this new elite was learning the distance-business techniques that in the 1980s would allow them to further attenuate their ties to the workplaces of downtown America. And as these residential transients severed their roots in the city, the suburb, and the land itself, they refashioned the natural symbols that undergird the environmental movement.[36]

The forms of nature appreciation that animated late-1970s environmentalism were diverse, ranging from those of the traditional Isaac Walton League outdoor enthusiast and the young back-packer to the nature consumer in Bean boots and an exurban home. This diversity was both a strength and a weakness in the environmental constituency. Indeed, old-style environmentalists found it difficult to achieve a balanced perspective on these new directions in nature appreciation. Like exurbia itself, the new exurban recreational resorts appearing in the deserts, mountains, and seacoasts of the West were perplexing. Recreation and leisure services were forms of low-impact economic activity that many environmentalists hoped would replace ecologically disruptive extractive industries like mining, drilling, and logging, and like the grand old Victorian-era summer hotels, these resorts helped to concentrate tourist activity in a single location, relieving development pressures in more pristine areas. Yet some, particularly the old-style Budweiser-and-beans environmentalists, saw this artificial nature as highly ironic. Tom Bell, editor of *High Country News*, described one Colorado vacation village as a "sort of place in which only Californians and John Denver fans could attend a conference on the environment and not feel hypocritical." Trees and mountains, he observed, were "gouged out to provide space for ski lifts, golf courses, softball fields, swimming pools, convention facilities, equestrian centers, and the rest of the prerequisites of wealth.... The place is a classic example of development that digs out all the hills, cuts down all the trees, drains all the water, and then sells subdivisions that tout the quality of their environment."

Bell saw Rocky Mountain resort villages like these as popular fantasies of the West, "never-never lands" made to imitate the last great enclaves of open space left in the continental United States. Florida and California, he observed, had once offered authentic open space, but runaway growth erased all but the memory. Disneyland and Disney World thus became "opportunist death symbols," commodified epitaphs to nature in areas now "too crowded and too cluttered to foster any grand dream of 'the great escape.'" The Rocky Mountain resorts likewise catered to the memory of open space, Bell noted ominously, and Disney Enterprises held options on land here as well. It was "too early in the growth history" of the region for another fantasy land, but someday Mickey Mouse would replace the elk in this Rocky Mountain ecosystem.[37]

Bell's take on the promise of knowing nature through consumption was echoed in an article by Edward Abbey, who expressed the concerns many felt as they watched familiar rural settings grow into upscale nature-friendly resorts. Telluride, Colorado, which Abbey claimed to have discovered during a "picnic expedition" into the San Miguel Mountains in 1957, had been an "honest, decayed little mining town of about 300 souls" before it was "Californicated" into a "bustling whore of a ski resort with a population of 1,500." California developer Joseph T. Zoline promised that Telluride's growth would be "controlled and orderly," but what Abbey found disquieting was not the loss of ecological balance, but rather the loss of local color and Telluride's organic sense of place—the folk in the natural landscape. "What some of us liked so much about Telluride was not the skiing but the quality of the town which Zoline and his developmental millions must necessarily and unavoidably take away: its rundown, raunchy, redneck, backwoods backwardness. That quality is one you cannot keep in a classy modern ski resort, no matter how much money is spent for preservation."[38] Clearly, this was not the environmental vision most activists had been weaned on, and the class biases in this carefully crafted vision of nature alienated many like Bell and Abbey, who were concerned about the class implications of consumption as a form of nature appreciation.

In a similarly critical article in *Audubon* magazine, Jack Hope noted that ski facilities from Maine to Oregon were "crammed to overcapacity," and he worried that to meet this burgeoning demand developers would build new resorts in even more fragile areas. In the mountains of New England, there were few long-range vistas left that did not include "the broad treeless swipes of ski trails radiating from mountaintops." The region's soils were thin, the slopes steep, and the growing season short; tree removal caused erosion, and gasoline, oil, and salt runoff from roads and parking lots polluted the small upland streams. Hope was impressed with the "ecological good sense" of some ski area developers, but "the ultimate ecological sensibility would have been never to build these facilities in the first place." Even more disconcerting to Hope was the obliviousness of the resorts' upper-class patrons. "There is nothing in a heated outdoor pool, a helicopter chauffeuring service, a fancy boutique, or a $66 room that encourages humility or respect either toward our natural environment or toward ourselves. Rather, a week at a place like Snowbird only reinforces the belief that if you have the money, the tops of mountains will be lopped off to provide you with entertainment."[39]

The concerns raised by Bell, Abbey, and Hope about the environmental insensitivity of the exurban consumer raise important issues about how these forms of nature appreciation translate into environmental values. Consuming nature images helps keep three out of four American nominally in the environmentalist camp, but the nature they defend is curiously abstract and, as these critics point out, often dangerously exclusive.

Still, for all their liabilities, nature consumption, exurban migrations, and nature resorts were expressions of America's ongoing love affair with nature, and when a hostile federal administration threatened this remote and sometimes abstracted construction of nature in the 1980s, Americans rallied to its defense. Swelling the ranks of environmental organizations in unprecedented numbers, they fought the Reagan administration to a standstill on almost every important environmental issue.

Despite the contradictions and the misgivings, the new forms of nature appreciation emerging in the late 1970s helped sustain mainstream environmental politics. The principles behind the movement—the right to clean air and water, the protection of endangered species, the sanctity of wilderness, the regulation or elimination of toxic materials—continue to receive widespread public endorsement. Americans see their connection to nature as inviolate, however conflicted their idea of nature might be. Commercialized regionalism and nature consumption blended with experiments in sustainable energy, a preference for organic foods, and a variety of landscape preservation programs as new forms of environmental enthusiasm embraced by Americans less engaged in civic activity, less rooted to a specific place, and more comfortable with the dictates of consumer capitalism. Like their predecessors, these environmentalists leave much undone, but they keep the door open for a more intensive engagement when new political and economic forces threaten the nature they associate so closely with their residential quality of life.

NOTES

1. This article is adapted from Chapter 7 of Richard W. Judd and Christopher S. Beach, *Natural States: The Environmental Imagination in Maine, Oregon, and the Nation* (Washington, DC: Resources for the Future Press, 2003).
2. Peter Walker and Louise Fortmann, "Whose Landscape? A Political Ecology of the 'Exurban' Sierra," *Cultural Geographies* 10 (2003): 469–491.
3. Adam Rome, *Bulldozer in the Countryside: Suburban Sprawl and the Rise of American Environmentalism* (New York: Cambridge University Press, 2001): 15, 37, 39, 40; Kenneth Jackson, *Crabgrass Frontier: The Suburbanization of the United States* (New York: Oxford University Press, 1985).
4. Elmer T. Peterson, ed., *Cities Are Abnormal* (Norman: University of Oklahoma Press, 1946); Terry H. Anderson, *The Movement and the Sixties* (New York: Oxford University Press, 1995): 16; Andrew Jamison and Ron Eyerman, *Seeds of the Sixties* (Berkeley: University of California Press, 1994): 39–44, 55–60, 82–85.
5. David Burner, *Making Peace with the 60s* (Princeton: Princeton University Press, 1996): 6, 220–221; Warren Susman in Andrew Jamison and Ron Eyerman, *Seeds of the Sixties* (Berkeley: University of California Press, 1994): 8; Paul Lyons, *New Left, New Right, and the Legacy of the Sixties* (Philadelphia: Temple University Press, 1996): 37–38; Raymond Williams in Lawrence Buell, *The Environmental Imagination: Thoreau, Nature Writing, and the Formation of American Culture* (Cambridge: Belknap Press, 1995): 62.
6. Leo Marx, *The Machine in the Garden: Technology and the Pastoral Ideal in America* (New York: Oxford University Press, 1964): 3, 5, 6, 7–9, 13, 21 28, 116; Lawrence Buell, *The Environmental Imagination: Thoreau, Nature Writing, and the Formation of American Culture* (Cambridge: Belknap Press, 1995): 35, 44, 62.
7. Rene Dubos in Robert Burns, "Cultural Change, Resource Use, and the Forest Landscape: The Case of the Willamette National Forest," (Ph.D. dissertation, University of Oregon, 1973): 205.
8. Henry L. Kamphoefner, "An Architect Protests," in *Cities are Abnormal*, ed. Elmer T. Peterson (Norman: University of Oklahoma Press, 1946): 128; Rutherford H. Platt, "From Commons to Commons," in *The Ecological City: Preserving and Restoring Urban Biodiversity*, eds. Rutherford H. Platt and Rowan A.

Rowntree (Amherst: University of Massachusetts Press, 1994): 31–32; James L. Machor, *Pastoral Cities: Urban Ideals and the Symbolic Landscape of America* (Madison: University of Wisconsin Press, 1987): 6–7, 10, 13, 168, 213; Mark Luccarelli, *Lewis Mumford and the Ecological Region: The Politics of Planning* (New York: Guilford Press, 1995): 28, 29, 48, 49, 51, 79.

9. In addition to Rome, see Bureau of Municipal Research and Service, *Procedure for Establishing County Planning and Zoning Under Chapter 537, Oregon Laws, 1947* (Corvallis: University of Oregon, 1947); Paula M. Nelson, "Rural Life and Social Change in the Modern West," in *The Rural West Since World War II*, ed. R. Douglas Hurt (Lawrence: University of Kansas Press, 1998): 38; Keith Montgomery Carr, "Changing Environmental Perceptions, Attitudes, and Values in Oregon's Willamette Valley, 1800 to 1978," (M.A. thesis, University of Oregon, 1978); Horace Sutton, "Land of Lumber and Lakes," *Saturday Review of Literature* 31 (May 15, 1948): 30.

10. "Quarterly Report of the Division of Sanitary Engineering, January, February & March, 1946," pp. 3–4, Health Board, Sanitary Engineering Division General Files, 1936–1950, B-14-1 (also file B-13-1), A-P; pp. 2, 5; "Memo," January 28, 1946, Report, Box 29, 2; "B-14-1 Nuisance complaints, A-P," Health Board, Sanitary Engineering Division General Files, 1936–1950, box 29, Oregon State Archives; "Selected Files, 1938–1961"; "Reports, 1936–1959"; "Reports (Quarterly) 1936–1947," box 35, ibid.

11. Lawrence Halprin & Associates, *The Willamette Valley: Choices for the Future* (Willamette Valley Environmental Protection and Development Planning Council, 1972): 7; Robert Straub in *Oregon Statesman* (December 9, 1969); Charles DeDeurwaerder, "The Fight Against Ticky-Tacky," *Oregonian Northwest Magazine* (November 16, 1969): 12; Peter James, "Ecotopia in Oregon?" *New Scientist* 81 no. 1135 (January 4, 1979): 29; "Quarterly Report of the Division of Sanitary Engineering, January, February & March, 1946," pp. 3–4, Health Board, Sanitary Engineering Division General Files, 1936–1950, B-14-1 (also file B-13-1), A-P: 5; "Memo," January 28, 1946, Report, Box 29, 2. See L.A. Helgesson Testimony to State Legislative Commission on Pollution, July 20, 1966, 1965–1967 Public Health Committee Minutes and Exhibits, Box 82, Oregon State Archives; Adam W. Rome, "Building on the Land: Toward an Environmental History of Residential Development in American Cities and Suburbs, 1870–1990," *Journal of Urban History* 20 (May 1994): 415.

12. Governor's Natural Resources Advisory Committee, *Water Pollution Control in the Willamette River Basin* (Salem: Oregon State Sanitary Authority, 1950): 4–9, 29, 32; George W. Gleeson, *The Return of a River: The Willamette River, Oregon* (Corvallis: Oregon State University, 1972); William G. Robbins, *The Oregon Environment: Development vs. Preservation, 1905–1950* (Corvallis: Oregon State University, 1975): 22, 25, 27; *Ninth Biennial Report of the Oregon State Sanitary Authority* (Portland, 1954–1956): 3; Courtland Smith, *Public Participation in Willamette Valley Environmental Decisions* (Corvallis: Oregon State University, 1973): 92; Richard Condon, "The Tides of Change, 1967–1988," in *Maine: The Pine Tree State from Prehistory to the Present*, eds. Richard W. Judd, Edwin A. Churchill, and Joel W. Eastman (Orono: University of Maine Press, 1995): 564–567; Neal R. Peirce, *The New England States: People, Politics and Power in the Six New England States* (New York: W.W. Norton, 1976): 404–405; Keep Oregon Livable, *The Fight to Save Oregon: An Environmental Progress Report* (Portland, 1972), n.p.; J. Herbert Stone to Pierre Kolisch, May 13, 1970, Governor's Committee for a Livable Oregon, Correspondence, 1970–1971, mss 2537, Oregon Historical Society Manuscript Collection.

13. Andrew Jamison and Ron Eyerman, *Seeds of the Sixties* (Berkeley: University of California Press, 1994): 93. See *Oregonian*, November 18, 1967; Keith Montgomery Carr, "Changing Environmental Perceptions, Attitudes, and Values in Oregon's Willamette Valley, 1800 to 1978," (M.A. thesis, University of Oregon, 1978): 16; "The Malthusian Dilemma Updated," in Robert C. Paehlke, *Environmentalism and the Future of Progressive Politics* (New Haven: Yale University Press, 1989): 41–75.
14. Richard Wescott and David Vail, "The Transformation of Farming in Maine, 1940–1985," *Maine Historical Society Quarterly* 28 (Fall 1988): 66–84.
15. E.B. White, "Letter from the East," *New Yorker* 36 (December 3, 1960): 238; James E. Sherow, "Environmentalism and Agriculture in the American West" in *The Rural West Since World War II*, ed. R. Douglas Hurt (Lawrence: University of Kansas Press, 1998): 59; Philip G. Terrie and E. Melanie DuPuis, "In the Name of Nature: Ecology, Marginality, and Rural Land Use Planning During the New Deal," in *Creating the Countryside: The Politics of Rural and Environmental Discourse*, eds. Melanie DuPuis and Peter Vandergeest (Philadelphia: Temple University Press, 1996): 103.
16. Anthony Netboy, "Cleaning Up the Willamette," *American Forests* 78 (May 1972): 15; Netboy, "French Pete for People," *American Forests* 76 (May 1970): 59; Carr, "Changing Environmental Perceptions, Attitudes, and Values." See John M. Findlay, *Magic Lands: Western Cityscapes and American Culture after 1940* (Berkeley: University of California Press, 1992): 277; Earl Pomeroy, *The Pacific Slope: A History of California, Oregon, Washington, Idaho, Utah, and Nevada* (New York: Alfred A. Knopf, 1965): 388; Edward P. Morgan, *The 60s Experience: Hard Lessons about Modern America* (Philadelphia: Temple University Press, 1991).
17. Daniel D. Chiras, *Beyond the Fray: Reshaping America's Environmental Response* (Boulder, CO: Johnson Books, 1990): 129; Riley E. Dunlap and Angela G. Mertig, eds., *American Environmentalism: The U.S. Environmental Movement, 1970–1990* (New York: Taylor & Francis, 1992): 4; D.B. Luten, "Fading Away?" *Western Outdoors Annual* (Federation of Western Outdoor Clubs; Spring 1973), n.p.; Richard N.L. Andrews, *Managing the Environment, Managing Ourselves: A History of American Environmental Policy* (New Haven: Yale University Press, 1999): 209, 238, 252–255; Mark Dowie, *Losing Ground: American Environmentalism at the Close of the Twentieth Century* (Cambridge, MA: The MIT Press, 1995): 59.
18. Stephanie S. Pincetl, *Transforming California: A Political History of Land Use and Development* (Baltimore, MD: Johns Hopkins University Press, 1999): 186–187, 236; Hal K. Rothman, *The Greening of a Nation? Environmentalism in the United States Since 1945* (Fort Worth: Harcourt Brace & Company, 1998): 5; David Burner, *Making Peace with the 60s* (Princeton: Princeton University Press, 1996): 217–219; Paul Lyons, *New Left, New Right, and the Legacy of the Sixties* (Philadelphia: Temple University Press, 1996): 101; William A. Shutkin, *The Land that Could Be: Environmentalism and Democracy in the Twenty-First Century* (Cambridge, MA: The MIT Press, 2000): 33–34.
19. See *Maine Environment* (Natural Resources Council of Maine), April, June 1977; *Wild Oregon: The Voice of the Oregon Wilderness Coalition*, July 1978, May/June 1979, January/February 1980; *Earthwatch Oregon* (Oregon Environmental Council), 1974.
20. Audrey Jackson in "The Oregon Eighties," *Earthwatch Oregon*, January/February 1980; Commission on Maine's Future, *Final Report* (Augusta, 1977): 3, 38–39; "North by East," *Down East Magazine* 21 (August 1974): 53; Lew Dietz, "Maine at the Crossroads," *Down East Magazine* 23 (May

1977): 67; *Maine Sunday Telegram*, February 5, 1978, January 4, 1981; Robert Deis, "Environmental Watch," *Down East Magazine* 25 (February 1979): 154–155; "Environmental Watch," *Down East Magazine* 26 (February 1980): 75, 124; Condon, "Tides of Change," p. 568.
21. Murray Bookchin, *The Limits of the City* (New York: Harper Colophon, 1974): 73; Evan Eisenberg, *The Ecology of Eden* (New York: Alfred A. Knopf, 1998): 158, 426.
22. Charles A. Reich, *The Greening of America* (New York: Random House, 1970): 152, 162–174, 194, 204.
23. Lawrence Buell, *The Environmental Imagination: Thoreau, Nature Writing, and the Formation of American Culture* (Cambridge: Belknap Press, 1995): 15, 21.
24. James T. Farrell, *The Spirit of the Sixties: Making Postwar Radicalism* (New York: Routledge, 1997): 11, 209, 215.
25. "L.L. Bean's Bonanza," *Down East Magazine* 1 (April 1955): 38–39 (from Tom Mahoney, *The Great Merchants*). See Leon A. Gorman, *L.L. Bean, Inc.: Outdoor Specialties by Maine from Maine* (New York: Newcomen Society, 1981): 7–8, 10, 13–14, 17–18, 20–21; M.R. Montgomery, *In Search of L.L. Bean* (Boston: Little, Brown and Company, 1984): 32, 38–39, 62; Richard Saltonstall, Jr., *Maine Pilgrimage: The Search for an American Way of Life* (Boston: Little, Brown and Company, 1974): 262; *Maine Times*, May 13, 1977; Michael L. Johnson, "A Vertical Ride: Cowboy Chic and Other Fashions of the New West," in *New Westers: The West in Contemporary American Culture* (Lawrence: University Press of Kansas, 1996): 14–54.
26. Joel Garreau, *The Nine Nations of North America* (Boston: Houghton Mifflin Company, 1981): 272–273.
27. Hal Rothman, *Devil's Bargain: Tourism in the Twentieth-Century American West* (Lawrence: University Press of Kansas, 1998): 14.
28. See, for instance, "The Vanishing Redwoods: An Album," *Audubon* 67 (November–December 1965): 359.
29. Michigan Society of Planning Officials, *Patterns on the Land: Our Choices—Our Future* (Rochester: Michigan Society of Planning Officials, September 1995): 21–22; Calvin L. Beale, *The Revival of Population Growth in Non-metropolitan America* (Washington, DC: USDA, 1975); Ray Rasker and Dennis Glick, "Footloose Entrepreneurs: Pioneers of the New West?" *Illahee: Journal for the Northwest Environment* 10 (Spring 1994): 34–39, in Regional List, Box # 1: The West, Samuel P. and Barbara Hays Environmental Archives, University of Pittsburgh. See Alexander Wilson, *The Culture of Nature: North American Landscape from Disney to the Exxon Valdez* (Cambridge, Massachusetts: Blackwell, 1992): 17; William Lucy and David Phillips, *Confronting Suburban Decline: Strategic Planning for Metropolitan Renewal* (Washington, DC: Island Press, 2000): 5; Robert Fishman, *Bourgeois Utopias: The Rise and Fall of Suburbia* (New York: Basic Books, 1987): 15; Murray Bookchin, *The Limits of the City* (New York: Harper Colophon, 1974): 73.
30. Samuel P. Hays, "From Conservation to Environment," in *Beauty, Health and Permanence: Environmental Politics in the United States, 1955–1985* (New York: Cambridge University Press, 1987).
31. Lloyd C. Irland, *The Northeast's Changing Forest* (Petersham, Massachusetts: Harvard Forest, 1999): 130.
32. Rothman, *Devil's Bargains*: 17.
33. Jack Lessinger, *Regions of Opportunity: A Bold New Strategy for Real-Estate Investment* (New York: Times Books, 1986): 76, 88–89, 93. See Rasker and Glick, "Footloose Entrepreneurs: Pioneers of the New West?"

34. Evan Richert, "Maine's Changing Population and Values," in *Changing Maine*, ed. Richard Barringer (Portland: University of Southern Maine, 1990): 30–36; *Maine Times*, December 13, 1974; Rasker and Glick, "Footloose Entrepreneurs"; Bookchin, *Limits of the City*: 72–73.
35. *High Country News*, August 2, October 25, 1974, March 26, 1976; Rothman, *Devil's Bargain*: 231, 280–281.
36. Harry W. Paige, "Leave if you Can," in *Rooted in the Land; Essays on Community and Place*, eds. William Vitek and West Jackson (New Haven: Yale University Press, 1996): 13–14; John M. Findlay, *Magic Lands: Western Cityscapes and American Culture after 1940* (Berkeley: University of California Press, 1992): especially p. 73.
37. Robert Fishman, *Bourgeois Utopias: The Rise and Fall of Suburbia* (New York: Basic Books, 1987): 147–148.
38. *High Country News* in *Maine Times* (July 29, 1983).
39. *High Country News* October 25, 1974. See Margaret Lynn Brown, *The Wild East: A Biography of the Great Smoky Mountains* (Gainesville: University Press of Florida, 2000): 304.
40. Edward Abbey in *High Country News* (October 22, 1976).
41. Jack Hope, "$160 Boots, $66 Suites, and Wild Canyon: The Beautiful Ski Crowd Discovers Utah's Mountains," *Audubon* 76 (March 1974): 85–92.

4 Airworld, the *Genius Loci* of Exurbia

Andrew Blum

EDITORS' INTRODUCTION

This chapter is about the contradictions in the contemporary experience of place and explores how the dream of exurbia as a retreat from the city and modernity is a clever lie. Andrew Blum is a writer and geographer interested in the construction and experience of modern places. In this chapter, Blum explores the infrastructure that enables exurbia. Starting with the premise that *here* is a confusing place, he traces some of the ways in which a sense of place is created in exurbia through iconography, material, and technology from distant places. Modern elements of exurbia are often effaced in the privileging of a sense of the local, the production of the trendy *here* and the seeming authenticity of nature that draws people to exurbia.

The growth of exurbia is supported by a supranational network of chains, airports, and communication technologies—an environment novelist Walter Kirn collectively terms "Airworld" in his 2001 novel *Up in the Air*, which has since been made into a movie of the same name. Airworld is often dismissed as a placeless, interchangeable, history-less landscape of diagrammed sandwiches and national newspapers. This paper attempts a re-conceptualization of place in the face of Airworld. It questions the idea of place as something exclusively and intensely local and historic, and examines an idea of place necessarily more human-centred and self-constructed.

Blum shows the links between exurbia and Airworld—two distinctive and yet linked ways of organizing experience of place—and argues that although the natural and local tend to provide a privileged landscape experience, this experience of place is no more real than the experience of traveling through the channels of air transportation and global communication networks. In fact, the ability to experience exurbia relies on these webs of transportation and communication, however much the privileged landscapes of exurbia seek to mask their presence. And despite the temptation to discount the experience of generic places, he argues that the absence of nature and the local does not invalidate the phenomenological experience of place.

The privileging of natural landscapes and the persistence of pastoral ideology in post-modern aesthetics raise some interesting questions. In a world where it is taken for granted that everything is connected to everything else, why do some people go out of their way to mask the privileges that these technological interconnections provide? One of the premises of this book is that exurbia may be a symptom of a selective rejection of modernity, at least symbolically. Blum asks what sort of reconciliation with modernity might be possible if experiences of Airworld could be embraced. Could the imagined escape that helps drive sprawl be foiled by an acknowledgement that *everywhere is here*, whether in a condo in the city or in an eco-home in the woods?

Using work on the philosophical meanings and functions of *place*, Blum points out that in interactions with everyday environments, people *decide* what will be part of their environments, to some degree; they are also shaped by the ways they have made places. Blum calls for a greater self-consciousness in noticing and shaping environments. Addressing a central tension of exurbia, this exploration of hybrid place helps us think through the dissonance between the natural, rural ideal of exurbia and the modern built environment. As Thomas Looker argues in greater detail in Chapter 9, Blum makes a case that the effort to engage with the duality of a landscape acknowledged as modern yet natural takes *work*, work Blum describes as acts of imaginative coherence—knitting together in the grounded experience of *here* the sometimes overwhelming and discontinuous array of inputs that constitutes experience of space and place.

Acknowledging this difficult and ongoing work is an important step in opening up dialogue on many issues of exurbia and sprawl that, like cell phone towers, are often hidden away behind the natural appearance of the exurban landscape. Turning to Walter Kirn's bestseller *Up in the Air* and Henry David Thoreau's *Walden* for two accounts of the work required to makes places into *here*, Blum makes a case for the need to address the reasons people escape from the landscapes of placelessness and also for the possibility of doing so. Exploring the imaginative effort already expended in reconciling hybrid places to the experience of *here*, wherever we are, Blum exposes an impact of exurban escape-to-nature narratives that often gets lost amidst the more common accounts of the problems of high-impact exurban living. Discomfort with making visible in the landscape the infrastructure of modern lifestyle makes it much more difficult to engage in conversations that could help reconcile modern lifestyles with the ideals of sustainable and satisfying living that often accompany them.

Drawing upon Henry David Thoreau's *Walden*, Blum shifts the emphasis of the Thoreauvian project away from notions of wilderness and solitude, and toward a conscious and deliberate attempt to live in the here and now. Using as an example Ryan Bingham, the frequent-flying protagonist of Kirn's *Up in the Air*, this chapter reflects on the experience of searching, like Thoreau, for experiences of place in a world of increasing speed and homogeneity.

—Kirsten Valentine Cadieux and Laura Taylor

■ ■ ■

"Don't tell me this isn't an age of miracles. Don't tell me we can't be everywhere at once."

—Walter Kirn, *Up in the Air*

"Why do precisely these objects which we behold make a world?"

—Henry David Thoreau, *Walden*

Walking recently on a heavily wooded ridge a dozen miles north of San Francisco, I came upon a small clearing filled with a handful of metal utility boxes painted a deep green and surrounded by a chain link fence. Above them, an antenna rose perhaps a dozen feet in the air, lower than the surrounding trees. Heavy wires led to a series of telephone poles that ran down the hill. Just a quick glance in my pocket confirmed my hunch: this was a cell phone tower, quietly buzzing with the communications of the valley below. A bird buzzed nearby as well, and a light fog swept over the hill from the ocean beyond. The tapping and sawing sounds of house construction permeated the forest. Poison oak spotted the ground.

Twenty minutes later, out of the woods and into town, the featured blend at the coffee place was from Sumatra. A guy in the window seat had his laptop open, and was emailing who-knows-where. My cup was decorated with a woodblock pattern of such elegant simplicity it must be ancient. As I tossed it in the recycling bin, my cell phone rang—no doubt summoned from those green boxes in the clearing up on the hill. It was my sister-in-law on a visit to New York, hoping I could remind her which subway would take her from Times Square to Union Square. I could hear the squeal of sirens in the background and the ruffle of wind on 42nd Street, and looking down at my feet to concentrate on that other place 3,000 miles away, my eye caught a rumpled copy of the day's *New York Times*—edited not a hundred yards from where she was calling from, but electronically transferred out here, then trucked up this valley.

By now, the fog had gotten thicker, carrying with it the smell of redwoods from the ancient stand nearby, protected by the federal government in the name of John Muir, champion of the wilderness. I crossed the street to grab a snack at a small restaurant selling "Indian Burritos" (Punjabi flavors wrapped in Nan) and washed it down with a "Brooklyn Cream Soda," bottled in Oregon. Around the corner was a small boutique named after a popular trail on the local mountain; in the window were hiking clothes from Italy and maps of the ground upon which I stood—of *here*. But—all these things kept insisting—*here* was a confusing place.

My immediate surroundings—the "here" of our day-to-day existences—are full of elements of faraway places. I may be eager to understand this place, but this place is not explicitly or even primarily about here. Instead, it is a "hybrid place," best characterized by the presence of other places. Its organic spirit, it *genius loci*, does not only bubble up from the earth over time but arrives nearly instantaneously on a wing or a wire, in a truck or with the breeze. And yet that makes it no less meaningful, no less rich an experience. It is neither "placeless" nor a "non-place;" the appearance of the remote or the faraway is not a "virtual reality" nor a "media-space" (although perhaps it has elements of both); and for all this confusion, this place is not entirely lacking in what could be called intrinsic character (although often what character there is seems incongruous). What results is, if not a paradox, then an incongruity in our understanding of the concept of place. This place, like most places today, is not purely local but a hybrid of the local and the remote. I call it a hybrid place.

Exurbia is a hybrid place. Inherent in the exurban ideal of being in the natural world is the parallel ideal of remaining connected to the wide world. As a result, while seeking a lifestyle closer to nature, exurbanites are often drawn deeper into the world they meant to escape—of Federal Express, technologically mediated communications, global supply chains, and the global marketplace. To engage with the processes of exurbia requires us to engage with both the "natural" processes that draw us there in the first place and the "technological" ones that allow us to stay—in other words, to re-imagine exurbia as a hybrid place. My hope is that if exurbia is a construct, an imagined landscape as much as a real one, then addressing the basis of those imaginings by re-thinking what we mean, and the meanings we summon, by the term "sense of place" can help us to better understand the things that make the landscape the way it is, and consequently allow us to begin a stewardship borne out of love for *that* landscape, rather than an imagined ideal.

A conceptualization of hybrid place relies heavily on the ongoing work to conceptualize "place" and "sense of place" in a way that is both coherent and respectful of the existential complexities that place entails—most notably the work of the philosopher Ed Casey and the geographer Ted Relph. As Casey is fond of quoting Archytas, "To be is to be in place," and as Relph has put it, "Sense of place is first of all an innate faculty, possessed in some degree by everyone, that connects us to the world."[1] There is an essential duality to places: they are both individually experienced and collectively held, and they consist of both figments of our imagination and the material world around us. Place is within and without; or, as Martin Heidegger writes, "it is neither an external object nor an inner experience."[2]

The enormity of all these statements illustrates the difficulty of grasping place in the abstract. And yet even more grounded discussions about place are often laden with a particular set of images. As Relph has written, "Somewhere behind most discussions about place lies an image of quiet simple landscapes where there are no great cities, no suburban tracts, no ugly factories, no money-based economies, and no authoritarian political systems."[3] As a geographical and philosophical concept, place is often constituted as a sense of the local blind to the manifestations of a remarkably global, telemediated world. For the most part our notion of place remains deeply rooted in traditional models of settlement and societies. It is often unashamedly anti-modern, in the association of place with local customs and characters, with a sense of continuous history, and with identity. In his book *Non-Places*, Marc Augé uses the term "anthropological place," which he defines with the slightly tongue-in-cheek equation "land = society = nation = culture = religion."[4] Augé's definition is deliberately overstated in opposition to the characteristics of "non-places," but the point remains the same: this is no longer the dominant structure of the world. Perhaps in aboriginal outposts one can find four out of those five, but in most places, the only synchronicity I would count on is that of land and nation. And sometimes even that too falls by the wayside. The traditional models of what defines places are no longer dominant, if they apply at all.

But I resist the temptation to throw out the baby with the bath water, to dismiss the idea of place altogether when faced with the absence of its more romantic characteristics. There is nothing in the definition of place—save certain particularities of their very emplacement in the world—that makes places obsolete when they no longer ascribe to the traditional characteristics of local distinctiveness. This is not to say that pre-modern, often traditionally "country" places are no longer powerful (although that very power sometimes makes them tourist attractions, with a whole other set of issues), but I seriously question the notion that modern landscapes do not have the same potential for meaning as pre-modern ones—in other words, that place is obsolete. On the contrary, in a complex modern world complicated modern places can be meaningful. In the end the feeling is the same: in a hybrid place, just as in a more traditional place (if we can find one), I may be profoundly aware of my being in the world, even if the characteristics that inspired that feeling diverge from what we have come to associate with the idea of place.

High among these associations is the requirement that places exist in locations. We are always, of course, *somewhere*, but today that *somewhere* is increasingly divorced from the meanings of places in our day-to-day experience. Instead, that meaning is often derived not only from our immediate location but from other remote locations. Even when standing still, even at home, our location is not necessarily our place—we are very possibly engrossed in somewhere else, whether staring at our smart-phone screens, reading the paper newspaper, or hiking through a wooded glade in the fog thinking of Yosemite. All perception may be local perception, but there remain aspects of our experience of the world that quite explicitly come from somewhere else, sometimes from faraway. As a result, we might often find ourselves living locally in a place that is not local at all. It is here and now, but only in our perception of it. In all other ways, it is from there, and even then—transmitted by any number of communications devices, shipped, carried with us as things or thoughts. We effortlessly incorporate into our lives technology that enhances our means of perception, allowing us to hear and see things faraway. This does not mean we experience places remotely, but we do incorporate the remote into our experience of place. We do it constantly, nearly unthinkingly: every time we talk on the phone, log onto a social network, eat an imported strawberry, or even acknowledge our individual places of origin, themselves likely complicated. Place is not exclusively local, and location is not necessary to place.

But faced with such a constant barrage of sensations from everywhere, how do we go about understanding them as any sort of coherent place, as a palpable *here*? If we are not only where we are, but elsewhere—or rather, if where we are *is* elsewhere—how do we understand it? I would argue that that sense of being in a place is achieved by bringing to our perceptions an act of imaginative coherence, a sort of temporary patch for the discontinuities of the world. When our perceptions are increasingly mediated through a variety of technologies, we rely more heavily on our imaginations to make them coherent, and even to learn new ways of understanding those perceptions. Without these leaps, the world can be a jumbled mix of incongruous stimuli in which nothing seems consistent to the time and place.

For example, literally walking (much less driving) while talking on a cell phone is an acquired skill; we must learn to hear the nuances necessary for conversation when they are utterly incongruous with the surroundings, especially when we must simultaneously react to those surroundings. It is not so simple as hearing the voice on the other end. Instead, we constantly bring to our perceptions an acquired understanding of their sources, and we learn the ability to incorporate those perceptions into everything else going on in the immediate environment. We suspend disbelief that the sounds we're receiving have actually "traveled," that they are not inherently here—even if, in some ways, there is no difference. When it comes to our experience of the remote, there may not be a "there there," but there is certainly a here here, some stimuli that our senses react to. Even sitting at my desk (wherever that may be) I am responding to the computer, the weather outside, the planes overhead, the pollen in the air, and attempting to organize them into some coherent sense of this place. These perceptions may be tricky and variable, but it seems increasingly unreasonable to preclude from our experience the extended reach that technology provides. Perception may not be exclusively local, but it remains the starting point for our being in place. But what then is the ending point? How, in Ed Casey's words, do we find ourselves "in the midst of an entire teeming place-world rather than in a confusing kaleidoscope of free-floating sensory data"[5]—especially when that sensory data is disembodied, incongruous, inauthentic or artificial? The easy answer might be because we want to—because we bring to objects of our experience what Relph describes as an "unselfconscious intentionality": "The basic meaning of place, its essence, does not therefore come from locations, nor from the trivial functions that places serve, nor from the community that occupies it, nor from superficial and mundane experiences—though these are all common and perhaps necessary aspects of places. The essence of place lies in the largely unselfconscious intentionality that defines places as profound centers of human existence."[6] Place depends on an infinite variety of characteristics and associations, but place is also what you make of it—and how it makes you.

But more and more, the places we inhabit are organized around and defined by a complicated mix of the local and the remote. In our contemporary experience dense with images, signals, people and things from everywhere, there remains the potential for meaningful places, which are often hybrid places—and are best understood as such. For example, I am continually awed at the ways in which we more fluidly construct our places, the ways we feel more at home anywhere. "Home" can be a café with a good cell phone signal, and the café itself (especially if it's a Starbucks) may likely mimic the material vocabulary of home. Granted, this idea of "home" may lack the profundity of the ancestral village, the house you were raised in, or a vivid mountain valley, but these are exactly the types of place-characteristics that have begun to slip away. If "home" itself is a construct, it seems difficult to discern the authentic from the inauthentic, the empty sign from the thing itself. And yet we still live our lives in place, we still engage in an "unselfconscious intentionality that defines places as profound centers of human existence," as Relph puts it. It is just that those centers need not be fixed in locations, and the elements of which they consist may be increasingly complex, and sometimes superficial.

And although I repeatedly say "increasingly" and "more and more" I want to emphasize that I do not think this idea of hybrid place is dependent on any particular narrative of technological development, nor is there a need to assert that it is something ostensibly new. Without a doubt, the remote has in some form been present in the local for as long as geography itself. (Why, after all, would we need to pay attention to our place if we did not know, or had not at least heard of, other places?) But I would argue that the seamlessness with which we incorporate elements of the local and the remote is characteristic of our era, whether you call it post-modernity, or late capitalism, or define it by the near-ubiquity of telecommunications. The writer and critic Pico Iyer has put it well when he describes the "present moment" as one in which "the world is spinning around us at the speed of light."[7]

And I also want to stress that I am not immune to the drawbacks, dangers, and consequences of hybrid places. Henry David Thoreau writes of his house in *Walden*: "with this more substantial shelter about me, I had made some progress toward settling in the world."[8] But in hybrid places, whatever progress we make toward settling in the world does not depend on shelter; it is not so much "substantial" as immaterial, reflective of a world spinning around us at the speed of light. Hybrid places are often confusing and choppy; hybrid places may be meaningful, but those meanings may not always be rich, sustainable, or just. But they are, however, often collectively held. The spinning world arranges itself in an intense and communally visible tension between the local and the remote.

Exurbia is one of these shared hybrid places. As other essays in this book explore, the myths, signs and symbols that sustain it do not arise out of the land itself, but are shaped by the broader zeitgeist—or rather, a reaction to it. It barely needs noting that we live in a "McWorld," a "Geography of Nowhere," "a world of multi-national corporations, universal planning practices and instantaneous global communications." In this "Airworld"—a term I borrow from the novelist Walter Kirn—the remote dominates, nothing is intrinsically local, and everything (it seems) comes from faraway, often at the command of corporate headquarters. Kirn uses "Airworld" to mean airports and their attendant spaces, such as chain hotels, convention centers, and Starbucks. But I also include in it more abstract and pervasive spaces, the networks and the mediascapes that define popular culture.

Taken all together, these spaces provoke an odd transcendence of place: an insistence on the possibility of being at home everywhere, if only because everywhere is like everywhere else. As Ryan Bingham, the protagonist of Kirn's 2001 novel *Up in the Air* puts it, "Don't tell me this isn't an age of miracles. Don't tell me we can't be everywhere at once."[10] Intense places (good or bad) have always inspired intense desires to escape, to abandon the bedrock of values that define quotidian life. But Airworld raises the possibility of a shift in this narrative. If everyplace is the same, if there is nowhere different to escape to, then the only option left (short of heading to a cabin in the woods) is escaping *to* the totality and attempting to make yourself at home everywhere, and therefore nowhere. This is to insist that place doesn't matter anymore because—for some people, at least—every place presents the same possibilities. In Airworld, we approach places with little regard for local specificity, characteristics, or location. We use technology and the centralizing and standardizing forces of capitalism, and our responses to them, to find meaning wherever we are—and often that meaning is empty, the places we construct houses of cards devoid of any psychic foundation. As the world becomes increasingly homogenized, it logically follows that the dissatisfied would seek out the unique—the special places, often the natural ones. But in a strange paradox, it also follows that the unique and the different become increasingly visible and transferable—and consequently not so unique, and not so different.

In Airworld, the belief that places are entirely fluid is often accompanied by the belief that all values are fluid. It is a distinctly postmodern stance, devoid of all absolutes, subscribing to a cosmology whose foundation lies in the cultural logic of global capitalism. As the marks of difference are illuminated against the bright and not always sympathetic light of sameness, they are also made ready for exclusion or appropriation—a tension that characterizes any speculation on the fate of place in a postmodern world. As David Harvey explains in *The Condition of Postmodernity*, "The less important the spatial barriers, the greater the sensitivity of capital to the variations of place within space, and the greater the incentives for places to be differentiated in ways attractive to capital."[11] There is the possibility that standardized places will form the backdrop for a new skyline of distinctiveness. But it is more likely that that distinctiveness will only be skin deep, and the new landmarks of locality will be funded by the same global forces, and cater to a fully commodified standard of regional difference. The complicated and subtle specifics of meaningful places do not translate as easily as they travel. It is a complicated set of paradoxes: the very fact of our mobility makes distinctive places more accessible, but consequently less distinct. And the increasing similarity of places means that where we are matters less, but also more.

This is the worldview exhibited by Ryan Bingham, the protagonist of *Up in the Air*. A thirty-five year old "career transition counselor" (he fires people for a living), Ryan has a quixotic mission: to accumulate one million frequent flyer miles. In the novel we accompany him as he travels around the American West on the last week of his journey to a million—which he calls with typical geographic acuity both a "horizon" and a "boundary."[12] The company he works for is headquartered in Denver, but Ryan lives, as he describes it, "in the margins of his itinerary," in "Airworld." "Planes and airports are where I feel at home," he says.[13] "Everything fellows like you dislike about them—the dry, recycled air alive with viruses; the salty food that seems drizzled with warm mineral oil; the aura-sapping artificial lighting—has grown dear to me over the years, familiar, sweet." As he explains:

> I call it Airworld; the scene, the place, the style. My hometown papers are *USA Today* and the *Wall Street Journal*. . . . Airworld is a nation within a nation, with its own language, architecture, mood, and even its own currency—the token economy of airline bonus miles that I've come to value more than dollars.[14]

It is, admittedly, totally at odds with what we expect from places. He craves homogeneity: "Unless a dish can be made to taste as good no matter where it's prepared, LA or Little Rock, it doesn't entice me," he says. Ryan flies enough to wave hello to shoe-shine guys and security guards, and to know the people at the hotel bar in any city in the West—not the locals, but his fellow-residents of Airworld, who are similarly at home nowhere and everywhere. Not that Ryan ignores the downsides. "A steady relationship with a good dentist is tough to maintain in Airworld," he acknowledges. And, "The cities don't stick in my head the way they used to."[17]

But Ryan puts up a convincing façade of loving it, if only because the "real" world he sees at its margins—in the American West, at least—offers no more distinctiveness. Throughout the novel people keep trying to get him to settle down and buy a house, but it is a source of pride with him that he is, quite literally, homeless. As he says at the beginning of the novel, recalling a seatmate, "the kid has a soft spot for homeless immigrants, which pretty much describes all of us out West, though some are worse off than others. We're the lucky ones."[18] Ever since Ryan's ex-wife left him while he was on a twenty-night business trip (leaving on the doorstep "a heap of rolled-up newspapers dating back to the morning of my departure"[19]), the idea of putting down roots in the real world means the disconcerting prospect of pulling up roots in Airworld. As he says, "Green grass is a losing battle in the West, which wants to go back to sage and prickly pear, and so is securing an outpost in the sprawl. I look down on Denver, at its malls and parking lots, its chains of blue suburban swimming pools and rows of puck-like oil tanks, its freeways, and the notion of seeking shelter in the whole mess strikes me as a joke."[20]

In this world he's avoiding he identifies a placelessness he successfully fends off in the world he lives in. Airworld is "familiar, sweet;" the exurban sprawl down there is just as superficial, but foreign, the food inconsistent. In Airworld's static antipode, the place Ryan would make for himself once he'd overcome what he calls "the necessary vigilance of nomadism" would be profoundly empty. As another character stresses to him over coffee in an airport lounge, pushing him to buy a house, "be aware of this, it *is* a community, not just a development."[21] In the context of this come-on, the notion of a brand new house providing "shelter"—at least the kind Ryan really needs—rings false, so fully is it submerged in the corporate marketing ideals that emanate from Airworld itself. Kirn—who himself left New York City to write novels on a farm in Montana—captures the rhetoric of exurban development perfectly, and the corporate consistency with which new neighborhoods are summoned from the earth. For Ryan, "securing an outpost in the sprawl" is "a losing battle"—so fully does it embody a sense of home so oppressive that it only begs to be escaped, even to nowhere. Reading *Up in the Air,* faced with a world of increasing absurdity and persistent reality, you are left constantly wondering if Ryan's affection for Airworld is an illusion of his wholeness or, in his imagining of it as a place, a fair match for the banality of sedentary America. If the sprawl is soulless, as Ryan sees it, why not place all value in frequent flyer miles?

But halfway through *Up in the Air* Ryan tells a story that helps to explain the psychic slippage inherent in Airworld's paradoxes. Traveling from Ontario, California to his sister's house in Salt Lake, he falls into a memory of the town he grew up in, Polk Center, Minnesota. It is a comically idyllic vision of place—the sort that fuels the growth of exurbia. Ryan remembers the cap he wore that said "Polk Center Gouda," and recalls the old traditions the town kept up, like polka bands and spinning cars on frozen lakes. It is this last one he remembers most vividly, especially the time his car broke through the ice and, unconscious, he was flown by helicopter to Minneapolis. Nearing the hospital, he woke up, looked out the window, and in a long and elegiac passage describes the world that opened up for him:

> I could see the western horizon, where I'd come from, and a dogleg of snowy river crossed by bridges sparkling with late-night traffic. The landscape looked whole in a way it never had before; I could see how it fit together. My parents had lied. They'd taught me we lived in the best place in the world, but I could see now that the world was really one place and that comparing its parts did not make sense or gain our town any advantage over others.
>
> Moments later, we landed. My stretcher jolted. As we waited for the helicopter's blades to slow, the medic said I would be home in a few days, not understanding that this was not the comfort it would have been had I never left the ground. He wheeled me out onto the roof under the moon, which had risen some since I'd seen it from the car. I lifted the oxygen mask so I could speak and asked how long we'd been flying. Just thirty minutes. To reach a city I'd thought of as remote, halfway across the state, a foreign capital. I told the man I was feeling lucky now.
>
> Tonight, in Salt Lake, I'm feeling lucky again. . . Three hours and thirty-five minutes, door-to-door, across the Great Basin to my sister's mansion in the foothills along the Wasatch Front. I slept, I woke, I hailed a cab, I'm here. Don't tell me this isn't an age of miracles. Don't tell me we can't be everywhere at once.[22]

It follows, in many ways, the familiar form of place epiphanies: "the landscape looked whole" to him, he is hyper-aware of both his body and the moon, it is "an age of miracles." But it is striking that the world opening up to him wasn't reassuring, because it wasn't the world he had known. Suddenly aware of the compression of space and time possible with flight, the old sense of place seems outdated, a "lie." Seeing Minnesota spread out before him from up in the air, the landscape of his youth wasn't so much placed in context as removed from context—not one part of many, but one part entirely. "I could see now that the world was really one place," Ryan says. And then his declaration—"Don't tell me this isn't an age of miracles. Don't tell me we can't be everywhere at once"—is defensive, but assertively so. Ryan doesn't merely pronounce that this *is* an age of miracles, and that we *can* be everywhere at once, nor does he suggest that he *believes* it so. Instead, in his insistent "don't tell me" there lies a deeper assumption at work that rejects any and all imposition of values, any "meta-narrative" that indicates Ryan's relationship to time and place. This seems a particularly, acutely, postmodern stance: one in which some people think that they can "act as, and become what, they please."[23] Ryan's statement neither invites nor makes judgment. If Ryan sees the world as whole, he can think of no grounds for anyone to disagree; and if he feels that wholeness is miraculous, so it is.

By this logic, the criteria by which we judge places has come unfixed and without it everywhere is as good as anywhere and anywhere is as good as nowhere. The paradox that where we are matters less but also more is resolved by the fact that nothing matters at all—or at least everything is debatable. It is akin to Robert Venturi and Denise Scott Brown's premise in *Learning from Las Vegas*: "The commercial strip challenges the architect to take a positive, non-chip-on-the-shoulder view," they write. "Las Vegas's values are not questioned here. The morality of commercial advertising, gambling interests, and the competitive instinct is not an issue."[24] No surprise then that Airworld, with Las Vegas as its capitol, is the land of nowhere and everywhere, the land of the strip, and of the particular relationship between capitalism and society that is best described as postmodernity. In Airworld, nothing is fixed, not places, people, or values. Whatever meaning we find in Airworld therefore seems ultimately precarious and ignoble, rather than illustrative of some great project of modernization. In Airworld, we are consumed by what Frederic Jameson calls "the problems of figuration," "conveyed by way of a growing contradiction between lived experience and structure, or between a phenomenological description of the life of an individual and a more properly structural model of the conditions of existence of that experience."[25] Having given in to the totality, we cannot grasp it. Jameson identifies this as "the so-called death of the subject," marked by its "fragmented and schizophrenic decentering and dispersion"—fair words to describe Ryan and, often, us.[26]

This is the world from which exurbia promises an escape. And yet, what are the terms of that escape? Especially when, as we've defined it, to be an exurbanite means to maintain connections with the world you've just left. But then, couldn't one engage with that duality, rather than suppress it? Couldn't one embrace exurbia as a hybrid place?

I think that, not for the first time, Henry David Thoreau offers some answers, or at least some good questions. Kirn's novel is a convincing portrait of ennui in America today, an ennui rooted in the embrace of exurbia and Airworld. But *Walden* offers another, more hopeful take on the connection between the natural world and Airworld as played out in exurbia. *Walden* is, among many things, Thoreau's attempt at understanding what he perceived as the encroachment of Airworld on nature. You could even call Thoreau the first exurbanite. His trip to the woods in *Walden* is explicitly framed as an escape from the hustle and bustle of Concord, Massachusetts, 1845. And yet that escape was not as bold as it is often presented. The daring part of Thoreau's experiment wasn't where, or even how, he chose to live, but his effort at observing and recording his experiences—"notching it on my stick," as he puts it. And, interestingly, many of those experiences were modern ones, deliberately so. He was not strictly a naturalist, but constantly concerned with the interface between the natural and human worlds, the modern and the pastoral. This, it seems to me, is his project: not to exclude all that is modern, fast, and new, but rather to examine it at arm's length. Overwhelmed by the changes in the world, Thoreau does not escape them by going to Walden Pond, but goes there to engage with them. Whether he likes them or not is beside the point. Thoreau's world is not black and white. *Walden* the book is full of contradictions. But what he is clear about is his method, his attempt at finding meaning in everything, especially the heralds of modernity—key among them the railroad.

The Fitchburg line, then as now, nearly touches the edge of Walden Pond, opposite his house. The tracks are nearly invisible until a train passes, momentarily dominating the landscape. Thoreau revels in this. "I watch the passage of the morning cars with the same feeling that I do the rising of the sun, which is hardly more regular. Their train of clouds stretching far behind and rising higher and higher, going to heaven while the cars are going to Boston, conceals the sun for a minute and casts my distant field into the shade, a celestial train beside which the petty train of cars which hugs the earth is but the barb of the spear."[27] Thoreau is searching for a way to incorporate the railroad—symbol of all his frustrations with the world, its speed, its silly commerce, its superficiality—into his world of celestial rhythms, and a landscape rich in metaphor and meaning. It is a strikingly contemporary attitude, and he follows it up with strikingly contemporary examples. In this same chapter, "Sounds," Thoreau goes on and on about where all the things have come from that the trains are carrying: lumber from the Maine woods, Manila hemp and coconut husks, salt fish from the Grand Banks, Spanish hides, molasses on its way to Vermont, the "cattle of a thousand hills."[28] He is in awe of the marketplace, in awe of the goods that pass by him, while he stays still at his little spot in the world. "So is your pastoral life whirled past and away," Thoreau writes.[29] Faced with all this, Thoreau does not give up on the place-world or the local, and yet he does not ignore the world as it is, trains and all. "I will not have my eyes put out and my ears spoiled by its smoke and steam and hissing," he hisses himself.[30] Thoreau is determined to see place in the midst of the passing trains. He is determined to live among the contradictions of the world, even as he lives alone in the woods. I take his constant use of natural metaphors to describe industrial things to indicate his eagerness to abandon their duality, and to find connections between them. Thoreau was deliberate in choosing a spot in the woods with a great view of a passing train. The train seems to me the driving force—the metaphor is appropriate—for Thoreau's endeavor to live in place, in that the trains' essential dynamism (they are places that move from place to place) stand in sharp juxtaposition to Walden Pond's immobility. The pond, Thoreau notes, has no outlet. It is literally not going anywhere, making it the perfect place for finding a way to live deeply in a world of increasing mobility.

Both Thoreau and Kirn draw upon both the natural and the technological, the local and the remote, as sources of meaningful experience. They understand and accept the world as a hybrid place. Can exurbanites too? Is it possible to live in a world of the local and remote in a way that neither ignores nor is subsumed by its tensions and contradictions?

This is the very tension that makes modernity so fascinating, a tension Hilde Heynen describes as "the relationship between all these divergent aspects, programmatic and transitory, pastoral and counter-pastoral."[31] She continues, "Marshall Berman argues that for the individual the experience of modernity is characterized by a combination of programmatic and transitory elements, by an oscillation between the struggle for personal development and the nostalgia for what is irretrievably lost."

This is also a way of understanding hybrid place: we struggle to incorporate the experience of the remote into our lives, while we bemoan the loss of the local as our primary way of being in the world. At times our experience is heightened by the presence of the remote and the contradictions of modernity, and the effort is satisfying. But at other times the effort is overwhelming, and the fissures are deep, and deeply personal. And yet in the indissolubility of this conflict exists a strange comfort. As Marshall Berman writes, "To be modern is to find ourselves in an environment that promises us adventure, power, joy, growth, transformation of ourselves and the world—and, at the same time, that threatens to destroy everything we have, everything we know, everything we are."[32] The key here lies in Berman's insistence that these forces run "at the same time." There will be no resolution of the local and remote, the pastoral and counter-pastoral, the human and the technological—not even in exurbia. Especially not in exurbia. Short of momentary transcendence, there will certainly be no resolution of the world outside and the world within, that basic duality of place. Mixed feelings are inescapable. Modernity will forever be a tense repose. "To be a modernist," Berman writes, "is to make oneself at home in the maelstrom, to make its rhythms one's own, to move within its currents in search of the forms of reality, of beauty, of freedom, of justice, that its fervid and perilous flow allows."[33]

In defining hybrid place, my hope has not been to resolve these contradictions, but to illuminate them. In our experience of place, we must be open to the elements of place that are local, intrinsic and timeless, *as well as* elements that are remote, technological and modern. As Heynen puts it, "To repudiate modernity as a monolithic whole that deserves to be censured is a conservative and reactionary attitude; not only does it ignore the fact that we are 'modern' whether we want to be or not; it also reneges on the promises of emancipation and liberation that are inherent in the modern."[34] And yet in the place-world, it is undoubtedly the modern that most frequently draws our disdain. That Heidegger's concept of "dwelling"—the holy grail of place philosophy (sought, but rarely found)—is rooted in just such repudiation indicates the depths of our struggle to incorporate the modern in our experience of place. To ignore the modern is to be profoundly disconnected from the world in which we actually live. The prevalence of remote experiences and of standardized places may not be good (and undoubtedly there are ways, if we are so inclined, to fight them) but they are here, quite literally. Acknowledging the potential for hybrid places—like exurbia—as meaningful sources of experience is an important step toward achieving a satisfying and sustainable relationship to the world.

NOTES

1. Edward S. Casey, *Getting Back Into Place: Toward a Renewed Understanding of the Place-World* (Bloomington: Indiana University Press, 1993); Edward Relph, "Modernity and the Reclamation of Place" in *Dwelling, Seeing, and Designing: Toward a Phenomenological Ecology*, ed. David Seamon (Albany: State University of New York, 1993): 208.
2. Martin Heidegger, *Poetry, Language, Thought*, translated by Albert Hofstadter (New York: Harper and Row, 1971): 156.
3. Edward Relph, (1993): 25.
4. Marc Auge, *Non-Places: Introduction to an Anthropology of Supermodernity*, translated by John Howe (London: Verso, 1995): 116.
5. Edward S. Casey, "How To Get from Space to Place in a Fairly Short Stretch of Time: Phenomenological Prolegomena", in *Senses of Place*, eds. Steven Feld and Keith H. Basso (Santa Fe: School of American Research, 1996): 17.
6. Edward Relph, *Place and Placelessness* (London: Pion, 1976): 43.
7. Pico Iyer, *Imagining Canada: An Outsider's Hope for a Global Future* (Toronto: Hart House, 2001): 27.
8. Henry David Thoreau, *Walden and Other Writings of Henry David Thoreau* (New York: Modern Library, 1992, originally published 1854): 81.
9. Edward Relph, (1993): 25.
10. Walter Kirn, *Up In the Air* (New York: Doubleday, 2001): 141.
11. David Harvey, *The Condition of Postmodernity: An Enquiry into the Origins of Cultural Change* (Oxford: Blackwell, 1990): 296.
12. Walter Kirn, (2001): 10, 11.
13. Walter Kirn, (2001): 5.
14. Walter Kirn, (2001): 7.
15. Walter Kirn, (2001): 75.
16. Walter Kirn, (2001): 21.
17. Walter Kirn, (2001): 150.
18. Walter Kirn, (2001): 2.
19. Walter Kirn, (2001): 27.
20. Walter Kirn, (2001): 25.
21. Walter Kirn, (2001): 161.
22. Walter Kirn, (2001): 141.
23. David Harvey, *The Condition of Postmodernity: An Enquiry into the Origins of Cultural Change* (Oxford: Blackwell, 1990): 5.
24. Perry Anderson, *The Origins of Postmodernity* (London: Verso, 1998): 21.
25. Frederic Jameson, "Cognitive Mapping" in *Marxism and the Interpretation of Culture*, eds. Larry Grossberg and Cary Nelson (Urbana: University of Illinois, 1988): 349.
26. Frederic Jameson, (1988): 351.
27. Henry David Thoreau, (1854): 110.

28. Henry David Thoreau, (1854): 115.
29. Henry David Thoreau, (1854): 115.
30. Henry David Thoreau, (1854): 116.
31. Hilde Heynen, *Architecture and Modernity: A Critique* (Cambridge, MA: The MIT Press, 1999): 13.
32. Marshall Berman, *All That Is Solid Melts Into Air: The Experience of Modernity* (New York: Penguin, 1982): 15; and Hilde Heynen, (1999): 13.
33. Marshall Berman (1982): 345–346.
34. Hilde Heynen, (1999): 14.

5 Rewilding Walden Woods and Reworking Exurban Woodlands
Higher Uses in Thoreau Country

Brian Donahue

EDITORS' INTRODUCTION

This chapter engages problems in valuing green sprawl for its aesthetic nature, and explores ways that the aesthetic of nature as pristine wilderness limits meaningful everyday environmental engagements that green infrastructure in urban and residential landscapes might otherwise make possible. The view of exurban landscapes as pristine nature—a view supported, in part, by a popularized version of Thoreauvian cultural heritage—undermines valuable possibilities for more sustainable and equitable land uses. Such land-use improvements might be encouraged by helping residents of green sprawl come to terms with the ways their landscapes could be managed to reduce their disproportionate resource appropriation.

Brian Donahue is an environmental historian and community farmer and forester, chair of the Environmental Studies program at Brandeis University in Massachusetts and author of *Reclaiming the Commons: Community Farms and Forests in a New England Town* and *The Great Meadow: Farmers and the Land in Colonial Concord*. Combining fascinating historical narratives with experience managing working forest and farmland in the Boston suburbs, Donahue presents an overview of land use in the exurban areas known as Thoreau country and explains how the ideology of wildness alone is not sustainable.

In "Huckleberries," Henry Thoreau wrote that "each town should have a park, or rather a primitive forest, of five hundred or a thousand acres. . .where a stick should never be cut. . ., but stand and decay for higher uses—a common possession forever, for instruction and recreation. All Walden wood might have been reserved. . ." A thousand acres surrounding Walden Pond has indeed today been preserved. This chapter examines the history of Walden Woods and of the relationship of people and nature in Concord as Thoreau encountered it, and as it stands today. It inquires into the higher uses of Walden Woods, and argues that there are no convincing ecological, historical, or even Thoreauvian grounds for preservation alone. It argues for cutting a stick. It lays out an approach to stewardship of Walden Woods that stresses cultural engagement with nature, including wildness, biodiversity protection, research, education, recreation, and sustainable harvesting of wood products for local use. Finally, it embeds this proposal for Walden Woods within a larger vision of sustainable community forestry and ecosystem management in the fragmented "Thoreau Country Forest" of the suburban towns surrounding Walden.

Appealing to Thoreau as an historical and regional authority on what should be done with the cultural landscape, Donahue critiques the passive sylvan aesthetic of exurbia. Calling for more actively engaged approaches to sustainable land use, he contrasts exurbanites' aesthetic fervor for "saving" trees with a pragmatic philosophy of engagement with resource production. Local forestry might well meet both goals, strengthening the ties of residents to their environments while at the same time reducing the consumption of forest resources from more distant environments.

This book asks questions about the role of ideals of nature in the production of green sprawl landscapes and in their subsequent management. One major use of nature is as an aesthetic backdrop for a particular lifestyle. One of the central tensions in the exurban landscape involves the replacement of pre-existing productive uses, such as farming and forestry, with landscapes valued for the amenity of their "preserved" nature. As amenity consumption transforms the landscape, residents and planners refer to the environmental impacts of productive processes as a threat to the landscape, rather than considering the role of these processes in shaping the landscapes they find desirable. Donahue's chapter cuts to the heart of the matter with his environmental history and future vision of Walden Woods, the landscape Thoreau made famous in his mid-nineteenth century book *Walden*. In this chapter, Donahue makes a case for harnessing the widespread reverence of nature-lovers for Thoreau and his famous Walden Woods to an active ethic of stewardship.

By situating the natural environment of the Walden Woods within specific ecological and human history, Donahue helps reverse the replacement of productive land uses by generic amenity lifestyle land uses in the geographic imaginary of green sprawl. He dispels a romantic imagination of Thoreau's iconic landscape as a wilderness, and of the exurban forest as a remnant of a "true" nature that, if untouched, might revert to some pre-contact, and consequently better, state. Reminding us of Thoreau's pleasure in "connecting to the forest through direct use of local natural materials" as a tonic to the call of the wild, Donahue traces the history of overlapping residential and productive land uses in the landscape familiar to many from Thoreau's writing and highlights opportunities for engagement with that landscape.

This material landscape, which sheltered Thoreau's transcendental nature experiences and which now welcomes half a million visitors per year, helps us question ideals of primeval nature by its very changes, both ecological and human—and by the possibility of an everyday working relationship with such a forest. Considering the material and symbolic implications of inhabiting a landscape as Thoreau did—not only revering landscape as wilderness, but also facing the processes of interacting with landscapes to meet everyday needs—helps to erode conceptual barriers to engaging with contemporary environmental issues. This embodied view of nature helps us begin to address the questions of where the natural resources we depend on come from, and how, and eases us away from an easily bruised, superficial environmentalism that makes us shield ourselves from such questions. How can we live in an environment that's supposed to look as if we're not there? Why do we seek our places within which, by our own ideals of nature, we cannot carry out the everyday tasks of inhabitation or display the modern trappings of such inhabitation honestly? Donahue's blend of environmental history and call to action helps make approachable these questions that are of particular importance for exurbia.

—Kirsten Valentine Cadieux and Laura Taylor

■ ■ ■

> I think that each town should have a park, or rather a primitive forest, of five hundred or a thousand acres, either in one body or several—where a stick should never be cut for fuel—nor for the navy, nor to make wagons, but stand and decay for higher uses—a common possession forever, for instruction and recreation.
>
> All Walden wood might have been reserved, with Walden in the midst of it, and the Easterbrooks country, an uncultivated area of some four square miles in the north of the town, might have been our huckleberry field... As some give to Harvard College or another Institution, so one might give a forest or a huckleberry field to Concord.
>
> —Henry David Thoreau, "Huckleberries," 1970[1]

Henry Thoreau recorded this passage in his journal on October 15, 1859, after returning from a walk to the Easterbrooks Country. It was not otherwise published, and for another century was read by only a few. But Thoreau's philosophy that "in wildness is the preservation of the world" was read by many, and continues to inspire the protection of natural areas throughout the world. By the time Thoreau's plea for a wild reserve in Concord finally *was* published in 1970, a thousand acres of woods surrounding Walden Pond had indeed been made "a common possession forever." But has this reserve within Walden Woods fulfilled Thoreau's vision? Over the same period, the social and ecological context of Walden Woods has changed: it is no longer embedded within an agrarian landscape, but within an exurban mosaic of reforested farmland and residential development. What can the complex history of Walden Woods and of the changing relationship between people and forest in Concord tell us concerning the "higher uses" of exurban woodlands today?

Higher uses can mean different things. For Henry Thoreau, higher uses were those that transcended mere utility. He said "instruction and recreation," but by that he clearly meant spiritual connection with nature as well. He suggested that the highest use for at least some tracts of forest was to let them "stand and decay"—preserving (or gradually recreating) what he called "primitive forest" and we now call "old growth," with its ancient trees, large standing snags, and woody debris slowly rotting on the forest floor. This is an ecological condition considered by many environmentalists to be most natural, healthy, and desirable.

But the phrase "highest and best use" is a term also applied to real estate appraisal. There it means the most profitable potential use of a piece of property, which is generally residential development. Exurbia, in this post-agricultural region, has become a battleground between these two conflicting impulses towards higher use—and the combatants are often the same people, contesting with themselves. They are affluent residents whose economic fortunes are tied to the highest and best use of property, but whose wish is to live surrounded by unmolested, wild nature, devoted to higher ecological and spiritual uses. Meanwhile, the other competing use of the forest that concerned Thoreau—that is, "cutting a stick" to make wagons or anything else—does not rank very high on either of these scales. Growing and cutting timber is less profitable than building a house, less "ecological" than leaving the forest alone. It is a low use, largely banished from exurbia. But is the conclusion that productive engagement has no legitimate place among the higher uses of an exurban forest supported by a careful reading of the work of Henry Thoreau, or the history of Walden Woods?

Walden Woods is a tract of about 2,000 acres of oak and pine forest surrounding Walden Pond in Concord and Lincoln, Massachusetts, some fifteen miles northwest of Boston (Figure 5.1).[2] It lies on a plain of excessively drained glacial outwash that encompasses a few projecting hills of rocky till, along with a few inscribed wetlands. Thanks to its poor agricultural soils, this stretch of country was almost immediately relegated by Concord's seventeenth-century English settlers to wood lots. It remained largely (although not entirely) forested even through the height of agricultural conversion in the nineteenth century, although it was frequently cut for timber and fuel. Walden Woods became famous not because it was a particularly outstanding piece of forest, but because it *survived* as forest and was the favorite walking place of the Concord transcendentalists, most notably Emerson and Thoreau. As a direct result of this legacy, a large part of Walden Woods was protected from development as the surrounding region both suburbanized and reforested during the twentieth century.

Figure 5.1 Walden Woods protected open space.

The resulting post-agricultural landscape is exurban: half forest and half suburb, with the lines between the two often blurred. It comprises a small remnant of farmland and a larger expanse of forest, interpenetrated with development ranging from dense villages, to tract housing, to scattered dwellings on large lots. Many of these large (some would say bloated) houses thus lie effectively within the woods. Roadways connect the residential areas and fragment the forest, while riparian zones and wetlands tenuously connect the forest and break up the developments. Among the larger tracts of relatively continuous, intact forest within this landscape is Walden Woods.

Walden Woods is special not because of its ecology, but because of its history. Geologically and ecologically it contains elements common to the greater forest of the Concord River basin, which I will call the Suasco woodlands. These woodlands are roughly coterminous with what some call Thoreau Country, though that is more a region of the mind than of the earth.[3] Because of its history, and particularly its association with Thoreau, Walden Woods has become an ecological icon. Although not particularly wild itself, it stands as the seminal place where the value of wildness was first articulated. The question naturally arises, what would be the most fitting roles, the higher uses for this symbolically extraordinary fragment of an ordinary exurban forest? To address this question fully requires more than a simple passion for wildness, in my view. It requires a deep understanding of the complex history of Walden Woods, and a broad vision of the cultural and ecological value of the encompassing Suasco woodlands. Such a vision might include not only provision for biodiversity and wildness, but also for human engagement and sustainable harvest of wood products. Can these purposes complement one another, and are they Thoreauvian?

Some followers of Thoreau would simply answer no. Many more might answer perhaps, but remain uneasy when it comes to logging anywhere near Walden Woods, whose protected status flows from Thoreau's call for wildness and fulfills his dream for a natural reserve on this very spot. But I am as passionate a Thoreauvian as any, and I think we need to address this question fully and fairly. By "we," I mean primarily those of us who live in this place and bear primary responsibility for its future. But I also mean all those who feel a legitimate interest in it. I have lived in these woodlands for four decades, in Concord, Wayland, and Weston, and I have played an active private and public role in protecting and managing them. So although I have conducted scholarly research on the land's history throughout this period, I can hardly speak as a detached scholar, but only as an attached citizen.[4]

I certainly believe it is fitting that a significant portion of Thoreau Country be left wild, including a substantial part of Walden Woods itself. But to me, striving for wilderness alone is no way for a Thoreauvian to walk—it amounts to hopping on one foot. To value only wildness is to ignore history, to trivialize Thoreau's vision, and to abdicate our responsibility as inhabitants of this forest today. Without restrained, responsible use of nature—the other part of Thoreau's message—wilderness will remain a luxury and a lie. By employing the binary vision of ecology and history, can we find ways to walk on two feet, on preservation and stewardship, in Thoreau Country?

THE HISTORY OF WALDEN WOODS

> *What a history this Concord wilderness I affect so much may have had!*
>
> —Henry D. Thoreau, *Journal*, January 1, 1858.[5]

Let us look first at the history of Walden Woods and the adjoining countryside, and then at its possible future. Beginning with the land, Walden Woods contains elements of the three major geomorphological forms of the surrounding region: glacial outwash plains, till highlands, and finer lake-bottom and alluvial lowlands. Colonial Concord farmers called these sandy lands, rocky lands, and moist lands. Walden Woods is dominated by coarse sand and gravel deposits that were washed southward into glacial Lake Sudbury (Figure 5.2). Such outwash soils are quite common in Thoreau Country, covering about half the landscape. They grade from exceedingly coarse and barren sands dropped at the retreating ice margin to sandy loams carried farther out into the departed lakes, where they can provide droughty but serviceable tillage soils. The ground at the northern edge of Walden Woods is so coarse as to be all but useless for tilled crops, whereas the land south of the Woods belongs to the same kame delta complex but is finer, and has been plowed since the middle of the seventeenth century. Much of it remains cultivated today.

Walden Woods also contains several famous heights composed of glacial till: Fairhaven Hill, Emerson's Cliffs, Pine Hill. These hills are mostly thin and stony, with bedrock outcrops. Till soils, composed of material deposited directly beneath the ice and compressed by its weight, occupy about a quarter of Thoreau Country. They grade from thin ablation till that is agriculturally worthless to thick boulder-clay that is hard to plow but makes admirable pasture and orchard ground, because it is well watered. Tills within Walden Woods are generally thin and poor. Some of the better tills, however—such as the north slope of Fairhaven Hill—were for a time cleared for apples and cows.

Finally, within Walden Woods there are a few wetlands, the landform that characterizes the remaining quarter of Thoreau Country. Wetlands in Walden Woods tend to be kettle holes that are either permanently flooded (such as the famous pond itself) or very small and isolated. By contrast, larger, more accessible wetlands in other parts of Concord were once an important agricultural resource, providing large supplies

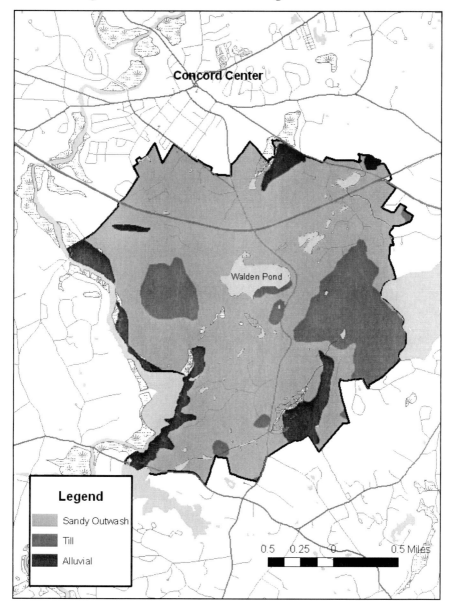

Figure 5.2 Walden Woods surface geology.

of native meadow hay. Half a mile south of Walden Woods, for example, Pond Meadow was divided into mowing lots by 1650, remained in hay production for centuries, and has only recently been abandoned and become a red maple swamp. In brief, Walden Woods comprises sandy, rocky, and wet lands typical of the surrounding countryside, but all lying toward the marginal end of the spectrum. For this reason, the Woods remained largely forested and intact even as the surrounding countryside was progressively "improved" for cultivation.

What did the native, pre-European forest of Thoreau Country look like? Its composition was very different from the forest we see today, but it was also different from the forest we *would* have seen at the time, had the land been completely wild and uninhabited. The Native people burned this landscape, perhaps for thousands of years. Historians and ecologists are still debating how widespread such fires were across southern New England as a whole, but there is good evidence that burning was pervasive in Concord, cooking each of the three parts of the landscape to varying degrees. The purpose of these fires would have been to increase the productivity of the forest and wetlands in providing nuts, berries, and game, by altering their species composition and ecological structure. The dry woodlands at Walden Pond and many other sandy places were described by the English settlers as "pine plains," composed of scrub oak, white oak, and pitch pine—a fire-type now confined mostly to Cape Cod and adjacent Plymouth County. This was once the dominant forest of much of Thoreau Country.

The fires probably spread less frequently onto the moister till uplands, but evidence that they occasionally did is provided by the nearly complete *absence* from our hills of fire-sensitive northern species such as sugar maple and beech. A small stand of beech in a valley east of Pine Hill (noted as unusual by Thoreau) remains today, and a few hemlock groves in steep ravines are probably also aboriginal. But in general, the till uplands of Walden Woods and Thoreau Country were dominated by oak, chestnut, and hickory, forming a taller and denser canopy than on the drier, fire-prone plains.[6]

Fire probably also swept from time to time across Concord's wetlands, which often become dry in the late summer and fall. But the pattern of disturbance there was complicated by the presence of beaver—hence "Pond Meadow," for example. As beaver colonies came and went many wetlands cycled from ponds, to grassy meadows, to wetland forests of species such as swamp white oak, black ash, and Atlantic white cedar, now greatly reduced or gone from the post-agricultural landscape. Where fires ran most frequently, especially in the broader river meadows, the wetlands may have had a stronger tendency to remain in grass, rather than reverting to swampy forest. Although patterns of flooding may account for some of it, the striking pervasiveness of grassy meadows across Concord's wetlands at the time of English settlement is hard to explain without fire.

This is not a forest mosaic we are likely to see again, in spite of the return of trees, deer, and beaver in the past century. On sandy, rocky, and moist lands alike, the forest that has re-grown across Thoreau Country is not only strikingly different from the eighteenth and nineteenth century farmscape of pastures, meadows, and wood lots; but also from the native forest and wetland ecosystems that preceded it. The forest we see today is the outgrowth of successive human disturbances that are much deeper than the agrarian and suburban irruptions of the past few centuries. Even left to itself, this forest seems unlikely to return to any specific pre-European condition. It will continue to change as it follows the ecological trajectory upon which it has been placed, and responds to fresh disturbances.

As they set about turning the native forest into an agrarian landscape more familiar to themselves, the English settlers of Concord quickly realized that the thirsty plain at Walden a mile south of the village would serve them best as woodland. Not that the wood growing there was particularly good, but the land wasn't much use for anything else. "Walden" is derived from Old English words such as wald and weld that *mean* wooded, so Walden Pond means simply "the pond up in the woods."[7] Walden Woods therefore means "woodsy woods," which suggests that the woods could hardly have been called this until everyone had time to forget what "Walden" meant. First they named the pond after the woods, and after a century or two they named the woods after the pond.

Concord was incorporated in 1635, and a commons system was established. The fifty or so proprietors laid out private lots for housing, mowing, and tillage, and left three-quarters of the territory the Massachusetts General Court had granted them in commons, for grazing, timber, and fuel. But at the coming of age of the second generation in the 1650s, the community changed course, and decided to privatize virtually all the remaining common land. One of the earliest of these proposals for new divisions was to lay out large private wood lots near Walden. However, this proposal was soon modified, and subsumed within a more sweeping "Second Division" of the entire town. Much of the southern portion of the proposed woodland tract was not divided into wood lots after all, but was instead granted to a family named Billings, who moved out beyond Walden and subsequently cleared the better land there for farmsteads. A parcel nearby was designated Ministerial plow land, which later became part of Baker Farm of *Walden* fame. The northern portion of the proposed woodland tract, by contrast, *was* subdivided among many owners and did indeed become permanent wood lots—it became Walden Woods. These men chose well: the boundaries they established between cultivation and woods have remained fundamentally intact now for three hundred and fifty years.

Throughout the colonial period, Walden Woods was composed of detached wood lots belonging mostly to yeomen living in and around Concord village. How often these wood lots were cut is hard to judge. They may have seen relatively little cutting during the early generations, as farmers kept themselves supplied with wood by clearing farmland closer to home. By the late eighteenth century, however, Concord was well stocked with farms and in fact straining at the seams. Yet some 25 to 40 percent of the town remained forested. Timber and fuel were essential to the colonial agrarian economy, and for the most part had to be cut locally. Yeomen typically owned twenty to thirty acres of woodland in poor or remote corners of town such as Walden Woods, and might have had to cut as much as one acre of that every winter for fuel alone. After twenty or thirty years, the regrowth would be ready for cutting again—a system of sprout-wood management was emerging. Many farmers did some additional woodcutting to supply the trade in ship timber, barrel staves, hoops, and oak tanbark. It would have been economically disastrous for Concordians to reduce their woodland much further than they had—they would then have lacked the fundamental resources they needed to live. Consequently, by the end of the eighteenth century, forest cover in Thoreau Country stabilized at from one-quarter to one-third of the landscape and headed into a regime of rotational harvesting. It is likely that by this time Walden Woods had seen at most one or two rounds of cutting and re-growth, and only a small portion of it would have been cut in any given year.[8]

This picture changed dramatically by the second quarter of the nineteenth century. Concord farmers adopted a more commercial approach to farming and greatly expanded their upland hayfields and pastures. This was the culmination of a period of what might be called "agrarian sprawl," rapid forest clearing across the state of Massachusetts that became most pronounced in the region around Boston. In its dynamics and distribution, this surge of agricultural development was remarkably similar to the sprawl of residential development that would take place a century and a half later. According to tax valuation data, by 1850 forest cover in Thoreau Country had been driven to its lowest level ever, about 11 percent—although there were no more *farms* in Concord than there had been in 1775, there was more farm*land*, and less forest. Concordians of the time read the same valuations as historians do now, which confirmed what they saw with their own eyes. Thoreau wrote in his journal in 1860, "the woods within my recollection have gradually withdrawn further from the village, and woody capes which jutted from the forest toward the town are now cut off and separated by cleared land behind."[9] This agricultural expansion was driven by urban demand for livestock and dairy products, but was *allowed* by the increasing availability of lumber, firewood, and coal, imported by canal and railroad. Concord farmers no longer needed to conserve local forests—they could cut them or even clear them as the market demanded. The character of the surrounding countryside began to be powerfully influenced by the growth of the city of Boston, and by an expanding national economy. The first wave of exurbanization was agricultural.

Agricultural clearing nibbled at the edges of Walden Woods, but most of these wood lots were excluded from conversion to farmland because their dry soils remained as marginal as ever for pasture or tillage. By 1845, when Henry Thoreau began his two-year sojourn at Ralph Waldo Emerson's lot by the pond, Walden Woods was one of the few surviving large expanses of forest in Concord. But Walden Woods was not immune from exploitation—while not much of it was cleared for farming, it was apparently more heavily cut than it had been in the past. Both local and urban demand for firewood remained high, in spite of the advent of coal. When the Fitchburg railroad passed close by the pond in 1844, it brought a new market for fuel, and for oak and chestnut sleepers. The wood lots in Walden Woods were cut less by farmers and villagers to supply their own needs, and more by commercial wood merchants in great swaths to supply the city and its lifeline, the railroad. As a result, during the middle decades of the nineteenth century a large part of Walden Woods was being cut over all at once, much to the dismay of the sauntering Transcendentalists.[10]

Henry Thoreau thus was born into an agrarian landscape that was undergoing a powerful commercial transformation, and that by his adulthood was much exploited and abused. Walden Woods was over-cut, and the surrounding countryside was over-cultivated and degraded.[11] Much of it was choked with scrubby pasture—the next forest in waiting, as it turned out. Thoreau responded to this wave of over-harvesting with trenchant criticism of both its shallow impulses and its destructive impact, and made his famous plea for wildness. His case was neatly laid out in the Beanfield chapter in *Walden*, which was set at the roadside end of Emerson's lot, within Walden Woods.

Thoreau's beanfield was marginal farmland, at best. The only folks who broke the huckleberry sod in Walden Woods tended likewise to be marginal people such as blacks, Irishmen, drunkards, and Transcendentalists. Thoreau, a renowned lifelong gardener at home in the village, knew exactly how poor his beanfield soil was. The whole bean exercise was in part a satire of the straitened accounts of many Concord farmers, and was designed to fail. It was also a pointed attack on the expansion of farmland at the expense of wildness. Thoreau tells us with great precision that he tilled two and one half acres of beans. He also lets us know that he lost one-quarter acre of those beans to a woodchuck—or one-tenth of his crop. Thoreau was a careful surveyor and a close reader of tax valuations, so when he made a case for resigning one-tenth of his crop to the woodchucks, birds, and weeds, he was also making a metaphorical plea for keeping that last tenth of Concord wild. Presumably, Thoreau would have been happy with more forest than that, but when he later called for a wild reserve in Concord of five hundred or a thousand acres he was by no means suggesting a *complete* return to wilderness. He was calling for the retention of sufficient wildness within an agrarian landscape of farmland and woodland—and of course the preservation of a corresponding spiritual wildness within ourselves. Thoreau never proposed that total wildness was any more desirable than total domesticity, and we might observe that he cheerfully ate the wild woodchuck that pilfered his beans. Was that Romantic, utilitarian, or both?[12]

Thoreau was deeply involved with the agrarian landscape, including its beleaguered forests. Throughout his life, Thoreau recorded the pleasure he took in working up firewood—often driftwood he collected in his boat—and how the burning of it recalled to him the many times it had already warmed him. But his family also bought their household fuel from a wood-dealer, in the ordinary way of village folk. Thoreau tells of paying the bill to Marshall Miles (a woodcutter who lived in the southwest part of Concord) one evening, and getting into a debate about the best way to split and stack cordwood. Miles ended the discussion by remarking that "I have handled a good deal of wood, and I think that I understand the *philosophy* of it." Thoreau was amused by the woodcutter's quaint use of the word "philosophy," but not nearly so amused (it appears) as was Miles by the philosopher's crooked genius regarding how to cut wood.[13]

Thoreau was more deeply implicated in the cutting of Concord's wood lots than as a consumer merely: he also laid the woods out for periodic slaughter. During the last decade of his life, Thoreau made his living as a surveyor, and much of his work involved lotting off woodlands. Woodcutters such as Miles would bid on the standing timber and then cut it clean. Other times, Thoreau resolved disputes by running lines between adjoining owners. Thoreau was an excellent surveyor, meticulous and precise as in everything he did. We still have many of his surveys, including several in Walden Woods. He did feel some ambivalence about being thus an agent in the felling of the woods he loved, which he expressed in his journal.[14] I don't mean to take a cheap shot at Thoreau or portray him as a hypocrite, but simply to point out that he was not a detached, idle critic, but a man who was intimately involved in the economic relationship between his community and its landscape.

Henry Thoreau did not object to cutting wood in principle. He did have several important objections to the way it was cut in practice. First, agricultural clearing had reduced forest cover too much overall. Thoreau called for the protection and restoration of forest, *some* of which—he proposed 1,000 acres—should be set aside to grow with time into untouched "primitive forest" once more. Second, too much of the remaining forest was being cut at one time, and often badly. Thoreau called for moderation in demand (particularly of industrial uses such as feeding the railway) and for improved methods of cutting. From his pioneering observations of the way trees disperse their seeds and grow, he believed that farmers were not working with nature to get the most from their wood lots. They did not see how nature prepared seedlings for the next forest; did not notice which species were becoming less common because of their practices. Thoreau expressed great admiration for farmers such as the old yeoman George Minott who *did* observe their woodlands closely and care for them faithfully. "Minott may say to his trees: 'Submit to my axe. I cut your father on this very spot.'" So we see that Henry Thoreau had a foot in both the preservationist and conservationist camps, in the wildlands and the woodlands, and took some of the first steps side-by-side in each. Minott's wood lot lay within Walden Woods, and is an important part of its heritage. In praising farmers like Minott, Thoreau called for direct, wholesome contact with nature that included both walking and working.[15]

When Henry Thoreau died in 1862, in his forty-fifth year, Walden Woods and the surrounding countryside were again changing dramatically, in complex and contradictory ways. Almost from that moment, the once-secluded pond has been heavily used for outings by urbanites. Had Thoreau only lived to age fifty he would have seen Walden transformed into an amusement park. Beginning in 1866, the Fitchburg Railroad established picnic grounds, a bathhouse, dancing platforms, a merry-go-round, ball fields, and a racetrack at "Walden Grove and Lake." Visitors wandering onto Emerson's adjoining property would sometimes meet the venerable sage of Concord himself, out for a stroll. Although somewhat troubled by these throngs of city dwellers who preferred to encounter nature in large, noisy groups—not exactly the Transcendental way—Emerson felt that any outdoor exercise and contact with the woods was better than none, and made the visitors welcome.[16] So the second exurban wave to hit Thoreau Country was recreational. The railroad facilities burned to the ground in 1902 and were not rebuilt, but Emerson's heirs gave 80 acres surrounding the pond to the state in 1922 to "preserve the Walden of Emerson and Thoreau, its shores and nearby woodland for the public who wish to enjoy the pond, the woods, nature, including bathing, boating, fishing and picnicking."[17] Another bathhouse was built and a beach improved on the opposite, eastern shore by the road, to better serve the onrushing age of the automobile. Today some 500,000 people visit Walden Pond every year, mostly to swim—although parking restrictions aim to limit the visitors to a mere 1,000 at any given moment. Some Thoreavians regard this as a desecration, but others see it as an educational opportunity—and for solitude, simply visit the Pond when the weather is miserable.

Meanwhile, the surrounding forest of Thoreau Country at large was enjoying a remarkable recovery. At first, this was not because of a decline in agricultural prosperity, but because of a change to market gardening and stall-feeding of dairy cows, which required less land. Concord farmers became, if anything, more specialized and more prosperous than they had been before the Civil War, even as agriculture throughout much of America slid deeper into depression and Populist revolt. By the end of the nineteenth century, the town of Concord ranked eighth in the state of Massachusetts in the value of agricultural production, even as forest rebounded to almost half of the countryside.[18] The wholesale abandonment of degraded, superfluous pastures led to a surge in the growth of white pine, and a better balance between farmland and forest was restored. Unfortunately, after about 1920 came a long decline in truck and dairy farming at the hands of Western competition and rising land values, and the forest continued to recover farmland. That is, in the age of oil the agrarian landscape was becoming more valuable for pastoral scenery even as it became less viable for agriculture itself. The third, most familiar wave of exurban development was underway, driving farmland simultaneously toward forest and houses. At first forest rushed ahead, but in recent decades residential clearing has begun to outpace reforestation, causing a net loss of forest once again. Forest cover in Thoreau Country today stands at about 60 percent, depending what you want to count. In many ways it is good that so much forest has returned to the region, but this was only made possible by the spread of industrial agriculture in other parts of the country. The celebrated "rewilding" of the East is nothing but the reverse image of a Colorado feedlot, seen through the looking glass.

Tentative agrarian incursions into Walden Woods such as the beanfield were among the first cleared patches in Concord to revert to forest. But subsequent residential development around the fringes of Walden Woods has left the acreage of contiguous forest there about the same today as it was in Thoreau's time. Meanwhile, the trees have been growing older. From the early twentieth century, woodcutting went into decline with the triumph of coal, gas, and oil for home heating and cooking, the chronically depressed state of the local timber market, and, in recent decades, preservationist antipathy to logging. Harvesting within Walden Woods has not completely ceased but the degree of cutting has greatly decreased, leading to a much more uniformly mature forest.

At the same time, forest composition throughout Thoreau Country has changed markedly. There has been a dramatic decline in fire since the late nineteenth century. Sparks from locomotives continued to set many fires in the immediate vicinity of Walden Pond for a time, including an 1896 blaze that destroyed the white pines Henry Thoreau had planted for Emerson at his beanfield in the 1850s.[19] But in general, fire became less common, greatly abetting the spread of white pine at the expense of pitch pine. Chestnut blight all but eliminated that beloved species in the 1920s, and gypsy moths have been especially hard on white oaks for more than a century now. Red maple has invaded almost all abandoned wet meadows, and in the absence of fire has spread onto uplands as well. As a result, a landscape once dominated by pitch pine, white oak, and chestnut has been replaced by one dominated by white pine, red and black oak, and red maple. Even deep in the woods, Henry Thoreau would hardly recognize Thoreau Country today.[20]

Many other changes are still underway in the Suasco woodlands, with uncertain implications for the composition of future forests. There is the invasion from residential and roadside plantings of many new species such as shade-tolerant Norway and sugar maple—one alien, the other native. There is the influx of pests and diseases such as hemlock woolly adelgid and ash decline, to name only two. There is the return of powerful agents of landscape change such as beaver and white-tailed deer, continued fragmentation by roads and residences, and the impact of air pollution and climate change. It is evident that the present forest of Thoreau Country is a product of its unruly history quite as much as any underlying geological or climatic determinism, and that the future is likely to be wilder still. This forest is apparently not on course for any pre-determined stable endpoint, and trying to distinguish what is "natural" from what is not is an exercise in frustration, if not futility. We should be thinking, instead, about what is feasible and desirable.[21]

The twentieth century and the automobile brought suburban development to Thoreau Country, which provoked a vigorous response. While the local conservation movement has not been able to preserve anything resembling an agrarian landscape, over the past half-century it has succeeded in protecting a significant amount of wetland and forest. In towns throughout the Suasco basin anywhere from one-quarter to one-third of the land has been set aside as open space, most of it wooded. One of many striking successes has been Walden Woods itself, where on the order of 1,000 contiguous acres are now protected. The Walden State Reservation surrounding the pond has expanded to over 400 acres. This reservation is loosely encircled by several hundred more acres of conservation land belonging to the towns of Concord and Lincoln, and to various land trusts and non-profit organizations. Walden Reservation is the direct legacy of Emerson and reflects a preservationist, recreational approach, but the encompassing town lands descend from a somewhat different, parallel conservation tradition. To the north, the Hapgood Wright Town Forest in Concord was purchased in 1935, and saw several decades of tree planting and wood cutting, in the town forest spirit of the early twentieth century.[22] To the west, Fairhaven Woods was donated by the Wright family, who had begun planting trees and carefully managing the woods in the late nineteenth century. Sustainable forestry is still practiced there by the Concord Land Conservation Trust. Southwest of the pond is Adams Woods, which has also seen a century of active management by the Adams family, and is now owned by the Lincoln Conservation Commission. Pine Hill, to the east of the pond, was thinned by the town of Lincoln as well, during the 1980s. In brief, the outer ring of protected land in Walden Woods reflects a tradition of active forest management, begun largely by wealthy estate owners and passed down through local town conservation culture. This aspect of Walden Woods is very much in the spirit of George Minott, so admired by Henry Thoreau.

The history of Thoreau Country reminds us that in most places the recent, vigorous expansion of exurban development seldom invades either pristine natural forest or a timeless, agrarian countryside. Instead, residential sprawl can be seen as simply the latest in a succession of transformations that the expanding commercial and industrial economy of the past two centuries has wrought upon both cities and their hinterlands. This is not to suggest that those who value the rural countryside, in all its ambiguous complexity, should simply stand aside and allow this latest, automotive residential wave to take its preordained course. It is to suggest that we ought to be defending not any particular idealized natural or pastoral past, but a vision of a better relationship to nature than we have yet seen. That vision might well incorporate some elements of the past that we *have* seen, but could also be guided by a full, critical awareness of the natural and cultural forces that have shaped the world as it has come down to us.

Henry Thoreau lived in an agrarian world, in which productive use of local resources was stimulated by strong market demand to what can fairly be called extractive excess, and little forest was left standing. His call for moderation of use and for preservation of wildness addressed the most glaring abuses of nature in his time. But today's Suasco woodlands belong to a different world, in which productive use of local resources is practically nonexistent. Instead, today's exurban residents fatten upon extractive industrial agricultural and forest practices far away, beyond their sight, while striving to preserve the woodsy view from their back windows. I don't think we need to waste much time wondering what Henry Thoreau would think of that kind of obfuscation—just what he thought about the New England manufacturing economy profiting by Southern slavery, one would suppose. Under today's circumstances, thoughtful exurbanites need more than the tonic of wildness. We need to combine the traditions of a wild forest, a walking forest, and a working forest into a single vision, if we can. Elements of that broader vision are already in place within Walden Woods and the surrounding countryside. They only need to be rediscovered, elaborated, and celebrated.

THE SUASCO WOODLANDS

It may be possible to make cogent ecological, historical, and even Thoreauvian arguments for artfully integrating working forests and wild reserves at various scales throughout the world, but putting such arrangements into place in actual ecological, economic, and political landscapes is more difficult. Exurbia presents its own peculiar conflicts, but also opportunities, for nowhere else (in the developed world) do so many people live so deeply immersed in forest. Thoreau Country—the fragmented exurban woodland of the Suasco basin, with Walden Woods at its symbolic heart—is a hyperexample. These woods on the outskirts of Boston are home to some of the best-educated, most environmentally aware people in the world, and it was here that Henry Thoreau walked. But there can be few places on earth where per capita consumption of natural resources, including wood products, is higher. There could hardly be a more fitting place to address the future of the human relationship to the forest.

The Suasco woodlands are not only ecologically fragmented, they are also subdivided among many public and private owners. Remaining private woods tend to be small parcels that are under intense development pressure. Virtually all of the larger forest tracts are owned by governmental agencies and non-profit organizations (which are legally private but act in the public interest): the National Park Service, the US Fish and Wildlife Service, state forests, state game lands, town forest committees, town conservation commissions, and a slew of local land trusts, to name just a few. This presents a complex challenge to regional forest stewardship.

Each conservation entity and landowner has its own stewardship policy, but the present tendency is toward little or no management of forestland. This is because of a widespread (though hardly unanimous) belief that unmanaged forest is ecologically superior, and that in any case wood harvesting is economically and politically difficult, if not impossible, in the exurban environment. But there are local conservationists who support a more active approach, for reasons that range from demonstrating sustainable use of natural resources to maintaining diverse habitats. Here and there, on both private and public woodlands, some rather limited logging continues to take place in Thoreau Country.

A comprehensive approach to the forest would benefit from a mechanism for coordination among disparate owners. The tenuously connected Suasco woodlands might be thought of as an ecological entity at the watershed scale. The long-term stewardship of this forest (and its associated wetland and aquatic ecosystems) might be collaboratively overseen by a "woodland council" of cooperating landowners and conservation organizations. Such a council would not have the power to determine the management of individual parcels, obviously, but could serve to facilitate communication among the owners so that a more coordinated stewardship policy could emerge. In particular, the council could keep track of lands devoted to wildness and those devoted to sustainable harvesting, allowing cooperation among these landowners. This would nurture the growth of an overarching strategy to secure the full array of "higher uses" of the woodlands. Those higher uses might include ecological stewardship, ecological research, environmental education, and sustainable harvest: the wild forest, the walking forest, and the working forest as one.[23]

Ecological Stewardship

Broadly speaking, the goals of stewardship are to maintain healthy functioning ecosystems to provide ecological services such as water quality, and to maintain habitat for viable populations of as many species as possible. These goals will inevitably be compromised in the fragmented, heavily humanized exurban environment. Some forest species are simply unable to flourish under these conditions (although others do quite well), and ecosystems suffer air and water pollution from roads, lawns, and waste disposal. Perhaps the best that can be done is to protect a connected natural landscape with as much forest as possible, and to attempt to identify, monitor, and maintain habitat for the entire range of existing species. History suggests that we cannot freeze present ecological conditions, and that both managed and unmanaged ecosystems will inevitably change. Managers face difficult choices among simply allowing nature to take its course, maintaining a generally diverse landscape, and trying to support species of concern by providing particular treatments. Wild reserves and managed woodlands can both play useful roles in this mix.

Wild reserves have a special place in the forest. When fully mature they form ecologically distinctive "old growth," they erect standards of comparison for managed areas, and they give many visitors a sense of enduring wildness that has become spiritually essential to modern life. The location of reserves may be determined partly by ecological considerations, but mostly by the policies of individual landowners. In almost any imaginable future, the Suasco woodlands will surely retain a wild core at Walden Woods, largely unmanaged except for crowd control and remediation of trampling. This small, symbolic wilderness will likely include the 411-acre Walden Pond State Reservation along with major parts of adjacent woodland parcels—how much will of course always depend on the management decisions of the towns, land trusts, and non-profit organizations that own the land. This honors the wild heart of Henry Thoreau's message. Given its numerous visitors, Walden will always be more a park than a true wilderness area, but as an *icon* of wildness it will more than serve.

It seems appropriate that every town have at least one such "Walden" of a few hundred acres within its boundaries, answering Thoreau's call for wildness. Beyond that, we have an opportunity to allow several larger tracts within the Suasco woodlands to age into variegated "old-growth" structure, with large trees, small gaps, and abundant standing deadwood and woody debris. Such reserves of a thousand acres or more would achieve a very special ecological purpose. They would bring to full maturity assemblages of dominant tree species that, while novel in their composition, might still provide the ecological structure to encourage groups of understory and wildlife species that would otherwise be rare or absent. Several tracts across Thoreau Country suggest themselves for reserves of this size, representing the major landforms of the region. One is Estabrook Woods, Thoreau's "huckleberry field" now grown up to forest and owned chiefly by Harvard University after all. This tract is geologically distinct from Walden, being mostly till and swamp, and also historically distinct, being mostly abandoned farmland. A sand plain forest located in the "Desert" area of Maynard, Sudbury and Stow is ecologically similar to Walden Woods but larger and less visited, and may be suitable for prescribed burning to restore fire-dependent species to one corner of the region. A large wetland forest is found in the Great Swamp in Acton and Stow, and the Great Meadows Wildlife Refuge along the Sudbury River provides extensive marsh and floodplain forest. Over time, these reserves could be expected to yield distinctive old-growth communities on our rocky lands, sandy lands, and moist lands.

But in exurban circumstances, reserves alone may not be sufficient to protect biological diversity and the delivery of ecosystem services such as clean, reliable groundwater and streamflow. Working in tandem with patches of old growth, careful ecosystem management of the surrounding Suasco woodlands can help maintain a healthy diversity of habitat, largely by mimicking natural disturbance patterns. Of course, natural disturbances themselves will continue to occur, but given a fragmented forest inhabited by a large human population we cannot rely on these sometimes catastrophic events to maintain diversity at the landscape scale. Exurbanites dispersed through the woods in expensive residences will not tolerate the full consequences of unbridled natural disturbances, even (or perhaps especially) at long intervals. We might not relish the impact of another storm with the fury of the category-three Hurricane of 1938 if it were to fall upon a uniformly mature unmanaged forest, and we would hardly accept the large-scale wildfires that might consume the dried windfalls a few years in the wake of such a storm. Ecosystem management in this situation might aim to substitute somewhat more frequent, smaller-scale, controlled disturbances—in other words, systematic logging. This would maintain a wider range of age-classes throughout the forest. Done right, this can moderate the impact of natural disturbances yet create a very similar landscape pattern that will provide diverse conditions hospitable to the vast majority of species inhabiting Thoreau Country, and acceptable to a settled exurban countryside.

Ecosystem management can also be used for more specific ends, for example to enhance populations of species of regional concern. Some species (such as towhees and golden-wing warblers) are declining not for lack of mature forest, but because younger, regenerating forests and shrub lands have grown up and vanished. Such conditions can be selectively restored— indeed, the small-scale patch-harvesting regime of sustainable forestry may in itself be ideal for some species. Beyond that, species that were once present in the native forest but have failed to pass through the gauntlet of agricultural clearing and reforestation can now be deliberately restored. Atlantic white cedar and black ash have become rare in our swamps, and many wildflowers have been slow to re-establish themselves on once-cultivated land that has returned to woods—if we want these species to be part of modern ecosystems, they need our help. In particular, we can assist the return of the chestnut, in the form of the blight-resistant 15/16 American chestnut hybrid being developed by the American Chestnut Foundation. Fittingly, a magnificent chestnut "mother tree" within Walden Woods has been enrolled in this worthy program.[25] In all these efforts our goal cannot be to restore "original" ecosystems, but rather to reintroduce lost species so that nature is playing with a full deck in establishing new ecosystems as conditions change. This follows Aldo Leopold's precept of intelligent tinkering, which retains (or restores) all the cogs and wheels, but recognizes that they may be assembled in new ways.[26]

Neither the complex ecological history of these woodlands, nor their current broken exurban condition supports the popular notion that only wild nature left to itself can reliably protect biodiversity and provide ecosystem services, and that any and all human management is a poor substitute. Instead, we might think of wild reserves as playing a special role within broader, sustainably managed woodlands that are every bit as ecologically valid and useful, and perhaps more suited to exurban reality. Unfortunately, ecosystem management is presently not so well suited to exurban *culture*, which tends to imagine these surrounding woodlands as timeless and pristine. That is why ecological stewardship, sustainable forestry, and environmental research and education need to be united in a comprehensive program, if any of it is to work.

Ecological Research

Undertaking such adaptive ecosystem management calls for a corresponding program of long-term ecological research and monitoring. Fortunately, Thoreau Country is rich in historical records, and also rich in research institutions that can collaborate on such an investigation. A long-term ecological study, focusing initially on Walden Woods, would be able to explore critical questions throughout the Suasco region. Because of its well-documented history and accessible location, Walden Woods presents one of the best opportunities for ecological and historical research in the world. A comprehensive study of the ecological functioning of a fragmented forest that has seen dramatic historical change would be important in its own right, but would also provide a solid foundation for stewardship of the surrounding Suasco woodlands, and a tremendous educational opportunity.

A long-term interdisciplinary study of Walden Woods has been initiated by Harvard Forest and the Brandeis Environmental Studies Program—I have only given the broadest preliminary outline of the story in the first part of this paper. The full study will eventually include biological inventory and monitoring through permanent survey plots; pollen cores of Walden Pond and smaller wetlands for a fine-grained record of vegetation change across Walden Woods; archaeological investigation of Native American sites; soil archaeology probing Native and European farming practices; comprehensive GIS mapping of historical land ownership and land use from town records, deeds, probated estates, tax valuations, maps, photographs, journals, and field evidence; and qualitative reconstruction of historical changes in populations of selected species using sources such as naturalists' journals (of which Thoreau's journal is only the most famous), ornithological, and botanical records. Gathered together, these strands of evidence ought to produce one of the most detailed accounts of historical ecological change ever assembled. This study will provide insight into the protection of species today, as well as a mechanism for monitoring ongoing change and for guiding adaptive management. The Walden Woods study could eventually be linked to comparative studies in other parts of Thoreau Country, in order to reconstruct the history of the landscape at a wider scale. Ideally, a network of study sites could be established to undertake ongoing research, and also to actively involve local school systems and help guide community forest programs.[27]

Environmental Education

A comprehensive environmental education program of unparalleled scope could be built upon these woodland initiatives, together with other projects already underway in Thoreau Country. Environmental education is a term that covers a lot of territory. First of all, there is the simple opportunity to walk in the woods. This remains perhaps the highest use for exurban forest, just as it was for Emerson and Thoreau. Many trails already exist through Suasco conservation lands, but they could be better maintained, better connected, better mapped, and more accessible—the exurban forest should be a metropolitan educational resource, not the private preserve of the affluent. And far more could be done with interpretive trails. A coordinated trail system with an interpretive center in Walden Woods would encourage visitors to the Pond to explore and understand more of the surrounding woodlands.

The exurban forest at large provides a splendid opportunity to engage young people with the woods that has scarcely been tapped. Thoreau Country is full of first-class educational institutions. Workshops in "place-based" education could help train teachers in every Suasco community to involve students with the forest. The pioneering Walden Woods Project "Approaching Walden" summer program for high school and middle school teachers could be expanded, and the lessons learned in Walden Woods applied to other towns. Community forestry programs following the model of Land's Sake in Weston could employ young people directly in forest stewardship, maintaining trails and even harvesting wood. The object of all these projects would be to raise children who have deep familiarity with and affection for the forest, a schooling that would mature into active engagement throughout their lives.

The value of woodland education in the exurban setting runs in two directions at once, involving both the opportunity and the utility of getting a lot of people into the woods on a regular basis. On the one hand, although these woodlands are broken into small tracts and hence lack both the ecological and economic wholeness of larger forests in more rural areas, because they are close to so many people they provide enormous educational possibilities that more remote places lack. On the other hand, only by actively involving exurbanites with the woodlands they inhabit will it be possible to muster a sufficient constituency to support politically difficult ecological stewardship and sustainable forestry practices that require the dreaded chainsaw. Comprehensive environmental education is both the end and the means, the main object and the mainspring, of the entire exurban forest enterprise.

Sustainable Forestry

Community forestry programs across much of Thoreau Country ought to be carefully considered for two reasons: first, because they provide a practical mechanism for ecosystem management, and second, because they help enact meaningful cultural engagement and ecological responsibility. The hallmarks of sustainable forestry are thinning stands of trees periodically as they grow, and then harvesting timber on long rotations. Thinning follows the natural process of maturing forests, concentrating growth on the most economically desirable trees and obtaining some usable firewood and woodchips along the way. Ecologically sensitive managers, however, always take care when thinning to maintain wide species diversity and to leave some "rough and rotten" trees in the stand for nesting sites. Timber harvesting then takes place on rotations of a century or more, ensuring that at any given time most of the forest is ecologically mature, but with plentiful areas of younger regeneration. Harvesting is carried out either by small group selection cutting that creates stands made up of trees of many ages, or by shelterwood cutting (removal of all the mature trees in two or three cuts spread over several decades) that creates a forest of many small stands of different ages. These techniques are designed to produce the highest quality rather than the greatest quantity of timber, and to mimic natural disturbance patterns such as hurricanes and northeasters. Through low-impact logging methods they can be carried out with minimal damage to the forest floor and remaining trees, and they can be used to maintain a forest of diverse structure and composition. Sustainably managed forests are far removed from the monoculture plantations, short rotations, and large clearcuts of industrial forestry. They do not replicate old-growth forests—that is the function of wild reserves—but they do maintain diverse conditions hospitable to the great majority of forest species. In my experience, most exurbanites (even if initially skeptical) find these practices quite attractive.

Such forestry has been going on here and there across Thoreau Country, and even within Walden Woods, for a long time. To establish these practices more widely and consistently would require actively recruiting public and private forest owners interested in sustainable management, which would be the function of the Suasco woodlands council. Some forest owners might want to go further and organize a regional woodlands cooperative. Similar groups are taking shape in several New England regions—for example, the Massachusetts Woodlands Cooperative and Vermont Family Forests.[28] The resulting substantial base of managed woodlands would allow long-term rotations to be established, resulting in reasonably steady yields of firewood, pulpwood, and timber of various species. This would in turn mean steady work for consulting foresters and logging contractors skilled in low-impact methods. It would then be possible to invest in technology to process and market local wood products—both low-value culls and top quality timber. A "Suasco Woodlands" brand could be certified among cooperating landowners, logging contractors, and processors who meet best management practices. This would allow local, sustainable wood products to receive a market premium from educated consumers. Without such comprehensive initiatives, restrained forestry will not be profitable, and profitable forestry will not be restrained. A Suasco woodlands council would enable productive use of the forest in an ecologically sound way, and would make the forest more visible to its residents: they would begin to see it inside their homes, as well as outside.

Perhaps the most appealing aspect of community forestry lies in its ability to engage exurbanites directly with the forest that surrounds them. Although most logging and wood processing operations are best carried out by highly qualified and well-equipped private contractors, there can also be useful roles for non-profit community forest organizations employing young people. The organization Land's Sake in Weston has shown that young people can be effectively employed maintaining trails, and splitting and stacking firewood.[29] A joint approach could be used in which professional contractors harvest mature timber using forwarders (articulated tractors with loaders that can remove logs from the forest with minimum disturbance), while community non-profit groups conduct thinning, work up and deliver firewood, and process smaller saw logs with portable bandsawmills. Together, involvement by community groups and local retailing of wood products could lead to a much more complex and rewarding interaction between people and trees.

From its own forest, Massachusetts presently produces only about 2 percent of the wood products (primarily lumber and paper) that it consumes. Yet much of the state—even in long-settled exurban areas such as the Suasco woodlands—is heavily forested. Currently, only about 10 percent of the sustainable yield of the Massachusetts forest is being cut. We are not living within our ecological means, and are hardly even trying. We choose instead to exhaust ecological capital elsewhere in the world while we "preserve" forest close to home, unused.[30] This does no credit to the memory of Henry Thoreau. Clearly, we cannot produce all our own wood products within Thoreau Country, nor should all woodlands be intensively managed; but we can at least make an effort to meet some of our needs in an exemplary way. If well practiced, such forestry could be consistent with other widely held ecological and educational goals of the Suasco woodlands.

Exurbia might be defined as a semi-rural place where the use of land is no longer driven by local consumption, and not even by commodity production at large, but by links to a nearby city. Historically, these connections have included agriculture, recreation, water supply, and of course residential development. There is an obvious contradiction between exurbanites' attraction to pastoral or natural surroundings, and the cumulative consequences of their own settlement within such landscapes. But this paradox need not be as intractable as it seems. The places where city and countryside meet could also provide models of more engaged, enduring interactions between people and land. What better place than Walden Woods and Thoreau Country to explore these possibilities?

Henry Thoreau began his sojourn in the woods at Walden Pond by cutting down six pine trees, to clear the site and frame his house. He afterwards declared that he felt closer to the forest for having cut those trees. A skilled carpenter and a frugal man, Thoreau hewed his own beams, sheathed his house with boards obtained from an Irish family's shanty along the recently completed railroad, and shingled it with pine slabs from a local mill. The solid little building thus embodied an instructive combination of locally harvested and recycled materials. It stood for both reducing our demands on nature, and for direct, responsible use of nature.

In June of 2001, the Thoreau Institute built an accurate replica of Thoreau's house at their headquarters on Pine Hill in Lincoln. The house was constructed of white pine beams felled and hewed by hand, in Walden Woods. It was sheathed with pine boards from a small local sawmill. This vital aspect of Thoreau's philosophy, the pleasure of connecting to the forest through direct use of local natural materials, is as alive in many of us today as is the call of the wild. We surely cannot find *all* the wood products we need, nor *all* the wildness the world needs, in our own exurban forest. But we can remind ourselves of the necessity and the pleasure of these things by finding both some wildness and some sustainable resources together, here at home.

Let us hope that this is only the beginning. In the coming years, a new visitor's center may be built at Walden State Reservation, where half a million people come every year to walk in the woods and swim in the pond. Let us first embrace the idea that it is a triumph, not a catastrophe that so many visitors can still come to Walden Woods today. Then let us envision a visitor's center that celebrates more than the worn clichés of passively appreciating wilderness and communing with nature. Let us present the ecology and history of Walden Woods in ways that incorporate all its irony and complexity. And let the building itself be built primarily of certified wood from the Suasco woodlands that surround that wild core. Let us imagine it framed with oak, like the houses of Concord in Thoreau's youth (those that were not framed with chestnut) before the railroad brought cheap pine studs from the Maine Woods. Let us imagine it sheathed with Suasco pine boards, and paneled and floored with a beautiful array of local hardwoods ranging from black birch to red maple. Let us think of the tables, chairs, desks, and exhibit cases crafted from these same woods, and of the passive solar building drawing supplementary heat from clean wood burning technology.

The construction of such a public building in Walden Woods would express not an unrequited reverence for nature merely, but a bold step toward practical reconciliation with nature. Henry Thoreau was a hard-edged man. His eye was keen, his pen incisive. We will deserve to think that the forest we inhabit is equal to Thoreau's vision when the statue we erect before his cabin has him holding not only a book but also his friend's ax, which he returned sharper than he borrowed it.

NOTES

1. Henry David Thoreau, "Huckleberries," in *The Natural History Essays* (Salt Lake City: Peregrine Smith, 1980): 259–260. This essay was compiled from portions of Thoreau's manuscript "Notes on Fruits" by Leo Stoller and first published as *Huckleberries* (University of Iowa Press, 1970). The entire

manuscript was subsequently edited by Bradley P. Dean and published in a more complete form as Henry David Thoreau, *Wild Fruits* (New York: W.W. Norton, 2000) in which the same passage appears (with slightly different punctuation) on page 238. The journal entry from which Thoreau rewrote the passage appears on October 15, 1859, *The Journal of Henry D. Thoreau* (Boston: Houghton Mifflin, 1906): vol. XII, 387.
2. I have drawn upon Thomas Blanding and Edmund A. Schofield, "Walden Woods" (Concord: Thoreau Country Conservation Alliance, 1989): and W. Barksdale Maynard, *Walden Pond: A History* (New York: Oxford University Press, 2004).
3. "SuAsCo" has since the 1950s been the acronym for the Sudbury, Assabet, and Concord River watershed.
4. My dual career appears in Brian Donahue, *The Great Meadow: Farmers and the Land in Colonial Concord* (New Haven: Yale University Press, 2004); and Brian Donahue, *Reclaiming the Commons: Community Farms and Forests in a New England Town* (New Haven: Yale University Press, 1999).
5. *The Journal of Henry D. Thoreau* (Boston: Houghton Mifflin, 1906): vol. X, 234. Thoreau meant human history within Walden Woods (which he was surveying at the time so that it could be logged), and the double meaning of the word "affect" sounds deliberate. Thoreau was mocking his own "affectation" of wilderness in a landscape that had been deeply "affected" by many generations of Concordians, including his own.
6. This description of Concord's forests is drawn from Gordon G. Whitney and William C. Davis, "Thoreau and the Forest History of Concord, Massachusetts," *Forest History* 30, 1986; and Donahue, *Great Meadow*. See also T. Parshall and D.R. Foster, "Fire on the New England landscape: regional and temporal variations, cultural and environmental controls," *Journal of Biogeography* 29 (2002): 1305–1317.
7. Blanding and Schofield, "Walden Woods," 9. "Walden Pond" appears often in the seventeenth- century land-division records of Concord; "Walden Woods," never.
8. Colonial use of wood lots in Concord is detailed in Donahue, *Great Meadow*. It is difficult to give a more precise account of forest cover because valuation returns included an ambiguous category called "unimproved."
9. October 20, 1860, *Journal* XIV, 161.
10. This impression of heavier cutting is given by many contemporary observers quoted in Blanding and Schofield and in Maynard. I hope to see it confirmed by quantitative study.
11. The ecological condition of this landscape is detailed in Brian Donahue, "'Dammed at Both Ends and Cursed in the Middle:' The 'Flowage' of the Concord River Meadows, 1798–1862," *Environmental Review* 13 (1989).
12. Henry D. Thoreau, *The Illustrated Walden* (Princeton, New Jersey: Princeton University Press, 1973): 155–166, 59; Robert Gross, "The Great Beanfield Hoax: Thoreau and the Agricultural Reformers", *The Virginia Quarterly Review* 61.3 (1985): 483–497.
13. December 15, 1859, *Journal* XIII, 29.
14. See Blanding and Schofield, 39–40. Thoreau spent much of December 1857 surveying in Walden Woods.
15. December 10, 1856, *Journal* IX, 177–178. For Thoreau's observations on woodlot practices see David R. Foster, *Thoreau's Country: A Journey Through a Transformed Landscape* (Cambridge, MA: Harvard University Press, 1999).
16. Blanding and Schofield, 24–25.
17. National Park Service, *Walden Pond and Woods: Special Resource Study: Reconnaissance Survey* (Boston: Boston Support Office, 2002): 9.

18. Horace G. Wadlin, *Census of the Commonwealth of Massachusetts 1895* (Boston: Wright and Potter, 1899): 257.
19. Blanding and Schofield, 37–38.
20. On expansion of red maple in North America, see Marc D. Abrams, "The Red Maple Paradox," *Bioscience* 48 (1998): 355–363.
21. For a full discussion of these issues, see D.R. Foster, "Insights from historical geography to ecology and conservation: lessons from the New England landscape," *Journal of Biogeography* 29 (2002): 1269–1275.
22. Edith Sisson, "The Hapgood Wright Town Forest", unpublished, 1972. On town forests in New England, see Robert McCullough, *The Landscape of Community: A History of Communal Forests in New England* (Hanover, N.H.: University Press of New England, 1995).
23. A statewide version of this dual vision of old-growth reserves and woodland matrix is laid out in the Harvard Forest paper David R. Foster, et al., "Wildlands and Woodlands: A Vision for the Forests of Massachusetts" (Petersham, MA: Harvard Forest, Harvard University: 2005). More information on the Wildlands and Woodlands vision for the New England landscape is available at http://www.wildlandsandwoodlands.org/. A "West Suburban Conservation Council" for the SuAsCo basin was formed in 2010, under the auspices of Sudbury Valley Trustees.
24. For a detailed blueprint see Frances H. Clark, "SuAsCo Biodiversity Protection and Stewardship Plan" (SuAsCo Watershed Community Council, 2000).
25. The American Chestnut Foundation program is at http://www.acf.org/.
26. Aldo Leopold, *A Sand County Almanac* (New York: Oxford University Press, 1949)
27. The Brandeis "Suburban Ecology Project" http://www.brandeis.edu/programs/environmental/sep/index.html provides a link to a "Walden Woods Webmap," as well as a detailed description of a similar research program in Weston, MA.
28. For more information about these projects, visit Massachusetts Woodlands Cooperative at http://www.masswoodlands.coop/, and Vermont Family Forests at http://www.familyforests.org/.
29. Sustainable forestry at the community level is detailed in Donahue, *Reclaiming the Commons*.
30. M.M. Berlik, D.B. Kittredge, and D.R. Foster, *The Illusion of Preservation: A Global Environmental Argument for the Local Production of Natural Resources*, Harvard Forest Paper No. 26 (Petersham, MA: Harvard Forest, Harvard University, 2002).

6 Sojourning in Nature

The Second-Home Exurban Landscapes of Ontario's Near North

Nik Luka

EDITORS' INTRODUCTION

"Place" and "landscape" are powerful conceptual tools for exploring interactions between people and their environments. This chapter focuses on an exurban analog and classic frontrunner of green sprawl: central Ontario's "cottage country." The appropriation of this territory of lakes and woods—chiefly by residents of nearby metropolitan centers—is increasingly characterized by processes of transforming landscapes of weekend leisure to full-time home settings, as summer cottages are converted to year-round dwellings. The material effect of the ideology of nature is strongly exhibited here.

This chapter starts by characterizing the landscapes of cottage country as they are commonly perceived and represented by people who live and/or holiday there, focusing particularly on the way that these landscapes are associated with a carefully constructed and yet relatively generic set of representations of ideal "natural" settings. The chapter then explores what a proactive planning and design perspective contributes to an analysis of these representations. Luka considers how planners might intervene to deal with issues in ways such as encouraging land stewardship and greater ecological responsibility on the part of users.

What does the widely shared phenomenon of "cottaging" reveal about aspirations toward relating to nature? As a short-term respite or escape into closeness to nature, how does cottaging get in the way of finding nature in the cities or the suburbs where most people reside or work? This chapter explores the idealized nature sought in the weekend and summertime retreats of Ontario vacation tradition, and raises question about this nature at the beginning of the twenty-first century. Is "nature" something we visit on the weekend or on holiday rather than something within which we actually live? Are urbanites unable to connect to "nature" in day-to-day life? This broad question is the first of two interconnected premises explored here. The second involves critical scrutiny of the social practice of "cottaging," focusing on central Ontario cottage country, and making the case that this landscape typifies and helps us to explore the motives for exurban growth.

Understanding more about Ontario's cottage country helps to illuminate themes that cottaging and exurbanization share, such as residential mobility and the preferential choice of natural landscapes. Using the conceptual framework of cultural landscape studies together with analysis of the literature, history, ecology, and landscape of cottage country, Luka describes the ways in which cottagers find landscape meaningful. Cottagers negotiate many issues that also arise in the exurban landscape—and many problems that are naturalized by ideals of single-family home ownership are hyper-evident in the seasonal context of vacation and second home communities.

This chapter provides insight into North American aspirations for the nature that is represented as part of the exurban dream. Exploring the environmental values and aesthetics held by cottagers, who are often an influential part of society, Luka examines how such values and aesthetics are written into landscapes culturally appropriated in both cottage country and "back home" in the landscapes of primary residence.

Understanding the intents expressed by many cottagers to engage with the natural environment enables us to ask important questions about their arguable failure to achieve their visions of ideal environments in their environmental practices both at the cottage and at home. Although the cottage is undeniably a place of nature experience for most Ontario cottagers, this experience of nature tends to degrade the natural environment, both through forms of lake-side development and also through the exacerbation through commuting of southern Ontario's dramatic smog problem—arguably a major environmental *push* toward what was once cleaner air at the cottage.

Despite his analysis of cottagers' and exurbanites' culpability in reproducing landscapes they are trying to escape and of which they are so critical, Luka does not simply blame them for their role in residential sprawl, but rather suggests that the tendency to vilify people who participate in green sprawl ignores the wider historical and structural context of which they are a part. Luka looks to planning traditions and to the discussions that have been taking place in many local cottagers' and residential associations to call for more informed and open discussion of the issues that characterize amenity residential landscapes. Bringing this conversation back to the metropolitan areas of primary residence from which cottage country extends, Luka offers valuable questions about how the cottage vision might be better integrated into the urban environment it critiques.

As an architect, planner, geographer, and cottager, Nik Luka brings to his academic study of Ontario cottage country a personal history there as well as a perspective informed by his professional experience as an urban designer. Exploring the impulse to sojourn, Luka articulates the ways in which a desire for a relationship with nature is met by cottage visits, which, although short in duration are long on meaning. In pointing out the paradoxes in the way this temporary relationship with the natural environment has come to define a lifestyle standard that is itself highly abstracted and idealized, Luka seeks to make sense of this impulse to escape and concludes by suggesting some key points for the negotiation of meaningful incorporation of nature into the residential environment. This chapter shows the reciprocal relationship between a cultural landscape and its inhabitants: culture produces landscape in a place where, in turn, the landscape has a powerful effect on its inhabitants, both short- and long-term.

—Kirsten Valentine Cadieux and Laura Taylor

■ ■ ■

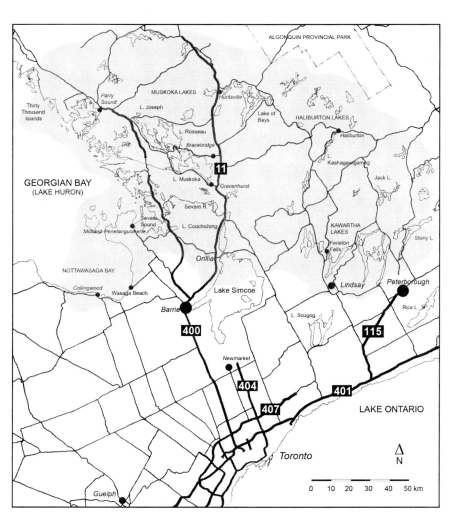

Figure 6.1 Central Ontario, showing the built-up parts of the major urban centers, major roads, expressways, and "cottage country" (shaded) to the north and east of the city, where pre-Cambrian rock of the Canadian Shield (as clearly indicated by the myriad of small water bodies) emerges from beneath the fertile plains of the Saint Lawrence Lowlands.

Wrapping around the western end of Lake Ontario is a continuously built-up urban area that is in many ways the economic, industrial, and service core of Canada. Centered on the city of Toronto, and with a total population of about eight million, this area is known as Ontario's "Golden Horseshoe" (Figure 6.1).[1] On any given summer weekend, a mass exodus occurs as tens of thousands of people head northward to what is called "the cottage." This can range from a small, rustic cabin to an elaborate lakeside mansion (Figures 6.2 and 6.3). Yet almost invariably, it is found in what is commonly referred to as the "cottage country" of the "Near North"—a land of lakes and woods where the rugged pre-Cambrian rock of the Canadian Shield emerges from the fertile plains of the St. Lawrence Lowlands, about 100 miles beyond the continuously built-up urban area radiating outward from Toronto. This "cottaging" activity is a long-established social practice; although access to some kind of cottage property is neither typical nor universal by any stretch of the imagination, the practice is certainly widespread. For instance, real-estate market studies conducted since 2000 suggest that cottage owners make up at least 10 per cent of the province's population, while another five to six per cent plan to acquire such a property within three to five years.[2] Invisible in these data

Figure 6.2 A modest cottage probably built in the 1930s (Crystal Lake).

Figure 6.3 An elaborate year-round "cottage" compound built in the late 1980s (Lake Muskoka).

sets, of course, are the relatives, friends, and colleagues of these cottage owners who visit, rent, or otherwise have access to cottage properties. Also centrally important is the fact that cottaging households tend to be based in the urbanized parts of the Golden Horseshoe. In other words, cottaging is a very popular activity among central Ontario residents of city and suburb alike.

Why should this Toronto-centric social practice of "cottaging" be of particular concern or general interest? After all, urban folk across North America and Europe often head when they can for more "natural" places in the mountains, at the beach, or in the woods.[3] This chapter suggests that the central Ontario case is an especially compelling example of how contemporary North American notions of "nature" are made manifest in the built environment. Drawing on William Cronon's work on the idea of nature, my aim is to demonstrate that cottaging in Ontario exemplifies how commonly-held notions of what constitutes "nature" may problematically hide their unnaturalness behind masks that are alluring, *precisely because they seem so very "natural."*[4] If cottagers seek a more meaningful relationship to the natural processes that sustain life in the "Near North," what does this tell us about how we relate "nature" to our everyday (sub)urban life settings? Must it always be "out there" relative to the metropolitan contexts in which most of us live, work, and play? Are we thus bound to merely *sojourn* in nature instead of really *dwelling* in it?

Defined by the *Oxford English Dictionary* as a "temporary stay at a place"—an act of tarrying or abiding only tentatively at some spot—the term "sojourn" aptly describes both exurbia and central Ontario cottage country, for in both cases, getting "close to nature" is arguably at the heart of the experience.[5] Yet it tends to be trite and fleeting, noncommittal and provisional—a glimpse, as it were; a mere visitation. Cottage country as settlement pattern may thus represent a sojourn amidst nature on the part of city dwellers who find themselves unable (or even unwilling) to bring nature into the fold of daily life *within* the more urbanized areas where much of contemporary life takes place. This is an extension of historian Allan Smith's argument that in nineteenth-century Ontario, while city dwellers could certainly indulge in sojourns amidst nature, these were to be strictly temporary experiences.[6] Popular publications at the time cottage country was being drawn into the metropolitan hinterland made this clear; for instance, one travel magazine writer advocated in 1874 for literally "stocking up" on "health and strength of body and mind" while asserting that this was best done through temporary summertime visits that would "serve for the rest of the year" in the city—where (and when) presumably one was to be industrious, productive, and societally useful, in keeping with the curiously Puritanical work ethic of "Toronto the Good" (as the city long was known).[7] Did cottaging in central Ontario thus have its inception in notions of sojourning amidst nature? Historical evidence suggests that its early growth was based on a strict dichotomy in which the urban spaces of industry and production were seen as quite distinct from—or perhaps irreconcilable with—the "Near North's" unspoiled, healthful, and regenerative spaces. Ironically, though, and as with similar such summer home settings, this cottage country really would not have come to exist were the major urban centers of the Golden Horseshoe and U.S. cities such as Pittsburgh or Cleveland not within relatively easy traveling distance. In many ways, then, the landscapes of central Ontario cottage country have always *themselves* been forms of scattered metropolitan growth.

The present-day issue, however, is this: striking changes are now clearly underway in central Ontario cottage country as rising numbers of "cottagers"—considered for official purposes to be "seasonal residents"—now live by their lakes on a year-round basis. At the same time, quasi-suburban subdivisions and elaborate permanent dwellings increasingly abound along lakeshores that have long been lined with smaller, more modest, and simpler summer cottages (Figures 6.4 and 6.5). Now more than ever, cottage country is in many ways an urbanizing edge where residents of city and suburb alike are appropriating and transforming the far reaches of the rapidly growing Toronto-centered urban region. We may think of these settings as important if underscrutinized pieces of the metropolitan puzzle, and indeed as a form of exurban growth, given the broad definition of exurbia as places inhabited by people who deliberately choose to leave urban settings for places where they feel they can be closer or more connected to landscapes they construe as being "natural" or even as the "wilderness."[8] This is not all there is to the process, of course, but it certainly appears to be a key motivating factor.

Figure 6.4 Typical configuration of modest early twentieth-century waterfront cottages (Severn River).

Figure 6.5 Lakeside subdivision in the Kawartha district (Pigeon Lake).

What follows here is the suggestion that cottage country in central Ontario reveals key questions of *how* Western culture is now increasingly scrutinizing its relationships with what we call "nature." At the risk of sounding cantankerous or arrogant, please be assured that I don't seek to pejoratively judge those people who "go to" cottage country (or any "exurban" setting), for they probably are seeking to genuinely deepen their connection and engagement with "nature." Certainly I myself am one of "those people"—I often spend my summer months on the rocky shores of central Ontario's waterbodies, as I have done almost every year of my life, and as my parents, grandparents, and great-grandparents did too. But in central Ontario cottage country, the results on the ground are not always ideal. Extensive concern is now being expressed over the "urbanization of cottage country"—but more on this in a moment. My case here is that if we are to figure out how we can really dwell *within* nature instead of merely sojourning in it, we should first come to terms with how we *don't* really live within nature—even where we explicitly intend to do so. On a more optimistic note, underpinning my argument is the belief that cottaging contexts present excellent opportunities to improve the ways in which we weave our settlement patterns into the natural processes that sustain life.

This chapter first looks at central Ontario cottage country as an imagined landscape, seeking to make sense of its rituals of sojournment—and by extension the relationships with "nature" thus represented. How does cottaging figure in landscape discourses in Canada, and how have its ideals infused cultural identity, both regionally and in pan-Canadian ways? Central Ontario's rich history of cottaging as a popular pastime has been matched by the development of a powerful cultural iconography, in ways ranging from the paintings of Canada's celebrated artists known as the Group of Seven to the ubiquitous beer commercial in which revelers are invariably shown escaping the city to enjoy life by the lake (usually in fine weather). Cottage country is thus in many ways considered a distinctly Canadian example of how people make sense of the landscapes in which they spend time. To unpack these images and meanings, this chapter considers how the cottaging phenomenon came about in the context of city folk seeking sojourns amidst nature. Drawing on historical narratives, I examine how cottage country images and meanings have been generated, with reference to the lakeside settlement forms of central Ontario, and how all of these are now changing. I also examine how cottage country is an intriguing landscape that may prove useful in making sense of how we tend only to sojourn in nature. How can we really begin to weave our settlement patterns into natural process? This is deliberately left as a lingering question for planning and design, as something that will only be possible once we move beyond mere "sojourning."

Before plunging into the argument, I should state that I am a fairly urban sort of person—in terms of my work in architecture, urban design, and planning, as well as in my day-to-day life. I live, work, and relax, for the most part, in city neighborhoods built up before the Second World War—in an era before cars became the predominant mode of travel in metropolitan areas. I stand with my feet firmly planted in the city, as it were, looking outwards to what lies beyond its boundaries, whether officially defined or otherwise perceived—all of which is in constant flux. But of course, my fascination with the cottage country experience is intense because I *am* a cottager, having grown up in the city while spending my summers in the "great beyond"—in quirky lakeside wooden houses that had no running water or much in the way of material comforts (but always at least one canoe!). My own early experiences were catalytic in shaping my current interests in how people think about space—and in how ideas like "nature" are key to understanding the motivation for people's behavior. For instance, as a researcher, educator, and practitioner, a key interest of mine is how to better integrate our own settlement systems with natural process: how to engage in ecological design.[9] My contribution to this book thus centers on the argument that the phenomenon of urban and suburban folk spending time in cottage country—and wanting to get closer to nature, but only for "sojourns"—merits critical reflection.

EXURBIA AND COTTAGE COUNTRY AS IMAGINED LANDSCAPES

To set the stage for the discussion to follow, we should first establish the importance of thought and feelings as they relate to the environments that surround us. From extensive work that has been done in the fields of environmental psychology and cognitive science, we know that people connect themselves with places and landscapes through the realm of the imaginary. They do so by creating meaning as they perceive, think about, and evaluate different settings in a fairly subconscious process. Landscape historians, theorists, and critics such as John Jakle, Edmund Penning-Roswell, Amos Rapoport, and Edward Relph have further demonstrated that what we apprehend within the landscape is a sophisticated function of our individual preconceptions, expectations, and cultural values, and indeed that landscape is something that enfolds our entire experience rather than something we behold in *tabula rasa* mode.[10] As concerns cottage country, for instance, the Canadian historian Claire Campbell has provocatively argued that, "the entire culture of cottaging and camping is predicated on a Romantic view of nature as a place of beauty and solitude, of restoration and spiritual communion."[11] A similar such argument can be made for exurbia, the growing literature on which includes texts by environmental historians, planners, designers, and other cultural observers. Many have taken their inspiration from the original work of the journalist Auguste Spectorsky. Now widely understood as a quasi-proper name for the region outside the suburbs of a city, Spectorsky's term "exurb" specifically described New York City's far-flung quasi-rural commuter settlements inhabited by well-to-do professionals and executives who had left the city, he argued, in search of a lifestyle more connected to nature.[12] Exurban settlement patterns can thus be seen as problematically complicated by their inhabitants' sense of being deeply connected to what they perceive as the natural landscape. Many of the central features of exurbia—extremely low-density albeit planned residential areas, usually consisting of so-called "McMansions" whose construction entails the conversion of forested or agricultural land—ironically contribute to the very problems that the social and environmental aesthetic of exurbia ostensibly attempts to avoid.[13] Yet the exurbs are imagined landscapes. Their growth has been shaped by ideas about nature that are based on attitudes, beliefs, and experiences, and which paradoxically often drive their subscribers to profoundly transform some of the very attributes that drew them into these settings in the first place. Clearly my argument here hinges on the importance of *representations* in our individual and societal understandings of what makes one landscape or setting different from another—something best supported by the discourses by which activity patterns, cultural idiosyncrasies, and societal norms are communicated.

The Imagined Landscapes of Cottaging

Given that the realm of the imaginary is so very rich, it is useful to start an appraisal of its manifestations by examining basic terminology. The quintessential central Ontario lakeside cottage has historically been a modest affair. Turning again to the *Oxford English Dictionary,* the very term "cottage" is commonly used in North American English to represent "a summer residence (often on a large and sumptuous scale) at a watering-place or a health or pleasure resort." This bespeaks its sojourning quality, for cottaging is thus an act of residential tourism: tourist activity made ritualistic and physically embodied in structures (in this case, housing).[14] Its first recorded use in this way dates to 1882, in reference to Bar Harbor in Maine.[15] Yet quite similarly, Partridge's Concise *Dictionary of Slang and Unconventional English,* first published in Britain in the 1960s, defines "cottaging" as "going down to one's cottage—often quite a largish house—in the country for the weekend."[16] The term could easily be applied to the analogous leisure settings found throughout the north-eastern U.S., central Europe (the Swiss have *chalets* or *Ferienwohnungen,* for instance), Norway and Sweden (where people go the *hytte*), and Russia (where people have *dachas*)—all part of the widespread trend across Europe, North America, and Australia in which urbanites keep second homes in non-urban settings.[17] The central-Ontario case is remarkable because its imagined landscapes are arguably a distinguishing icon of cultural identity in middle Canada. In this regard, the term "cottage country" refers to a discrete, easily-identifiable set of central Ontario landscapes—the products of growth and development factors that are arguably city-based.

Sojourning in Nature 131

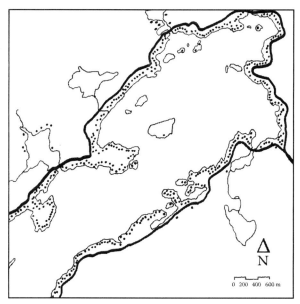

Figure 6.6 The typical pattern of cottage buildings and roads: as in much of central Ontario cottage country, the lake is lined by private dwellings on relatively small parcels of land, which make it inaccessible to the general public (Kennisis Lake).

Figure 6.7 Aerial view of a typical central Ontario cottage setting (Parry Sound).

What does central Ontario cottage country look like? Typically, its settings are lakes and rivers lined by private properties that make the water bodies virtually inaccessible to the general public, although the extensive tracts of land behind the single row of private lots are usually Crown land to which the public has right of access (Figure 6.6; Figure 6.7).[18] While eclectic in their architectural detail, the general way in which space is organized on each of these lots tends to be quite consistent, even homogenous (Figure 6.4). As a settlement pattern, then, cottage country in central Ontario is typologically well-developed and lends itself surprisingly well to generalization—but does the same hold true in terms of its imagined landscapes? How do bonds between people and places develop amidst the rugged contexts of lakes, rocks, woods, and cottages? The *Nelson Canadian Dictionary of the English Language* defines a "cottage" as "a recreational property with a house, especially for summer use." A more humorous definition comes from a best-selling popular book, *At the Cottage*, published in the late 1980s:[19]

> Whether it is a cottage, cabin, shack, or lodge, or whether it is camp, it is probably near a body of water, usually a lake. It has fewer creature comforts than its urban, suburban, or even rural counterpart. It has more bugs, less lawn, at least one boat, at least one mouse, a smaller kitchen, a larger birdhouse. Neighbours are farther away. So are stores. There may be a road to it; it may be accessible only by water. Either way, it is harder to get to than the place people live in the rest of the year. That may be its charm. It is hard to get to. It is hard for other people to get to.

This definition is evocative, if not romantic and wholeheartedly sentimental (ironically, given that it points to the romance and sentiment of cottaging by speaking of some aspects that people probably "love to hate"). Focusing for the moment, however, on some basic data, there seem to have been around 28,000 households maintaining summer holiday "cottages" or "camps" across the province of Ontario in 1941; this number increased to approximately 164,000 by 1973, reaching 216,000 in 1991 and 374,000 in 2003, corresponding to about eight per cent of the provincial population in that year (the most recent for which comprehensive data can be obtained).[20] It must be noted, however, that precise counts are nearly impossible to come by, as households can opt to declare their "cottage" as their primary dwelling and because Canada's federal data-collection agency, Statistics Canada, collects data only on "cottages" as household amenities or "perks" (and using methods that have varied over time). Ongoing studies by Royal Lepage indicate continued rates of demand and ownership.[21]

To grasp more fully the cottage phenomenon, we must turn to popular culture. One popular author jokes that "everyone goes, at some point, to something called 'Our Summer Place'"—claiming that cottaging is "Canada's summer obsession" from coast to coast.[22] Other examples are found in the many books published by popular Canadian presses, such as one entitled *Escape: In Search of the Natural Soul of Canada*.[23] Released only a few months following the conference that gave rise to this collection of essays, this book is a fine example of how we in North America too often associate "nature" with an "escape" from the city context. Even if the term "cottage" is highly evocative, though, it represents a geographical concept whose nomenclature varies regionally: "In Northern Ontario, around Thunder Bay, people don't go to The Cottage. Cottages are effete eastern creations, for sissies who eat quiche on screened verandas and sip white wine before retiring indoors to watch something on the VCR. Around the head of Lake Superior, people go to Camp. Albertans near the Rockies go to The Cabin. In Saskatchewan, people go to The Lake."[24]

A curious point is raised here. The preponderance of the very term "cottage"—a characteristically central Ontario name for a summer home—may be indicative of how this cottage phenomenon is analogous to exurbia. Spectorsky argued that many of the original "exurbanites" in the New York City region were employed in media and cultural industries and were therefore particularly adept at creating and maintaining cultural images, meanings, myths and narratives—in a word, representations, both individual and collective.[25] Indeed, Toronto's own Marshall McLuhan agreed with Spectorsky shortly after *The Exurbanites* was published. Exurbanites, in McLuhan's view, were symbol-manipulators who had mastered the grammar and rhetoric of the "new" media of the postwar boom years, which were so vitally important to the image-driven society in which we still find ourselves.[26] Similarly, central Ontario arguably controls much of the imagined landscape of Canada. The country's marketing, advertising, and broadcasting media are largely controlled from Toronto, as are many of the country's somewhat less market-driven cultural institutions. It is not an exaggeration to say that a great many movers and shakers in these circles are apt to frequent the lakeside settings of central Ontario cottage country. Consider, as an example, *Cottage Life* magazine—an enormously popular publication which, according to its publishers, now has a circulation of 70,000 with 800,000 readers for each of its six annual issues.[27] The magazine is only one part of a business venture that includes an annual trade show attracting almost 40,000 people and a popular television show—yet all of this is unquestionably dominated by images and narratives of cottage life in central Ontario, reflective of the fact that the magazine and television show are based in Toronto. Another more esoteric indicator is "Rune Arlidge," a critically acclaimed three-act drama by Michael Healey, which unfolds entirely on the front porch of the ubiquitous Ontario cottage. These examples bespeak the ways in which representation so dominantly can overlay direct experience: the importance of the "imagined landscape."

The main question to ask of the popular culture of cottaging is this: What exactly is being imagined by the many thousands of people who subscribe to cottage life?[28] Evidence suggests that a prominent place is given to experiencing nature, howsoever it may be perceived. But cottage country landscapes can be seen as indicators of the complicated yet ultimately ambivalent relationship that Canadians seem to have with natural process—ecosystems in all their diverse, messy, and uncertain complexities.[29] Could it be that the recognition of the many charms and undeniably positive effects of the fresh air, water, and woods that characterize much of central Ontario seems to invariably coincide with the knowledge that this "wilderness" is a thick, unrelenting, and unforgiving forest—a deep, dark space in which one can easily stray from the comforts of "civilization" and perish? For the early pioneers from Europe who colonized its southern reaches, we can only imagine how rudely it must have contrasted with the long-settled landscapes of Britain, Ireland, Germany, and the other places from whence they came.

We can turn here to the themes that dominate Canadian literature, as an indication of important ideas in the collective imagination of its people. Two particularly celebrated Canadian analysts are Margaret Atwood and Northrop Frye. Atwood is famous for her assessment of how a "survival" mentality has scarred the collective identity of Canada, recurrently appearing in our literature, in both French and English: "Our stories are likely to be tales not of those who made it but of those who made it back, from the awful experience—the North, the snowstorm, the sinking ship—that killed everyone else. The survivor has no triumph or victory but the fact of his survival; he has little after his ordeal that he did not have before, except gratitude for having escaped with his life."[30] In Atwood's view, most Canadian literature up to the 1970s deals with this kind of "grim survival" as its central theme, and as experienced by various individuals or groups. Frye similarly argued that Canadian thought has been characterized by a "garrison mentality" in which ideas were formulated in the context of the vast, terrifying emptiness of Nature—a *modus operandi* that functioned conceptually but not poetically and thus bogged down our very ways of thinking.[31] These ideas are rooted in historical narrative: consider, for instance, the experience of nineteenth-century Norwegian and Icelandic settlers who floundered in the undeniably severe central Ontario winters, and died out or moved away to greener pastures.[32]

Marked as it was by tragedy and hardship, pioneer life in Ontario may have helped give rise to a "grim survival" imagination of the "bush" as representing the dense, unforgiving back country that seemed to remain largely untouched by human hands. It's a notion well worth considering here, for it is within or adjacent to this proverbial "bush" that we find cottage settings—and the terms are often used somewhat interchangeably, as they both impart a sense of the rugged "backwoods" of central Canada (that is, somewhat removed from the lands that are arable in the conventional sense).[33] Included in the *Oxford English Dictionary* is a definition of "bush" as land "more or less covered with natural wood"—a term specifically applied "to the uncleared or untilled districts in the former British Colonies which are still in a state of nature, or largely so."[34] It would thus seem quite justifiably applicable to the context in which central Ontario cottage country has come to exist. Observers such as Roy MacGregor argue that "our" idea in Canada of the "bush" stands in rather stark contrast to the (ostensibly analogous) notions of the English "countryside" and the American "woods."[35] Certainly, both these concepts refer to very particular settings that are hardly "wild."

Sorting through this semantic puzzle is made easier by referring to a compelling argument made by Michael Bunce in his book *The Countryside Ideal: Anglo-American Images of Landscape*. Bunce argues that in both English and American contexts, certain archetypal "rural" landscapes serve as the focal point for elaborate cultural mythologies or aesthetic and social ideals that have come to be bound up in the very lay of the land; the countryside ideal is thus a culturally-constructed nostalgia for life in carefully-cultivated pastoral settings (but definitively *not* the "bush" and its "wilderness" analogues in America and England), as expressed by countless observers from Blake to Jefferson to Thoreau.[36] He thus refers to the English "countryside" and American "woods" as "universally domesticated rural landscapes" found across the settled parts of those countries.[37] At the same time, though, Bunce suggests in his book that a specifically Canadian version of the countryside ideal seems to be poorly developed. If we believe esteemed observers such as Atwood or Frye, this gap in collective narratives of Canada most certainly cannot be filled by the "wilderness"—as this is just not a suitable setting in which to lead everyday life. And so, my case here is that to some extent, in Ontario, at least, cottage country fills this gap of the imagination: as Ontario's "Near North," it represents a process of "making do" with a narrow belt of in-between landscape that is wild enough to be "away from the city," but "close enough to home" to be safe and secure. To make sense of this case, we can turn to a short summary of how this ritual of cottaging came about, including a more detailed description of cottage country landscapes and their lakeside settlements, with attention paid to what is what is happening there today.

COTTAGING IN ONTARIO: ORIGINS AND EARLY GROWTH

Historically, much of central Ontario cottage country was still untouched by Europeans until the middle of the nineteenth century, when government land grants brought immigrant farmers to its rocky shores. It did not take long to realize that the rocky soil was hardly arable except in small pockets—one observer noted that he had known "instances where a mound of earth has been sought for, and looked upon as a treasure."[38] Frustration and heartbreak may have awaited would-be farmers, but loggers recognized the potential value in the old-growth forests—that "bush" that loomed so large in the imagination—and this brought about a lumbering boom in the latter part of the nineteenth century.[39] Once the land had been literally stripped of its virgin forest (Figure 6.8), the loggers moved on, leaving the door wide open for tourism and leisure. A handful of hunters and sportsfolk, many of whom were wealthy Americans, had pushed their way into the land by the 1870s, establishing hunting and fishing camps or lodges for seasonal use (Figure 6.9). But this did not last for long, as savvy locals and tourists soon realized the potential for organized recreation geared to city dwellers from the growing industrial centers to the south. Here, at last, was something that could sustain a local economy—as illustrated by historical narratives of the life of several Canadian families on Browning Island on Lake Muskoka, considered by many to be in the very heart of cottage country.[40] For instance, "My father, Captain Charles Wesley Archer, came to the Island as a young man [in the 1900s]. He loved nature and the peace and quiet there. My mother came from Lucknow and my father was born in Bracebridge. They first came to the Island, cleared a campsite, and recognized the future of the Island. In time they purchased much of it and built cottages, docks, and the like, as well as operating a series of steam and motorboats."[41]

Figure 6.8 A late nineteenth-century postcard showing a steamer picking its way through a log boom in the Muskoka Lakes at the tail end of the logging boom.

Figure 6.9 A proudly posed late nineteenth-century postcard image of the kill at a Muskoka hunt camp.

The target market generally comprised well-heeled residents of the industrial cities of the Great Lakes basin, particularly Toronto, but also Pittsburgh and Cleveland. One popular history argues that in the late nineteenth century, central Ontario cottage country quickly became "one of the premiere holiday regions in North America" because of its antidotal effects as a "summer escape from heat, congestion, smoke, and the pressures of business."[42] To be sure, central Ontario cottage country is at higher elevations and further north than the industrialized Great Lakes lowlands, and its average summer temperatures are comfortably lower relative to the stifling humid heat of places like Toronto and Pittsburgh. With brisk westerly winds bringing fresh unpolluted air, it is not difficult to see the appeal to city dwellers, especially before the advent of mechanized air-conditioning. Those with the means thus headed northward, often for the whole summer. There were of course strong physiological and psychological benefits associated with doing so: "by leaving an atmosphere tainted with sewer gas to inhale the tonic perfume of the pine bush," the "weary, over-worked toiler[s] of the city" would find "their cheeks flushing with freshened tints of purified blood."[43] The sojourn was seen as a necessary coping strategy for regimented, over-structured, and perhaps "unnatural" city life. Cottage country offered freedom, space, and peace, and a healthful environment, particularly for children growing up in the city. One advocate of summer sojourns into the Muskoka region thus wrote in 1886 that, "the prudent man flies from all artificial conditions and yields himself to the soothing influences of nature on the shores of the lakes and rivers in the depths of our primeval forests."[44]

Figure 6.10 A 1908 advertisement for Canadian Pacific Railway's new express service to the Muskoka Lakes.

Direct rail links to cottage country from Toronto had been established by the 1870s, and accessibility increased during the initial boom period of the early 1900s as the Canadian Pacific and Grand Trunk Railways competed fiercely to bring the "soft waters" of the Muskoka Lakes within relatively easy reach of major population centers. It was still quite a journey from Toronto, however, especially when contrasted with the two- or three- hour drive of today. At the turn of the twentieth century, for instance, trains left Toronto at 9.30 a.m., reaching Muskoka Wharf in the village of Gravenhurst at about 2.00 p.m. (Figure 6.10).[45] Awaiting travelers at the wharf were steamers that would shuttle travelers to various resorts that had been built in great numbers and, in later years, to private waterfront cottage properties. At the height of this "golden period" of the grand cottage country hotels in the years before the First World War, there were over 100 resorts that together provided accommodation for tens of thousands of tourists (Figure 6.11).[46] The resort industry was helping to sow the seeds of its own demise, however, as guests increasingly sought private properties on which they could build their own retreat.

Figure 6.11 An early advertisement for the Grand Trunk Railway service to Muskoka, boasting of the "splendid service" afforded by hotels "set in fragrant pines."

Following the First World War was a moderate boom in cottage construction. This early growth was often characterized by clustered groups of families and friends; the following account, again from Browning Island, is quite typical.

> In 1904, Alexander Stark bought the point later occupied by the Greey family. He had a cottage built on this property on the northeast part of the Island, opposite the tip of Chief Island. Here he and his wife spent several summers. Unfortunately, he became ill in 1907 and could no longer travel to Muskoka. . . . In 1905, Frank J. Stark, who had visited his uncle Alexander Stark the summer before, and who had been greatly impressed by the beauty of the country, had a cottage built immediately to the south of his uncle's property. . . . Mr William G. Francis and his family rented Alexander Stark's cottage . . . They liked the Island so much that they bought land immediately to the north of that property. They moved into their own cottage in the summer of 1907.[47]

Reflecting the "sojourn" mentality of cottage country, the bulk of settlement form was rarely permanent. Most of the wooden houses built near the

Figure 6.12 A "classic" nineteenth-century cottage compound nestled amongst the trees on an island in Lake Rosseau.

lakes were not at all for use in the winter, although some of them were quite large and elaborate (Figure 6.12). The sparse seasonal use of the lake country was reflected in the fact that shops and services were usually provided on a somewhat ad-hoc basis by locals. For instance, until well into the 1920s, provisions were delivered by boat on many cottage country lakes: "The vessel servicing Browning Island was the *Nymoca*. It included a butcher shop and grocery. The *Nymoca* landed at all the larger docks three times a week. Cottagers with smaller docks gathered at the larger ones nearby to do their shopping. The gangplank was bridged to the dock from the freight deck, and one walked into the fully stocked grocery area, produce section, and butcher shop. Fresh fruit, vegetables, and regular staples such as flour and sugar were always available."[48]

Architecturally, the cottage structures were eclectic but they made extensive use of verandahs or porches from which genteel city dwellers could enjoy the view and the fresh breezes—suggesting that there were rarely any pretenses about "roughing it" or being on the land. To some degree, this was grounded in pragmatism: as pleasant as it may have been, cottage country was infamous for its black flies, mosquitoes, and other bugs that made being outdoors at times unbearable, especially in early summer. Sitting on a raised porch that was well ventilated by lake breezes was far more enjoyable. The building-up of cottage country in fact gave rise to an architectural feature now known by many as a "Muskoka room"—a fully-furnished screened porch in which one could be closer to the outdoors in comfort (Figures 6.13 and 6.14). While

Figures 6.13 and 6.14 The "Muskoka Room"—a furnished screened porch in which one can be closer to the outdoors without worrying about mosquitoes and other insects (Balsam Lake).

some cottagers brought in architects and builders from further afield, in many cases local carpenters were hired to design and build these structures. There are parts of cottage country where clusters of structures are marked by their remarkable similarity—not because they were prefabricated, or based on an off-the-shelf design, but simply because they were all built by one carpenter who saw no reason to vary his design. The result: pockets of curious vernacular architecture with a distinct local variability.[49]

Postwar Growth and Contemporary Changes

Through the first half of the twentieth century, cottage country remained a perk for the very wealthy, but the postwar years brought a huge increase in the number of (sub)urban households owning a second home by a lake somewhere in central Ontario. While very much driven by the cultural value associated with summer sojourns in cottage country, this postwar growth was facilitated by rising real and disposable incomes in Canada and more widespread access to private automobiles, which coincided with the construction of much better roads into cottaging areas. Another catalyst in this regard was extensive selling off by the government of lakefront Crown land at rock-bottom prices. It is hard to say whether or not this was motivated by an ideological motivation concerning the appropriateness of the citizenry having more direct access to the bountiful "natural" heritage of Ontario, or whether the government was merely looking to make money. Earlier in the twentieth century, as historian Allan Smith suggests, there seems to have been a desire to celebrate, honor, and elevate pioneer life, especially in what was considered a more authentic, innocent, and pure relationship with the land.[50] By the postwar years, however, the opening up of the land by the government may have been driven more by the wish to spur economic development in parts of the province that tended to be chronically impoverished.[51]

Regardless of what brought it about, the extent of this growth was astonishing: from 1941 to 1971, as the population of the Toronto metropolitan region rose rapidly from around 900,000 to almost three million (a growth of 221 per cent), the number of households owning cottages grew by a whopping 671 per cent, to about 164,000.[52] These data also show how metropolitan population expansion and cottage country growth were mutually reinforcing in the Toronto region, as undoubtedly in many others.[53] The net result of so many people having easier access to a greater number of cottages through the 50s and 60s—no doubt amplified by the workings of the media—was that cottaging thus came to be a well-known and characteristic pastime in the Toronto-centered metropolitan region. While it may have only been practiced by a relatively small slice of the population (rarely more than 20 per cent of Ontario residents have historically owned a cottage property, although an additional proportion of Toronto-area residents have rented and/or visited with friends and family having access to a cottage), it became a widely-shared vision of the good life. Moreover, it spread in a very particular way. Cottage country was resolutely a landscape in which one sojourned amidst nature, and perhaps even more so than had been the case with the wealthy, who had often spent long periods of the summer at their cottage. For the working middle class, the cottage was a place for weekends and holidays at a moment in history when leisure time had become much more plentiful than it had been in the past.[54]

What makes central Ontario cottage country so compelling as a settlement pattern has to do with the ways in which it reveals what happens when so many thousands of people give material expression to their images of an ideal environment—when they attempt to reconcile *what they wish for* with *what exists* on the ground.[55] Meaning is generated in environmental experience through processes of perception, thinking, and evaluating. This meaning is then related to preconceptions, which function as a real layer of experience in the landscape, even if the preconceptions bear no relationship to what is perceived.[56] While this disparity is true of most residential settings, it is especially vivid in cottage country. The desire to reconcile *what is perceived and evaluated* with *what is desirable* is a huge part of what drives people to build, to move from one dwelling to another, to renovate, to plant—to effect many kinds of environmental change. In central Ontario cottage country, we see the results of these processes in a landscape context that is highly charged for its users in terms of culture, individual meanings, and the ideology of "nature." The results are often architecturally eclectic, reflecting many individual choices (especially as cottages have often been owner-built), and bespeaking an imagined landscape of leisure—a vision of the good life amidst the lakes and woods. Figure 6.15 is a good example of this. In it is seen the standard spatial arrangement of the cottage and waterside boathouse, both having a carefully positioned and subtly framed vista over land and water. Rather than the savage, impenetrable "bush" is a pleasantly thinned copse of trees that help nestle the principal structure into the hillside, while keeping it on show (certainly in this example). A handful of "harsh" elements such as select boulders add a romantic flair to the scene, but the overall atmosphere is one of aesthetic refinement, space, and comfort.

Driven very much by the surge in conspicuous consumption of the postwar years, the cottage-building boom continued until the recession of the mid-1970s, when cottaging seems to have declined at least temporarily— partially because there was little vacant lakefront land left within easy driving distance of Toronto, but also because people just weren't going up to the cottage as much anymore. The Baby Boomers who had enjoyed waterfront life as children were now teenagers and young adults, preoccupied for the time being with other things. The oil crisis, too, had its effect; driving to the cottage suddenly became a pricier way to pass leisure time. Things picked up again in the 1980s, for a range of reasons, in many ways because of demographics—many of those Baby Boomers now had young families and were seeking to return to the holiday spots they knew when they were younger. But it was still done with a sojourn mentality: cottage country for most was a place to go on the weekend and on holidays.

Figure 6.15 An overstated example of a standard spatial arrangement in cottage country (Lake Joseph).

Figure 6.16 View in the "Port 32" lakeside subdivision in the village of Bobcaygeon (Pigeon Lake).

A sea-change began in the early 1990s. Although the familiar patterns of cottage life continued, people were progressively spending longer periods of time at their cottages, converting them for year-round use (known as "winterizing") if this hadn't already been done. Some—many of whom were nearing retirement age—decided to just live there year-round. As a result, by 1996, Canadian Census data already suggested an exodus of sorts with (sub)urbanites moving further afield from older urban centers. Many of these migrants were Baby-Boomers retiring to live at least semi-permanently in the leisure settings of cottage country.[57] This tendency has continued apace, marked by the widespread conversion and/or replacement of second homes at the water's edge, resulting in a highly dispersed landscape of year-round dwellings (Figures 6.3 and 6.5).[58] But it is not merely a matter of a graying population. Metropolitan-area housing developers have picked up on the trend by building other forms of housing on choice lakeside sites, and the archetypal "cottage" thus now increasingly vies for space with lakeside subdivisions such as "Port 32" near the village of Bobcaygeon (Figure 6.16) and condominium clusters inhabited by a range of demographic groups. The mortage crisis that dramatically transformed the U.S. housing market did not have commensurate effects in Canada, where the population moreover continues to enjoy increasing life expectancies and stable rates of early retirement. Given the importance of cottage country as a meaningful cultural landscape for its users, these latter-day transformations have caused no little uproar as images of summer holiday settings clash with burgeoning residential realities that in many ways constitute exurban growth. Many of the changes have been nicely scrutinized by geographer Greg Halseth in his detailed studies of how "cottage conversion" plays out in different settings across Canada. His conclusion is that while images of quaint cottages nestled beside wooded lakes are a significant part of the Canadian geographic imagination, "the idyllic calm of that lakeside scene belies a more complex and often somewhat contentious landscape" marked by controversy and debate.[59] Many cottage landscapes have recently become hotly-contested environments: spaces in which the priorities and expectations of different groups (locals, cottagers, cottagers-cum-locals, and others) now conflict more than ever before. The fiercest disputes are traceable to the clash of representation and reality—arguably the transformation of the sojourn into something else as people make their "first" or primary dwelling in cottage settings, such that these landscapes of sojournment become places of permanent settlement.

More broadly speaking, a familiar story in the history of metropolitan growth in North America now underlies changes in central Ontario cottage country: the migration of (sub)urban residents further afield in search of pleasant settings. Urban expansion now stretches well beyond the generally agreed-upon functional and administrative limits of the Toronto metropolitan region, as more and more of the relatively long-established summer cottage settings at or near the water's edge have come to be quite fully assimilated into the Toronto-dominated housing market—that is, the effective area that can be considered part of a real-estate search area for a household seeking to move while still retaining everyday ties to employment, services, and social networks across its metropolitan area.[60] But *why* is this happening? Ever-improving communication technologies have brought cottage country within everyday reach by virtually eliminating the friction of geography and distance. Inhabited as it increasingly is by (sub)urban folk who deliberately choose to move far afield to where they feel they can be closer or more connected to what they construe as "nature" or the "wilderness," cottage country is, much like exurbia, an urbanizing edge of the growing metropolitan region. Perhaps *unlike* exurbia, however, the choice to move into cottage country permanently is also often bound up in deeply-held visions of the good life that are in many regards unique to central Ontario—the myriad ways in which cottage country is literally a meaning-filled environment for its users. Bits and pieces of the imagined landscapes of cottaging, including ideas and ideologies associated with nature, the wilderness, and the perceived inhospitability of cities, are all key elements. From the Muskoka Room (Figures 6.13 and 6.14), for instance, one might see distinctive windswept pines and oaks, the rugged rock of the Canadian Shield, the blue waters, and in the far-off distance the forest primeval, the perhaps terrifying "bush" of Ontario folklore. One can easily stare into the face of the wilderness, but only from the comfort and calm of the cultivated lakefront landscapes of cottage country that are, for lack of a better term, rather urbanoid.

GETTING COTTAGERS (AND OTHERS) TO TAKE A CRITICAL PERSPECTIVE

On top of the social impacts of cottage country transformation, there are other significant impacts that may turn out to be highly problematic in short order. Where the swelling metropolitan housing market collides with cottage-country settings, there are bound to be troublesome effects. Ecologically, there is the unavoidable problem of the increase in impervious surfaces with the building of so much permanent housing directly adjacent to water bodies, as seen in Figures 6.3, 6.4, 6.7, 6.15, and 6.16. Massive amounts of energy are needed to travel through the expanded life-spaces of the city-which-includes-cottaging, to say nothing of the road infrastructure needed, and pollution generated. Economically, there have long been questions raised about the structural dependence of rural economies on the cottaging population based in the city, as with all tourism-based economic "development" (what happens if demand suddenly shifts, and locals are left without what have become their primary sources of income?). There are problematic social impacts—for instance, as the "sacred" spaces of cottage country are being transformed in ways that are unexpected (perhaps because they are "too urban"). More fundamentally troublesome is the socio-economic division in which the local resident population is often a service provider for wealthier (ex-)urban users, who, in turn, have pushed land values far beyond the means of many if not most "locals." With dramatic non-inflation-related increases in cottage country real estate values in the past two decades,[61] these landscapes are once again becoming the privileged haunts for only the wealthiest of their devotees—a process of landscape gentrification.[62] In early 2004, for instance, the unthinkable happened: the average cost of a cottage in Ontario for the first time exceeded the average cost of a house in Toronto.[63] The growing sense of anxiety is palpable in the popular media, as exemplified by a feature in *Cottage Life* magazine: "The 1990s gave birth to an economic boom in Canada of historic proportions, putting big money in newly deep pockets and big dreams in many minds. One of those dreams was to acquire finally a refuge by the lake in Ontario cottage country and to enjoy its timeless experiences of space, freedom, and peace. But the good times also gave birth to a powerful development spiral that, 10 years on, threatens the very cottage *Zeitgeist* that was so alluring to so many."[64]

The message here is that cottage country seems to be threatened by urban growth. Yet ironically, it is a form of urban growth! Central Ontario's lakeside cottage settlements have always been tightly bound up in the ebb and flow of the Toronto-centered metropolitan housing market. While it may not clearly present itself as an extension of the more spatially contiguous urban form of the Toronto region, this cottage country is undeniably an important, perhaps integral part of the metropolis. This is true in more "objective" terms—as would be indicated by patterns of residential mobility, for instance—and more "subjectively" in that growth is arguably being driven by representations, meanings, or ideologies that also motivate people to live in suburban or exurban settings. In other words, central Ontario cottage country, as a sojourn-driven form of urban growth, is simply a variation on the theme of "escape" from urban life.[65]

That people are motivated to expand their life-spaces to cover an area hundreds of square miles in size is but another parallel between cottage country and exurbia. Certainly, cottage country was first opened up to European settlement primarily during the resort era of the late nineteenth and early twentieth centuries, and thus represented an effective and crowd-pleasing way to reconcile a European sense of the tamed landscape with the menacing and untamed wilds of the Canadian "bush." The lodges and later "grand hotels" gave genteel city folk opportunities to see the "wilderness" without having to get too close for comfort. In short, the genesis of cottage country was as an ideal setting for sojourns amidst nature. Yet we now see in cottage country a problematic cultural tension between city life and a desire to dwell within nature, one that is painfully obvious in popular publications:[66]

> If there is one thing that links cottagers, it's this: The cottage is much more than just a building where we go to spend weekends and holidays—it's a place full of memories and traditions, and when we're not there, we just can't wait for that next trip to the lake.
>
> Cottage Man lives close to nature—maybe not right there, exactly, but closer than he usually gets. Elements of his closeness to nature matter to him. . . . By city terms, Cottage Man is 'away from it all'. By his own terms, Cottage Man is not away from it; he is *at* it. City stuff takes on, in his mind, an irrelevance. . . . Cottage Man watches, from a world that seems real to him, the so-called Real World go by.
>
> . . . it is the promise of escape, however temporary, that makes everything from black ice to conference calls bearable for the rest of us in our everyday lives. . . . There are millions of us who fill out a street address and a postal code, but this tells only where we do what is necessary to get to where we truly live: the lake, and the bush around it.

This tension bespeaks what is arguably also the very crux of the exurban phenomenon. In many ways, these landscapes are examples of the "trouble with wilderness" as stated by William Cronon—that is, the way that "wilderness" hides its unnaturalness behind a mask that is beguiling because it seems so very natural.[67] Cottage country is seen by its users as a "natural" foil to the "unnatural" everyday landscapes of the more densely built-up metropolis, a setting in which reluctant city dwellers can sojourn amidst nature without having to get too close for comfort. Like exurbia, then, it is a way to reconcile an ideological attachment to the *idea* of nature while avoiding committing to its messy processes, which are distastefully complex, diverse, and uncertain—not unlike what Richard Sennett describes as the useful "disorder" of the city.[68] Indeed, empirical studies of how nature is perceived by Americans and Canadians suggest that residents of cities and their metropolitan suburbs tend to view it as "scary, disgusting, and uncomfortable,"[69] the same way many of them also describe city life—particularly those who choose to live in suburban settings.[70] In this world-view, nature is something foreign to be kept at bay, somewhere "out there." But by effectively getting its users to continue thinking of nature in this way, and to merely sojourn *amidst* it instead of *coming to terms with it* so as to better integrate our very ways of dwelling into natural processes, cottage country may be just another kind of "sprawl"—which can be understood as settlement patterns that do little to respond to their natural and cultural context.[71]

As elements of metropolitan-regional growth and development, then, both cottage country and exurban settings are genuinely problematic in terms of sustainability and biodiversity because they too often are rife with "dumb design"—a term used to describe the unresponsive, template-based postwar urban form that spills across North America, which ecological-design advocates Sim van der Ryn and Stuart Cowan argue is literally "dumb" because it does not speak to its natural and cultural context.[72] More to the point, though, attitudes toward these settings reveal what is wrong with our cities. In this is a great paradox, as Cronon puts it: "wilderness embodies a dualistic vision in which the human is entirely outside the natural. . . . To the extent that we celebrate wilderness as the measure with which we judge civilization, we reproduce the dualism that sets humanity and nature at opposite poles. We thereby leave ourselves little hope of discovering what an ethical, sustainable, honorable human place in nature might actually look like."[73]

Compounding the problem is the collective difficulty we North Americans seem to have with accepting change in nature. There seems to be in both cottage country and exurbia a profound reluctance to admit the possibility for change, which is linked in many ways to our unintentionally uncritical reproduction of a nostalgic and romantic representation of "nature." Cottage settings in particular are thought to be unchanging and unchangeable, as in this account by Edward Bartram, a noted painter of Canadian landscapes: "Each of us who has been captured by the spirit of the Bay has, in a sense, developed his own mythology of place. I, for example, continue to imagine the Bay in its natural state as I experienced it by canoe, exploring as a youth wonderful lakes and bays named Spider, Moon, Twelve Mile, and Moose."[74]

The sense of timelessness associated with cottage country landscapes is nonetheless being upset by current transformations—hence the great uproar over its threatened *Zeitgeist* articulated in the quote from *Cottage Life* magazine cited above. Yet uncertainty and change are both inevitable in human settlements, as with the functioning of healthy ecosystems.[75] In this regard, urban history in the Anglo-American world abounds with examples of how cottage settings have come to be integrated into the fabrics of the metropolitan region to which they are appended, often transforming themselves dramatically in the process.[76] The cottage phenomenon is in many ways a foreshadow effect of urbanization, most clearly seen in districts that once were cottage hinterlands such as the Toronto lakefront neighborhood now known as the Beach (Figures 6.17 and 6.18). Originally a summer cottage colony to which city folk would migrate on weekends and holidays, this area was gradually integrated into the metropolitan fabric as a suburban district early in the twentieth century. Ironically, it is now considered quite "urban"—one among many of Toronto's much-loved "downtown" neighborhoods. While such a dramatic change is not to be expected of more far-flung cottage country settlements, the point remains that change is often dramatic and inevitable—but not necessarily bad except where it clashes with preconceptions of what a given landscape ought to be, now and forever.

Figure 6.17 View of the main commercial strip in the east-end Toronto district known as The Beach, which evolved from being an early metropolitan cottage hinterland in the 1880s into a lakeside urban neighborhood.

Figure 6.18 A scene in the Beach district of Toronto showing in its built form how the metropolitan landscapes of cottage country can evolve over time. The wooden house in the center of this photo was originally a summer cottage that was made into a year-round dwelling early in the twentieth century, with other houses built on what were originally its private grounds (the one on the left in the 1920s, the one on the right in the 1980s).

As a final comment before concluding, I note that cottage country seems strangely similar to another increasingly important kind of sojourn landscape in metropolitan regions. There are curious parallels between what is happening in cottage country and recent scholarly analysis of what is happening to many city-center contexts that are being rebuilt as what urban planning theorist Susan Fainstein and others have termed "spaces for play." In a growing number of North American cities, large parts of the "downtown" area are being transformed into entertainment-oriented "destinations."[77] A good example is the 1990s rebuilding of Times Square in New York City, which is now geared to day-trippers from the suburbs as much as the ubiquitous tourist visiting the Big Apple from much further afield.[78] City cores, like nature, are now subject to the powerful "tourist gaze" to which John Urry has famously referred.[79] And insofar as they are predicated on the sojourn and the notion of "escape" from the monotony of everyday life—in a word, tourism—cottage country settings and city-center tourism destinations may seem benign. Yet an unfortunate result is that these "destinations" tend to be quite solipsistic or enclave-like, as is much tourism-based development.[80] In this regard, central Ontario cottage settings have always been exclusive and exclusionary; the myriad lakes and rivers of cottage country—held by the Crown but ringed by individually-owned properties that make them virtually inaccessible to the general public—are nothing if not private enclaves. Dealing with this issue nevertheless raises fundamental questions about rights of access and land ownership, and forces us to confront some very deep-rooted cultural ideologies in this regard, having to do with ideologies of homeownership, land rights, and the like. Add to this the challenge of acknowledging a powerful cultural iconography suggesting that central Ontario cottage country may be perceived as archetypically Canadian (as expressed in literature, music, and painting as well as in the more popular discourses mentioned here), situated as it is on the rocky foundation of the North American continent that is the pre-Cambrian shield. Tempering a cultural entity this central to the identity of so many—even if it is a troublesome culture of sojournment—is a daunting task to say the least.[81]

CONCLUSION: FROM SOJOURNING TO . . .

I have made the case here that many of the people who sojourn in cottage country are seeking to deepen their connection and engagement with what they perceive to be the "natural world." We have considered some of the ways in which these places are deeply meaningful to the people who frequent them, but this bears a summary repetition. In many ways, central Ontario cottage settings are *home landscapes*—what Ted Relph has described as those irreplaceable centers of significance: the spaces in which the act of being occurs at scales above and beyond that of the dwelling unit, as has been observed in many other landscape contexts. This helps to explain the intensity of the debate over the so-called (sub)urbanization of cottage country. Apparently sweeping across central Ontario cottage country is the unwitting (if not inadvertent) replacement of relatively "untouched" places—often characterized by tremendous diversity and richness—by urbanoid landscapes that appear detached from their natural and cultural contexts.[82] Somewhat ironically, this transformation is squarely part and parcel of the environmental crisis in which we now find ourselves. Here, I draw on arguments made by ecological design advocates Sim van der Ryn and Stuart Cowan for rethinking how we place ourselves within natural process.[83] Charging that what we have done in the past fifty years to the North American landscape reflects an impoverished and perhaps even entrepreneurially bankrupt industrial imagination, they bemoan how we too commonly and unquestioningly reduce the myriad complexities of North American landscapes into an asphalt network stitched together from a dozen or so crude design templates that have set the patterns of our everyday experience, "insinuating itself into our own awareness of place and nature." Cottage country in central Ontario is no exception.

Here's the rub: in most cases, cottagers, like other exurbanites, may be driving the very processes of change they so dread and despise. Cottage country may represent and indeed embody a commitment to place that gets stuck in a holding pattern of symbolic and sentimental interaction with natural process. While powerful, the web of thoughts, memories, feelings, and social practices that make up landscapes of home do not necessarily equip people with the kinds of "instructions" they need to take care of these landscapes, and to live lightly within them. Blasphemy? Consider again William Cronon's claim that commonly-held notions of what constitutes "nature" problematically hide their unnaturalness behind masks that are alluring, *precisely because they seem so very "natural."*[84] In other words, good intentions for "fitting natural process into everyday life" seem to go awry—but how? In cottage country, we increasingly see another kind of "sprawl"—that is, urban form linked together, in the words of van der Ryn and Cowan, by "an environmentally devastating infrastructure of roads, highways, storm and sanitary sewers, power lines, and the rest."[85] We must therefore seek to transform how we make sense of how we fit into ecosystems and landscape processes.

How, then, might we tap into the images evoked so strongly to go from sojourning to something more profound and sustainable? This is the task of learning "how to come home" to nature: "If the core problem of wilderness is that it distances us too much from the very things it teaches us to value, then the question we must ask is what it can tell us about home, the place where we actually live. . . . we need to discover a common middle ground in which all of these things, from the city to the wilderness, can somehow be encompassed in the word 'home.'"[86]

The "sojournment" aspect of cottage life may contain a strange but useful symmetry. While indicative of what is wrong with many of our cultural constructions of what constitutes "nature," it could very well be a useful portal into the "common middle ground" to which Cronon refers. Certainly there seems to be amongst cottagers a heightened sensitivity to landscape and urban form, an awareness of the impacts, both positive and negative, that settlement patterns can have on the natural processes that draw people into these pleasing settings in the first place. For instance, organizations promoting environmental stewardship have developed a loyal following amongst cottage country inhabitants, even if the results on the ground remain somewhat disappointing in terms of actually getting people to practice environmental stewardship—taking care of a landscape that is used for only a short time before passing it along to subsequent generations. A potential energy seems to be expressed in the activities and architecture of cottage country. This can be converted and harnessed but only if there is an acknowledgement that current practice may be missing the mark.

The challenge faced in central Ontario cottage country, which exemplifies the one faced in metropolitan regions across both Canada and the United States, is therefore quite significant. How can we get beyond *sojourning* in nature and begin really *dwelling* in it? In learning to do so we may also find ways that enable us to embrace nature within the cities and suburbs that define our metropolitan settlements. As William Cronon has argued, only if we can stop thinking of nature as "(just) out there" and instead see it as "(also) in here" can we get on with the struggle to "live rightly" in the world.[87]

The challenge may ultimately be this: in a proactive planning and design perspective, how should we capitalize on the images and meanings associated with the "natural" settings of exurbia—as made especially clear in central Ontario cottage country—in ways that enable us to establish stronger links between the human-cultural and natural components and processes that make up these landscapes, such as encouraging land stewardship and greater ecological responsibility on the part of users? How, in other words, might we come to terms with planning and designing for the deeply meaningful landscapes of home?

ACKNOWLEDGMENTS

Thanks are extended to the editors of this volume as well as to Emily Gilbert of the University of Toronto for useful comments on previous incarnations of this chapter.

NOTES

1. Defining the extent of the Toronto metropolitan region is a challenge, as it is the largest urban area in Canada and one of the fastest growing and most diverse settings on the continent. The standard definition is the census metropolitan area (CMA). Statistics Canada and Ontario Ministry of Finance data current as of 2012 indicate a population of 5.6 million in 2011 on a

territory of 1,930 square miles (5,000 km²). This delimitation, however, is under-bounded given the continued dispersion of growth and the increase in urban-initiated activity in the rural and recreational fringe (which is the focus of this chapter). The provincial government now deploys the more pertinent concept of the Central Ontario region in an attempt to capture the full extent of urban dispersion for strategic planning. This region covers over 14,500 square miles (37,000 km²) with an estimated population of 9.1 million in 2011. See L.S. Bourne, M.F. Bunce, L. Taylor, N. Luka, and J. Maurer. "Contested Ground: The Dynamics of Peri-Urban Growth in the Toronto Region" in *Canadian Journal of Regional Science/Revue Canadienne des Sciences Régionales 26*, no. 2–3 (2003): 251–270.

2. Annual studies using robust research methods are commissioned by the Canadian realty firm Royal LePage. While detailed results are not published every year, the 2001 and 2002 studies suggested that Ontario was home to almost half the Canadian cottage-owning population (43 percent) when the sample was weighted by regional population distribution—with 155 of 342 cottage-owners and 89 of 234 aspiring owners interviewed. See Ipsos-Reid, *Mansfield-LePage Cottage Study—Detailed Tables* (Toronto: The Ipsos-Reid Group / Royal LePage Real Estate Services, 2002); Royal LePage Real Estate Services, *Royal LePage Recreational Property Report* (Toronto: Royal LePage Real Estate Services, 2001, 2010, 2011, 2012).

3. A collection of essays documents this practice in the U.S., Canada, Ireland, Australia, New Zealand, South Africa, Spain, Scandinavia, and the Mediterranean. See *Tourism, Mobility, and Second Homes: Between Elite Landscape and Common Ground*, edited by C.M. Hall and D.K. Müller (Clevedon [England]: Channel View Publications, 2004).

4. W. Cronon, "The Trouble with Wilderness; Or, Getting Back to the Wrong Nature" in *Uncommon Ground: Rethinking the Human Place in Nature*, edited by W. Cronon (New York: W.W. Norton, 1996), 69–90.

5. *Oxford English Dictionary, 2nd edition*. Prepared by J.A. Simpson and E.S.C. Weiner (New York: Oxford University Press, 1989).

6. A. Smith, "Farms, Forests and Cities: The Image of the Land and the Rise of the Metropolis in Ontario" in *Old Ontario: Essays in Honour of JMS Careless*, edited by D. Keane and C. Reade (Toronto: Dundurn Press, 1990), 71–94.

7. A. Fidelis. "The Thousand Islands" in *Canadian Monthly and National Review* (July 1874): 44.

8. See A. Blum, K.V. Cadieux, N. Luka, and L. Taylor. "'Deeply Connected' to the 'Natural Landscape': Exploring the Places and Cultural Landscapes of Exurbia" in *The New Countryside: Geographic Perspectives on Rural Change*, edited by D. Ramsey and C. Bryant (Brandon MB: Brandon University, 2003), 104–112.

9. This approach is reflected and articulated in my own work. See N. Luka, "Reworking the Canadian Landscape Through Urban Design: Responsive Design, Healthy Housing and Other Lessons" in *Industrial Ecology: A Matter of Design?*, edited by A. Dale, R. Côté and J. Tansey (Forthcoming); N. Luka and N.-M. Lister, "Our Place: Community Ecodesign for the Great White North Means Re-integrating Local Culture and Nature" in *Alternatives 26*, no. 3 (2000): 25–30. See also I. McHarg, *Design with Nature, 2nd edition* (New York: Wiley, 1995); N.-M. Lister and J. Kay, "Celebrating Diversity: Adaptive Planning for Biodiversity" in *Biodiversity in Canada: Ecology, Ideas, and Action,* edited by S. Bocking (Toronto: Broadview Press, 1999); A. W. Spirn, *Granite Garden: Urban Nature and Human Design* (New York: Basic Books, 1985); S. van der Ryn and S. Cowan, *Ecological Design* (Washington DC: Island Press, 1996).

10. J. Jakle, *The Visual Elements of Landscape* (Amherst: University of Massachusetts Press, 1987); E. Penning-Rowsell, "Themes, Speculations, and an Agenda for Landscape Research" in *Landscape Meanings and Values*, edited by E. Penning-Rowsell and D. Lowenthal (London: Allen and Unwin, 1986); A. Rapoport, *Human Aspects of Urban Form: Towards a Man-Environment Approach to Urban Form and Design* (New York: Pergamon Press, 1977); E. Relph, *Place and Placelessness* (London: Pion, 1976) and *The Modern Urban Landscape* (Baltimore: Johns Hopkins University Press, 1987).
11. C. Campbell, "'Our Dear North Country': Regional Identity and National Meaning in Ontario's Georgian Bay" in *Journal of Canadian Studies* 37, no. 4 (2003): 68–91.
12. Based on *Oxford English Dictionary*, 2nd edition and A.C. Spectorsky, *The Exurbanites* (New York: J.B. Lippincott Co., 1955). See also Blum et al., "'Deeply connected'"; J.R. Crump, "Finding a Place in the Country: Exurban and Suburban Development in Sonoma County, California" in *Environment and Behavior 35*, no. 2 (2003): 187–202; J. S. Davis, A.C. Nelson, and K.J. Dueker, "The New 'Burbs: The Exurbs and their Implications for Planning Policy" in *Journal of the American Planning Association 60* (1994): 45–59; R. Keil and J. Graham, "Reasserting Nature: Constructing Urban Environments after Fordism" in *Remaking Reality: Nature at the Millennium*, edited by B. Braun and N. Castree (London: Routledge, 1998), 100–125.
13. Blum et al., "'Deeply Connected'"; R. Keil and J. Graham, "Reasserting Nature."
14. See K.V. Cadieux, Engagement with the Land: Redemption of the Rural Residence Fantasy? in *Contrasting Ruralities: Changing Landscapes*, eds. S. Essex and A. Gilg. (Cambridge: CABI, 2005): 215–229.
15. *Oxford English Dictionary*, 2nd edition.
16. P. Beale, ed., *Partridge's Concise Dictionary of Slang and Unconventional English* (New York: Macmillan, 1989), 106.
17. Useful overviews are found in J. T. Coppock (ed.), *Second Homes: Curse or Blessing?* (Oxford: Pergamon, 1977), N. Gallent & M. Tewdwr-Jones, *Rural Second Homes in Europe: Examining Housing Supply and Planning Control* (Aldershot: Ashgate, 2000), and more specifically to Russia, R. Struyk and K. Angelici, "The Russian Dacha Phenomenon" in *Housing Studies 11*, no. 2 (1996), as well as in Hall & Mueller, *Tourism, Mobility, and Second Homes*.
18. In Canada, Crown land is technically owned by the monarch currently sitting on the British throne. It is controlled and administered by the provincial or federal governments on behalf of the monarch and the people of Canada (in turn). For instance, under Ontario law, Crown land is generally accessible to citizens of Canada who wish to travel through it for recreational purposes, and camping is allowed free on any one site for up to 21 days per year except where posted.
19. C. Gordon, *At the Cottage: A Fearless Look at Canada's Summer Obsession* (Toronto: McClelland & Stewart, 1989), 6–7.
20. Data compiled from reports published by from Statistics Canada. *Canadian Statistics*—online database, available from www.statcan.ca/english/Pgdb/ (2003), accessed May 2003 and September 2004, and from G. Halseth, *Cottage Country in Transition: A Social Geography of Change and Contention in the Rural-Recreational Countryside* (Montréal & Kingston: McGill-Queen's University Press, 1998) and R.I. Wolfe, "Summer Cottagers in Ontario" *Economic Geography* 27.1 (1951): 10–32. See also N. Luka, "Le «cottage» comme pratique intergénérationnelle: narrations de la vie familiale dans les résidences secondaires du centre de l'Ontario" in *Enfances, Familles, Générations 8* (2008): 86–117.

21. See note 2.
22. C. Gordon, *At the Cottage*.
23. R. MacGregor, *Escape: In Search of the Natural Soul of Canada* (Toronto: McClelland & Stewart, 2002).
24. C. Gordon, *At the Cottage*, 6–7.
25. A. C. Spectorsky, *The Exurbanites*.
26. M. McLuhan, "The Old New Rich and the New New Rich" in *Explorations Magazine* (1957): 23. See also N. Klein, *No Logo: No Space, No Choice, No Jobs: Taking Aim at the Brand Bullies* (Toronto: Knopf Canada, 2000).
27. Cottage Life. *Cottage Life 2012 Media Kit*—available from http://www.cottagelife.com/ (2012), accessed July 2012.
28. One might also wonder what others are getting these people to imagine. For instance, Roger Keil and John Graham ("Reasserting nature," p. 117) argue that urban growth now increasingly occurs through nature—in their words, "symbolic natures cement the foundation of a pervasive sales pitch in which garments of 'green' environments are articulated with post-Fordist urban lifestyle needs."
29. This articulation of a healthy ecosystem as characterized by complexity, diversity, and uncertainty is based on N.-M. Lister, "A Systems Approach to Biodiversity Conservation" in *Environmental Monitoring and Assessment* 49 (1998): 123–155.
30. M. Atwood, *Survival: A Thematic Guide to Canadian Literature* (Toronto: Anansi, 1972): 33.
31. N. Frye, "Conclusion," in *Literary History of Canada*, ed. C. F. Klinck (Toronto: University of Toronto Press, 1965).
32. See D.E. Gislason, *The Icelanders of Kinmount: An Experiment in Settlement* (Toronto: Icelandic Canadian Club of Toronto, 1999); W. Kristjanson, "John Taylor and the Pioneer Icelandic Settlement in Manitoba and his Plea on Behalf of the Persecuted Jewish People" in *Transactions of the Manitoba Historical Society* 3 (1976); O. Øverlund, *Johan Schrøder's Travels in Canada, 1863* (Montréal: McGill-Queen's University Press, 1989).
33. Indeed, this may reflect Anglo-American notions of the wilderness, which Olwig argues are derived from a differentiation between land that was arable and that which was not—traceable to the origins of settled agricultural life. See K.R. Olwig, "Sexual Cosmology: Nation and Landscape at the Conceptual Interstices of Nature and Culture; Or, What Does Landscape Really Mean?" in *Landscape: Politics and Perspectives*, edited by B. Bender (Oxford: Berg, 1993), 307–343.
34. *Oxford English Dictionary*, 2nd edition.
35. R. MacGregor, *Escape*.
36. See M. Bunce, *The Countryside Ideal: Anglo-American Images of Landscape* (London and New York: Routledge, 1994); he nicely summarizes the ideological obsession with rural life in Anglo-American culture, which has also been well documented by scholars in urban studies, history, environmental psychology, and related fields. See also W. Cronon, "The Trouble with Wilderness"; J.K. Hadden and J.J. Barton, "An Image that Will Not Die: Thoughts on the History of Anti-urban Ideology" in *New Towns and the Suburban Dream: Ideology and Utopia in Planning and Development*, edited by I.L. Allen (Port Washington NY: National University Publications—Kennikat Press, 1977), 23–60; P.C. Jobes, *Moving Nearer to Heaven: The Illusions and Disillusions of Migrants to Scenic Rural Places* (Westport CT/London: Praeger, 2000); R. Williams, *The Country and the City* (New York: Oxford University Press, 1973).
37. These settings might even be described by paraphrasing Geert Mak's description of rural Holland: landscapes whose austere and synthetic quality is

much-loved, as is anything to which we have grown accustomed. See G. Mak, *Jorwerd: The Death of the Village in 20th-Century Europe [Hoe God verdween uit Jorwerd]*, translated by A. Kelland (London: Harvill Press, 2000 [1996]), 113.
38. Quoted in C. Taylor, *Enchanted Summers: The Grand Hotels of Muskoka* (Toronto: Lynx Images, 1997), 2.
39. For more detailed accounts of logging in cottage country, see F.B. Murray, ed., *Muskoka and Haliburton 1615–1875: A Collection of Documents* (Toronto: Champlain Society / University of Toronto Press, 1963); G. Wall, "Recreational Land Use in Muskoka" in *Ontario Geography 11* (1977): 11–28.
40. Browning Island is one among many other examples; see for instance the Ontario histories compiled by C.E. Campbell, "'Our Dear North Country': Regional Identity and National Meaning in Ontario's Georgian Bay" in *Journal of Canadian Studies 37*, no. 4 (2003): 68–91; P. Jasen, *Wild Things: Nature, Culture, and Tourism in Ontario, 1790–1914* (Toronto: University of Toronto Press, 1995).
41. Quoted in R. Attfield, ed. *Browning Island, Lake Muskoka: Cottagers Remember the Good Old Days* (Huntsville ON: Fox Meadow Creations, 2000): 80.
42. C. Taylor, *Enchanted Summers*, vi.
43. W.B. Varley, "Tourist Attractions in Ontario" in *Canadian Magazine* (1900): 29; J. Hague, "Aspects of Lake Ontario" in *Canadian Magazine* (1893): 263.
44. A. Stevenson, "Camping in the Muskoka Region" in *The Week* (13 May 1886): 382.
45. R. Attfield, ed. *Browning Island*.
46. C. Taylor, *Enchanted Summers*; R.I. Wolfe, "Summer Cottagers in Ontario."
47. Quoted in R. Attfield, ed. *Browning Island:*, 25–26.
48. Quoted in R. Attfield, ed. *Browning Island:*, 38.
49. In the U.S. and Canada, vernacular architecture is an increasingly rare yet vital part of the cultural landscape. See J.B. Jackson, *Discovering the Vernacular Landscape* (New Haven CT: Yale University Press, 1984); A. Rapoport, *House Form and Culture* (Englewood Cliffs NJ: Prentice-Hall, 1969); A. Rapoport, "A Framework for Studying Vernacular Design" in *Journal of Architectural and Planning Research 16*, no. 1 (1999): 52–64.
50. Smith, "Farms, Forests, and Cities."
51. See R.I. Wolfe, "Summer Cottagers in Ontario" as well as U. Mai, *Der Fremdenverkehr am Südrand des Kanadischen Schildes: Eine Vergleichende Untersuchung des Muskoka District und der Frontenac Axis unter Besonderer Berücksichtigung des Standortproblems* (Doctoral dissertation, Philipps-Universität Marburg am Lahn, im Selbstverlag des Geographischen Institutes der Universität Marburg, 1971), and A.P. Hammer, *The Distribution and Impact of Cottagers in Toronto's Urban Field* (Toronto: Centre for Urban and Community Studies, University of Toronto, 1968).
52. In 1941, the total population of the Toronto metropolitan region (including outlying suburbs) increased from 909 928 in 1953 to 2 923 082 by 1971. Population data compiled by J.T. Lemon, *Toronto Since 1918: An Illustrated History* (Toronto: James Lorimer & Company, 1985). Estimates suggest that there were some 28 000 households maintaining summer holiday "cottages" or "camps" across the province in 1941, a figure that had increased to about 164 000 by 1973—a growth rate of 671 percent over 40 years. Data taken from G. Halseth, *Cottage Country in Transition*; R.I. Wolfe, "Summer Cottagers in Ontario."
53. See for instance B. Boyer, *The Boardwalk Album: Memories of The Beach* (Elora ON: Boston Mills Press, 1985); N. Brais and N. Luka, "De la ville

à la banlieue, de la banlieue à la ville: des représentations spatiales en évolution" in *La banlieue revisitée*, edited by A. Fortin, C. Després, and G. Vachon (Québec City: Éditions Nota Bene, 2002), 151–180; M. Campbell and B. Myrvold, *The Beach in Pictures, 1793–1932* (Toronto: Toronto Public Library Board, 1988); R. Fulford, *Accidental City: The Transformation of Toronto* (Toronto: Macfarlane, Walter & Ross, 1995), ch. 6.
54. See e.g. W. Rybczynski, *Waiting for the Weekend* (New York: Penguin, 1991).
55. For elaboration, see A. Rapoport, *Human Aspects of Urban Form*, ch. 1; A. Rapoport, *The Meaning of the Built Environment: A Nonverbal Communication Approach, revised edition* (Tucson: University of Arizona Press, 1990).
56. See N. Luka and L. Trottier, "In the Burbs: It's Time to Recognise that Suburbia Is a Real Place Too" in *Alternatives 28* (2002): 37–38; A. Rapoport, *Human Aspects of Urban Form*.
57. F. Dahms, "The Greying of South Georgian Bay" in *Canadian Geographer 40*, no. 2 (1996): 148–163.
58. G. Halseth, *Cottage Country in Transition*.
59. G. Halseth, *Cottage Country in Transition*, 3.
60. See L. S. Bourne et al., "Contested Ground."
61. See G. Halseth, *Cottage Country in Transition*.
62. See especially G. Halseth, "The 'Cottage' Privilege: Increasingly Elite Landscapes of Second Homes in Canada" in *Tourism, Mobility, and Second Homes: Between Elite Landscape and Common Ground*, edited by C. M. Hall and D.K. Müller (Clevedon, Buffalo, and Toronto: Channel View Publications, 2004), 35–54.
63. Average house cost in Toronto was C$325,000; average cottage price in Ontario was C$349,000. Reported in T. Wong, "Renovation Nation: Price of Cottage Serenity Keeps Rising" in *The Toronto Star* (23 May 2004), C1–C3. Annual market studies confirm that cottage prices have remained high across the province; see Royal LePage Real Estate Services, *Royal LePage Recreational Property Report* (2010, 2011, 2012).
64. D. Lees, "Subdivide and Conquer" in *Cottage Life* (March 2001): 40–46, 108–113.
65. This relates to a potentially touchy issue: the ways in which central Ontario cottage country does not at all reflect the ethno-cultural diversity that makes Toronto one of the world's most culturally diverse cities. Although not explored here, this needs to be examined in much more detail—what does this say about differential sorts of identification by certain groups with nature vis-à-vis environmental justice? See M.N. Philip, H. Mistry, G. Chan, and K. Modeste, "Fortress in the Wilderness: A Conversation about Land" in *Borderlines* (1997): 20–25.
66. Quoted (in order) from Cottage Life, *Cottage Life*; C. Gordon, *At the Cottage*, 186; R. MacGregor, *Escape*, ix–x.
67. W. Cronon, "The Trouble with Wilderness," 69.
68. R. Sennett, *The uses of disorder: personal identity and city life* (New Haven CT: Yale University Press, 2008 [1970]). The core characteristics of self-sustaining ecosystem processes are compellingly described by Nina-Marie Lister as complexity, diversity, and uncertainty. See N.-M. Lister, "A Systems Approach to Biodiversity Conservation."
69. R.D. Bixler and M.F. Floyd, "Nature Is Scary, Disgusting, and Uncomfortable" in *Environment and Behavior 29*, no. 4 (1997): 443–467.
70. See N. Brais and N. Luka, "De la ville à la banlieue" as well as R. Feldman, "Settlement—Identity: Psychological Bonds with Home Places in a Mobile

Society" in *Environment and Behavior* 22, no. 2 (1990): 183–229; D.M. Hummon, *Commonplaces: Community Ideology and Identity in American Society* (Albany NY: State University of New York Press, 1990).
71. The definition of "sprawl" used here is based on the concept of ecological design. See N. Luka and N.-M. Lister, "Our Place" as well as S. van der Ryn and S. Cowan, *Ecological Design*.
72. *Ibid*.
73. W. Cronon, "The Trouble with Wilderness," 80–81.
74. Quoted from Bartram's introduction to W. Harris, E. MacCallum, and J. Fraser, *Mad about the Bay* (Toronto: Key Porter Books, 2004).
75. In multiple ways, this refers to the dynamic, unpredictable qualities of ecosystem health. See N.-M. Lister, "A Systems Approach."
76. For instance, useful work has been done by James Borchert on urban districts of this sort, which he terms residential city suburbs. See J. Borchert, "Residential City Suburbs: The Emergence of a New Suburban Type, 1880–1930" in *Journal of Urban History* 22, no. 3 (1996): 283–307.
77. These ideas are discussed in S.S. Fainstein and D.R. Judd, "Cities as Places to Play" in *The Tourist City*, edited by D.R. Judd and S.S. Fainstein (New Haven CT: Yale University Press, 1999), 261–272; S.S. Fainstein and R.J. Stokes, "Spaces for Play: The Impacts of Entertainment Development on New York City" in *Economic Development Quarterly* 12, no. 2 (1998): 150–165; and P. Eisinger, "The Politics of Bread and Circuses: Building the City for the Visitor Class" in *Urban Affairs Review* 35, no. 3 (2000): 316–333. See also E.S. Ruppert, *The Moral Economy of Eities* (Toronto: University of Toronto Press, 2005).
78. See A.J. Reichl, *Reconstructing Times Square: Politics and Culture in Urban Development, Studies in Government and Public Policy* (Lawrence: University Press of Kansas, 1999) and L.B. Sagalyn, *Times Square Roulette: Remaking the City Icon* (Cambridge MA: MIT Press, 2001).
79. See J. Urry, *The Tourist Gaze: Leisure and Travel in Contemporary Societies* (London: Sage, 1995, and in its revised second edition, 2002).
80. See P. Eisinger, "The Politics of Bread and Circuses" and S.S. Fainstein and D.R. Judd, "Cities as Places to Play" as well as J. Hannigan, *Fantasy City: Pleasure and Profit in the Postmodern Metropolis* (London: Routledge, 1998) and C.C. Hinrichs, "Consuming Images: Making and Marketing Vermont as a Distinctive Rural Place" in *Creating Countryside: The Politics of Rural and Environmental Discourse*, edited by E.M. Dupuis and P. Vandergeest (Philadelphia: Temple University Press, 1996), 261–278.
81. On the cultural significance of cottage country, see e.g. C.E. Campbell, "Our Dear North Country."
82. E. Relph, *Place and Placelessness*. See also N. Brais and N. Luka, "De la ville à la banlieue" and C. Després and P. Larochelle, "Modernity and Tradition in the Making of the Terrace Flats in Quebec City" in *Environments by Design* 1.2 (1996): 141–161, as well as R. Feldman, "Settlement—Identity."
83. S. van der Ryn and S. Cowan, *Ecological Design*, 9–10. See also N. Luka and N.-M. Lister, "Our Place: Community Ecodesign for the Great White North Means Re-integrating Local Culture and Nature" in *Alternatives* 26 (2000): 25–30.
84. W. Cronon, "The Trouble with Wilderness," x.
85. S. van der Ryn and S. Cowan, *Ecological Design*, 9–10.
86. W. Cronon, "The Trouble with Wilderness," 87–89.
87. *Ibid*., 90.

7 Design and Conservation in Québec City's Rural-Urban Fringe
The Case of Lac-Beauport

Geneviève Vachon and David Paradis

DESIGN TEAM:

Érik Aguila, Carlos Aparicio, Redouane Bagdadi, Frédéric Bélanger, Suzanne Bergeron, Marie-France Biron, Vickie Desjardins, Claude-Bernard Lauture, David Paradis.[1]

EDITORS' INTRODUCTION

This chapter summarizes results of an advanced urban design studio on urban/rural design and conservation in Lac-Beauport, a small mountain community located to the north of Québec City's metropolitan territory. The design studio was intended to test participatory design methods, including roundtable discussions and a charrette with local experts. The built and natural landscapes of this exurban zone are in transition from pre-war *villégiature* resort settlements into generic postwar subdivision patterns. Modest summer cottages have been replaced by a heterogeneous amalgam of single-family housing types ranging from 1960s bungalows to million-dollar villas.

    The design studio explored how local officials and residents negotiate landscape meaning in their community, taking into consideration their desire to live within a forested landscape of escape and leisure. Despite an emphasis on making sustainable design decisions, the planning process made it clear that it would not be easy to resolve tensions that emerged from the municipality's desire to encourage urban-style growth in an otherwise undeveloped area on the edge of the provincial capital. Drawing on approaches to conservation and ecological design, the main objective was to support local officials in guiding future development within the broader perspective of sustainability. Many questions are raised by this ongoing action-research project: What can be learned from morphological analysis to understand local identity and guide design strategies? Which form and density can be adopted to preserve natural landscapes and resources? How might participatory design methods enhance local decision-making for sustainable community development?

In their "quest for a view," exurbanites in Lac Beauport appear to seek the same type of engagement with nature evident elsewhere globally. This chapter is not as much about the design approach to dealing with green sprawl, but about the challenge to the ideology of nature that accommodating growth presents. The researchers found that ideals of nature created meaning for residents in the landscape, but that these residents found it challenging to reflect and articulate those values in the design process. The explicit expression of landscape ideology is challenging to do in substantive ways and is so often subsumed by the urban growth machine. As discussed by Laura Taylor in (Chapter 2), the inability to compromise between urban and rural, trading natural landscapes for urban ones, results in the reproduction of sprawl.

This chapter on Lake Beauport illustrates how urban design produces knowledge in planning research. GIRBa is Laval University's Interdisciplinary Research Group on Suburbs (Groupe interdisciplinaire de recherche sur les banlieues). In the last several years, GIRBa's transdisciplinary research program has adopted a collaborative planning strategy to help many partners (governing bodies and community groups) orient the redevelopment of Québec City's first ring of suburbs and think about the future of its exurban territories. GIRBa's research program essentially consists of an iterative process between empirical, action and design research.

The urban design proposals emerging from that back-and-forth movement are not considered as definite, even though they are realistically anchored in collaboratively derived diagnoses and objectives. The projects nourish a collective reflection and fuel new questions to constantly challenge the program, its goals and strategies. The program also derives strength from the presence of architects, urban designers and planners whose training is based upon the development of abilities for solving multidimensional problems through design solutions. This multidisciplinarity encourages participants to integrate and translate multiple approaches into design objectives and proposals.

After working on Québec's first ring of post-war suburbs, GIRBa focused on exurbia while continuing to put forward design research as a rich complement to other types of investigation. Their goal is to understand better the dimensions that underlie urban sprawl in a context of demographic stagnation, population ageing, employment precariousness, family transformations, and increased mobility. How do exurbanites perceive and experience urban sprawl and its impacts? What reasons motivate their choice to live in the periphery in spite of compromises imposed by mobility to access services? What are their representations of city, suburbs, and countryside? How do they perceive exurban landscapes that form a confluence of nature and culture? GIRBa's latest book, *La banlieue s'étale*, proposes a few answers, including a morphological portrait of Québec's exurban landscapes' transformation, as well as in-depth analyses of residential choices from angles such as mobility, social trajectories, relationship to nature, etc. This design exercise on Lac-Beauport's sustainable development, although not conducted as a GIRBa project at the time, nevertheless constitutes a first foray into Québec's urban edge by tackling a few of these issues.

—Kirsten Valentine Cadieux and Laura Taylor

■ ■ ■

As the urban realm enters a phase of profound change, built settlements are more and more scattered, forming sprawling landscapes that invade the countryside. The limits between urban and rural territories, long considered two distinct entities separated by suburbs or middle landscapes, have become blurred.[2] They form ambiguous environments. Some refer to them as "urban countryside," "rural metropolis," or, as in this book, "exurbia."[3] Increasingly mobile, urban dwellers are appropriating and experiencing this territory at different scales—from the local to the regional. They construct multiple and complex territorial identities that transcend such categories as "urban," "suburban," or "rural."[4] Urban territories are subjected to forms of dispersal and de-structuring that are rooted in the individualization of urban practices, and in cocooning (insulating ourselves in our private space) or *repli domestique*.[5] Young house buyers in search of their first new home find exurbs particularly enticing. Longer distances to work, services, or leisure seem less and less determinant of their residential choices. Although individual household choices differ, their patterns of consuming remain the same: exurbanites shop at malls and big box stores and go to the mega multiplex cinema. Changing mobility patterns compounded by the ageing of the population have had an impact on the built and natural environment along a city-suburbs-exurbs continuum, creating all-too-evident ruptures in scale among neighborhood public spaces (such as parks and shopping streets), spaces of mobility (such as highways and spaces designed for car-oriented lifestyles like "power centers") and the urbanizing countryside (such as "strip" villages). Such ruptures reinforce the de-structured aspect and feel of metropolitan landscapes. They become a patchwork of "not quite urban, not quite rural" areas of segregated uses and piecemeal development, auto-oriented urban forms and generic types of residential and commercial buildings.

This effect of dislocation is compounded by urban sprawl which continues apace while ecological threats seem to multiply daily despite regulatory controls and political commitments that prioritize environmental issues. These phenomena of sprawl and ecological threats also exist in Québec

City's newly amalgamated metropolitan region. Coarse-grained developments of big box stores, factories, office campuses as well as a technological park, all surrounded by immense parking lots, have sprouted up along highways. Oftentimes, they cover up wetlands and replace wooded areas which until recently acted as natural filters for hydrocarbon-laced runoff water, for instance. Roads, power transmission lines, and haphazardly sited pockets of low-density housing developments eat up the few remaining cultivated lands and menace the integrity of ecosystems found along rivers and in ever-receding forests, thereby rendering watershed planning impossible.[6] Areas that form the third and fourth ring of Québec's suburbs—the exurban ring—are thus expanding to form a seemingly limitless landscape of dispersion.

Lac-Beauport is such an example of the exurbanization process. Located in Québec's exurban hinterland, it is neither a city, suburb, town, nor village. It is, however, widely known as Québec's great playground and an important economic center of recreational tourism. The challenges facing Lac-Beauport's future development are numerous and complex. They mainly center on the problem of allowing residential and economic development that is respectful of the natural character of place, as well as its vernacular qualities. These qualities are reflected, for example, in the way cottages and houses, built according to the construction "know-how" and aesthetic sensibilities of local builders, artisans, and residents, forge some of the landscape's essential traits. This ambiguous relationship between ecological concerns and development imperatives is indeed troublesome to designers. The notion of place-identity also poses difficulties in view of Lac-Beauport's role as a regional holiday destination and its status as a booming residential community protective of a "suburban" quality of life. Furthermore, residents seemingly consider nature and leisure as other "consumable" elements of their daily exurban experience. How can these issues inform the work of designers whose methods are mostly oriented toward urban and suburban intervention?

As places like Lac-Beauport with high natural value and, some argue, strong development potential continue to face an on-going exurbanization process, the issues of place design become increasingly relevant and timely. This chapter engages a reflection about practical ways in which to address design in exurbia through an exploratory process that takes into account the needs as well as the common *and* conflicting values of local exurbanites. The merits of this "applied" example of collaborative design lie less in the "feasibility" of the proposals than in the actual process of accompanying residents and local officials while they grapple with varied assumptions about the natural and historic character of the landscape. More specifically, this chapter highlights how designers and local participants confront concepts like place attachment, landscape preservation and built heritage, among others, with the challenges posed by tourism, real estate development and sustainability. Therefore, this book's main themes, which center upon considering exurbs from the residents' standpoint as well as exploring ideologies of nature, have a strong resonance in the Lac-Beauport case.

The exploratory design project presented here stems from a 2002 Urban Design Laboratory held at Laval University's School of Architecture, where students pondered issues of development and conservation in Québec's exurban fringe. This initiative builds upon the work of the Interdisciplinary Research Group on Suburbs (*Groupe interdisciplinaire de recherche sur les banlieues, GIRBa*),[7] a group of scholars who, through a program of empirical, action and design research, promote the sustainable development of ageing postwar suburbs through their revitalization,[8] as well as a better understanding of the reasons underlying the transformation of "rurban" territories.[9] The case of Lac-Beauport provides a concrete (if not realized) example of intervention in exurbia, with an accent on collaboration among designers and local participants to propose responsive and sustainable design solutions to help the community better address the challenges of development. The first part of the chapter presents the context underlying Lac-Beauport's planning challenges, followed by a portrait of the area, its residents and their representations of place. A briefly sketched morphogenesis of the area sheds light on the major influences that have helped shape the landscape since the early nineteenth century. The second part discusses how the collaborative design process and the ensuing design proposals for three very different sites raise questions about place-identity and the nature/culture dichotomy shaping the "rural urbanity" (or "*paradis verts*"[10]), which is the subject of this book.

PLANNING CONTEXT AND CHALLENGES

Lac-Beauport is a community of about 7,000 residents (5,400 in 2002) located 18 kilometers north of downtown Québec City, at the foot of the Laurentian Mountains (see Figure 7.1). As early as the 1830s, it had become a summer holiday destination with cottages and farms settled in the forest near and around one of the region's largest lakes.[11] Today, the municipality of Lac-Beauport is a regional recreational center. It houses a ski resort (*Le Relais*) and other leisure amenities. It is also a gateway to nature trails and snowmobile paths that crisscross the province. Only ten to fifteen minutes away by car from the center of one of UNESCO's world heritage cities, Lac-Beauport, not surprisingly, advertises itself as the provincial capital's backyard, a "playground" where its residents claim to have the best of both worlds. As a mélange of "suburb-by-the-lake," holiday destination, and natural reserve, Lac-Beauport dons the qualities of an exurb: it resembles neither town, village, nor resort. For

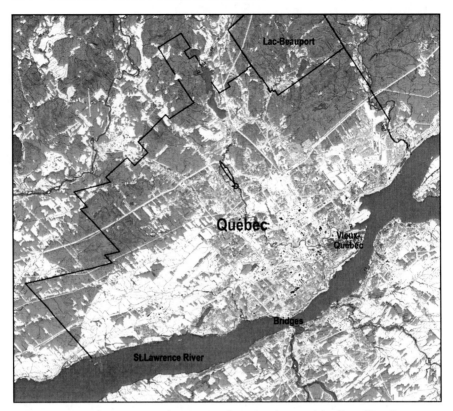

Figure 7.1 Québec's metropolitan region locating the municipality of Lac-Beauport, just outside Québec's city limits. Source: Photocartothèque Québécoise, Ministère des Resources Naturelles du Québec—MRN.

instance, there is no clearly defined center in Lac-Beauport, although older parts west of the lake bear traces of a rich colonial heritage. A new public library building functions as a community center, and together with the primary school forms a civic node. The community is surrounded by forest but is also linked to the Québec-bound highway by a strip-like commercial boulevard. Mostly residential, Lac-Beauport resembles a collection of small neighborhoods—some are older and mix houses with cottages, others are generic post-war suburban tracts while the newer clusters look like exclusive communities of large houses. Summer cottages around the lake are now part of a growing community and are being converted into year-round dwellings. In fact, only 18 percent of Lac-Beauport's residential units remain cottages. Apart from such physical clues, Lac-Beauport's exurban character owes much to its residents' idea of a green paradise that literally bridges the forest with the city, a stone's throw away, to form a different residential landscape.

Despite the fact that twelve other municipalities around Québec City were amalgamated to form a new, larger agglomeration in 2002, Lac-Beauport avoided being included in the new metropolitan entity. The reasons invoked at the time had a lot to do with the fact that Lac-Beauport belonged to the Jacques-Cartier *Municipalité régionale de comté* (MRC), a rural governance body with a different policy framework compared to the other twelve "urban" municipalities. Lac-Beauport and the Jacques-Cartier MRC are now part of the Québec Metropolitan Community (*Communauté métropolitaine de Québec*), which includes other MRCs and cities (such as Lévis) located south of the St. Lawrence River. Nevertheless, at the time of the 2002 amalgamation, Lac-Beauport officials were asked to prepare a strategic plan that would "prove" the uniqueness of Lac-Beauport's character as well as the necessity for local governance, in order to remain "outside" the newly amalgamated city of Québec.

According to local officials, the challenges facing Lac-Beauport's future mainly center on the issue of local identity, which is strongly defined by the community's leisure amenities and even more so by the surrounding natural landscape. Consequently, the municipality has fairly recently realized the need to ensure the preservation and sustainability of its ecologically sensitive natural assets (mainly lakes and rivers). In this regard, local officials in the last few years have tightened their environmental regulations regarding the use of herbicides, pesticides, and de-icing salts to protect waterways. They also regulate deforestation for residential development, to protect rivers and lakes from the impacts of erosion (such as eutrophy). Since only parts of the residential areas are serviced by a conventional sewer infrastructure, many house lots are required to have wells and septic tanks (or fields), quite an onerous challenge in such a mountainous area. In the 1990s, Ducks Unlimited, a non-profit Canadian wetlands conservation group, launched a restoration program for parts of the Rivière Jaune, which flows though the municipality, to bring back indigenous fauna near the urbanized area.[12] Such efforts, combined with more stringent regulations regarding construction near the river's flood plain, minimum house lot dimensions, and boat noise control, have contributed to increase residents' consciousness of the ecological value of sensitive areas near their homes. Nevertheless, while residents' and officials' preoccupations with the sustainability of waterways and woodlands have become more obvious in the last decade or so, the municipality paradoxically takes a "pro-development" stance toward residential and economic growth; indeed, it faces strong pressures from the construction and the tourism industries. Among other projects, local officials propose creating a town center not unlike the *Mont Tremblant* ski resort north of Montreal, but at a smaller, mountain village scale.

In this context, during public meetings in early 2002, Lac-Beauport's mayor, local officials, and developers presented residents with sketches for two new housing complexes. Residents mobilized to oppose the projects and voice their concerns about the impact of development on the area's integrity as well as its ecological balance and the overall sustainability of their community.[13] For instance, citizens raised issues such as water supply insufficiencies (in serviced areas), noise incurred by traffic increase, as well as the impact of introducing multi-family dwellings and parking in wooded areas. In the face of such pressures and struggles, Lac-Beauport has become a contested environment. This is not surprising since nearby Québec City, in line with provincial planning legislation, has enjoyed a tradition of public consultation as part of municipal policymaking, which has given a stronger voice throughout the region to specific interest groups, including local environmentalists.

In brief, Lac-Beauport officials hold a planning vision that seeks to reconcile the potentials offered by development with the omnipresence of nature. In their view, the natural environment is a resource to exploit in order to attract new residents and tourists, rather than a limiting construct in terms of constraints imposed by ecological sustainability. Taking these considerations into account, we asked ourselves: How can the needs of the municipality's highly mobile residents looking for a residential lifestyle intertwined with nature be harmonized with those of tourists seeking an ideal landscape of leisure? How can the ecological impacts of development be minimized in this type of exurb? Can the creation of a village center reinforce local identity? Can the collective memory of this community yield clues toward designing "responsive" places respectful of the environment, local needs, and heritage? The next section attempts to anchor this line of questioning with a portrait of the residents, the area, and its built form.

FORM, POPULATION, AND SPATIAL REPRESENTATIONS

Lac-Beauport's 63 square kilometers is a rugged landscape of woods, lakes, and rivers. Its geography is composed of three major lakes (Beauport, Neigette, and Parent), three mountain peaks (Mont Cervin, Mont Tourbillon, and Mont St-Castin-Le Relais[14]), and the Rivière Jaune. Lac Neigette is the only completely undeveloped lake remaining in the area. Although there are deforested areas where some of the neighborhoods now stand, our first impression of Lac-Beauport is of a forested valley centered upon a lake, which is surrounded by an eclectic collection of low-density clusters of houses built at different periods. Considering the way that the dramatic topography and the deep forest dominate the residential setting, Lac-Beauport's overall image is quite different from other rural areas near Québec where fields and farms give the landscape a very different character (see Figure 7.2).

Design and Conservation in Québec City's Rural-Urban Fringe 167

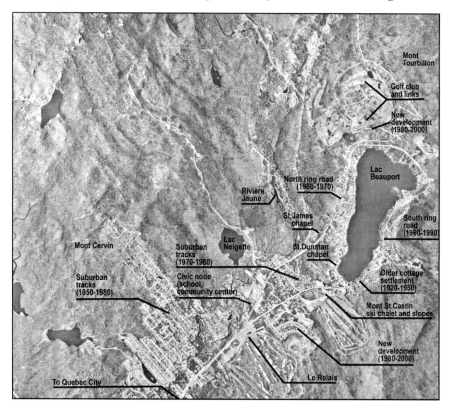

Figure 7.2 Aerial photograph showing Lac-Beauport's different areas and their construction periods, as well as major landmarks, circa 2002. Source: Hauts-Monts, Membre du Groupe Alta.

The ring road surrounding the lake (*Chemin du tour du Lac*), where the first cottages were built, is indeed the most prominent physical feature linking all recreational sites and tourist attractions (hotels, spa, ski centers, golf club, water slides, nautical club, etc.) built in the valley. It is connected to the *Chemin du Brûlé*, which dates from Lac-Beauport's founding, and to the entry boulevard (*Boulevard du Lac*) (see Figure 7.3). Although parts of this wide boulevard are not unlike typical suburban strips, dotted with houses and box-like stores fronted by parking, other parts, closer to the lake, convey a different image. For instance, some of the small cluster of shops and restaurants were built or renovated to look like brightly colored Victorian houses. The nearby pitched-roof gas station *cum* post office and small grocery store looks more like a Vermont village version. Signage regulations enforce a certain "village" image by encouraging the use of gold-etched wooden signs along public roads. The Relais ski center sits at the municipality's entrance where a green stretch of the Boulevard du Lac becomes the lake's ring road. Fronted by a large parking area, the relatively modest ski chalet (compared to other Québec ski resorts such as Mont Ste-Anne) fits rather well in the wooded landscape, as does the now vacated Mont St-Castin ski lodge, a few dozen meters away nearer the lake. Overall, Lac-Beauport is considered to be one of the Québec area's upper scale residential environments. Many large new houses with indoor garages are creeping up the hillsides or they are built on the few remaining lakeside lots with sprawling "landscaped" lawns (see Figure 7.4). The sense of exclusivity that permeates certain areas, particularly around the main lake, contrasts with clusters of generic bungalows and cottages turned into more modest yearlong houses. Finally, there are only a few buildings of heritage value in Lac-Beauport: the Protestant and Catholic chapels near the lake, the first school building (now abandoned) and the Simons house, home to a pioneer Scottish family who immigrated in the eighteenth century and later founded the province-wide Simons department store chain.

In morphological terms, Lac-Beauport's rather diffuse settlement form is without a clear structure. It is the product of a dual process: the densification, around the lake, of a cottage settlement having vernacular qualities mainly derived from "non professional" building traditions. It is also the product of three subsequent phases of suburbanization, which echo Québec City's stages of development. As a result, Lac-Beauport is

Figure 7.3 Aerial photograph of Boulevard du Lac, leading to the ring road and the lake, circa 2004. The Relais ski resort appears in the foreground Source: GIRBa and Pierre Lahoud.

Design and Conservation in Québec City's Rural-Urban Fringe 169

Figure 7.4 Typical residential developments built in the 1980s and 1990s, creeping up the mountain slopes, circa 2004. Source: GIRBa and Pierre Lahoud.

roughly subdivided into three parts: 1) the old cottage settlement and civic amenities which extend around the lake; 2) a postwar suburb to the west; and 3) a more recent mountainside development around a golf course to the east.

According to the 2006 census, Lac-Beauport's population is relatively young compared to that of the province of Québec and of Québec City. The majority of its residents are under 44 years of age and they have a relatively higher level of education (41.8 percent of adults have a college degree or the equivalent compared to 31.7 percent for Québec City). Lac-Beauport residents also have a higher mean household income of about C$87,919 (Québec City's is C$45,770). About 95 percent of the working population travel to jobs outside of the municipality, by car in about the same proportion, mainly in the services sector. Lac-Beauport's population growth rate, which peaked at 20 percent between 1986 and 1991, has been the highest of the Jacques-Cartier MRC.[15] Between 2001 and 2006, the rate remained steady at 10.2 percent, whereas Québec City's has been more modest at 3.1 percent after near stagnation (0.5 percent) between 1996 and 2001.[16]

Interviews conducted in Spring 2002[17] confirmed that Lac-Beauport's residents are highly mobile. Almost every household owns a car since there is no public transportation. They shop locally for daily needs and services on their way to and from work in the city. They also shop in suburban malls, big box stores as well as in various urban and suburban neighborhoods in Québec City. Patterns of sociability among Lac-Beauport residents are seemingly linked to their residential trajectories. For instance, since many residents come from other Québec suburbs, there are few Lac-Beauport "natives" with family living nearby. Combined with the geographical distance, they rely on mobility to structure a network of intergenerational and social links.

Mental maps drawn by residents, primary school students (see Figure 7.5) and a few visitors to Le Relais ski center we interviewed revealed an image of Lac-Beauport strongly associated with leisure activities (summer camps, restaurants), culture (yearly crafts exhibition) and, of course, sports (ski, golf, and water activities). In fact, their discourse alluded to many past and present renowned local athletes (ski, kayak) who hold a significant place in the collective memory. However, few interviewees mentioned or located historical buildings as markers of the past. No one identified a town center or even a neighborhood as significant elements of the landscape. Not surprisingly, the lake and its lakeshore, the mountains, ski slopes and forests are prominent features of Lac-Beauport's "imageability" (the quality of a place that vividly evokes a clear mental image) as are the entry boulevard and the small cluster of shops.

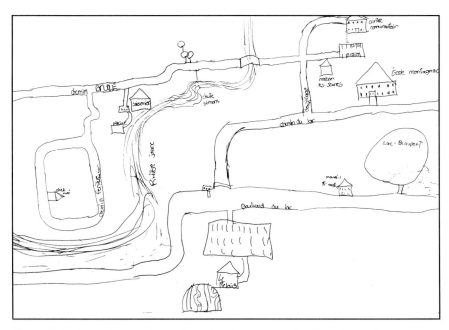

Figure 7.5 Mental map drawn by a local primary school student, showing Lac-Beauport's main roads, civic amenities, Le Relais, the Rivière Jaune and, to the extreme right, the main lake. Source: GIRBa.

HISTORY AND MORPHOGENESIS

Lac-Beauport has a rich colonial history as part of the great Beauport *seigneurie* established in 1634. To this day, there are no traces of a permanent Native North American settlement near the lake, although Hurons from the Wendat Nation who had settled further south near Québec may well have crossed this part of the forest en route to hunting grounds along the Jacques-Cartier River, further north.[18] During the colony's first decades, few French settlers were interested in that remote part of Québec. Instead, Lac-Beauport's founding history owes much to the British. In the early nineteenth century, soon after the Napoleonic wars, English, Irish and Scottish immigrants arrived en masse in Québec only to find that all of the land along the St. Lawrence River was already settled. When they instead founded a community near the lake to the north, the French *seigneur* Duchesnay named the area "Waterloo Settlement" to attract more British settlers. The St. Dunstan Catholic chapel built on the lakeshore in the 1830s took the name of an English archbishop. The St. James Protestant chapel was built much later in 1890. These pioneers faced difficult living conditions further aggravated by a rugged terrain ultimately inappropriate for agriculture.

By the end of the 19th century, Lac-Beauport had gained a reputation as a summer holiday destination, as evidenced by its large lakeside hotels. In fact, Québec's bourgeoisie regularly journeyed up north during the summer months for fishing excursions, picnics and hikes in the woods. A few stayed in lakefront cottages built on *pilotis* up off the ground, completely

Figure 7.6 Typical Lac-Beauport summer cottage. The Lavigueur house, circa 1908. Source: Duval 1983.

unsuitable for winter (see Figure 7.6). Between 1920 and 1937, three ski centers opened to make Lac-Beauport a winter holiday destination as well. Around that time, the proprietors of *Château Frontenac* bought ads in the *New York Times* to promote Québec as a worldly wintering city. They used photos of skiers coming down Lac-Beauport's mountain slopes to attract Americans (see Figure 7.7).[19] After the war and up until the 1960s, more recreation amenities were added to the year-round landscape of leisure: the

Figure 7.7 Montage of postcards from the 1940s showing different "wintering" sites in Lac-Beauport, including Le Relais (bottom image). Source: Duval 1983.

Club Nautique, now a training facility for international kayak athletes, and the *Mont Tourbillon* golf club, which spurred an important residential boom in the 1970s and 1980s. Today, Lac-Beauport's local economy is reportedly healthy: hotels and recreational centers report stable business with seasonal peaks.

After 1950, Lac-Beauport developed quite rapidly. Unsuitable agricultural land was finally abandoned and subdivided for single-family housing. In the frenzy of this building boom, the Rivière Jaune's flood plain was also built-up. During the 1960s, with less land remaining, construction started to creep visibly up the mountainsides surrounding the lake, transforming the landscape. Development continued unabated during the 1970s. Ignoring topography and natural elements, the same suburban subdivision pattern was repeated everywhere. To preserve the forest, many conservation areas were designated as public parks. Around the same time, the new highway to Québec City was inaugurated, spurring another wave of development on the last remaining land. This population increase soon translated into demands for new services such as a kindergarten and elderly housing. New hillside enclaves of huge houses with indoor garages are now worlds away from the post-war bungalows and winterized cottages. Their siting shows little regard for topography, orientation, or the eclectic character of the surrounding vernacular landscape, which reflects the local material culture. Nevertheless, recent residents seem paradoxically more aware of the negative impacts of this kind of construction on the natural environment. For instance, according to local officials, fewer trees are now cut down during construction. This attitude coincides with the relatively recent restrictions and regulations discussed previously. In sum, Lac-Beauport residents' awareness of their impact on the environment and the overall landscape is a very complex issue. It was made even more confusing when meetings with members of the community, addressed in the next section, revealed conflicting values among newcomers and established residents.

MISSION, APPROACH, AND DESIGN ORIENTATIONS

For the case study of Lac Beauport, the mission of the Université Laval School of Architecture's Urban Design Laboratory was to assist local officials and residents in identifying ways to ensure the sustainable development of Lac-Beauport. The intention was not to produce a master plan for the entire community but rather to exemplify a few sustainable design principles with proposals for specific sites.

The Laboratory is a place where students, professors, and practitioners meet to consider complex design problems involving such issues as identity and sustainability. As part of this academic requirement, students are expected to: 1) identify the contributing factors to the built environment's potential for transformation; 2) define the problem in as many dimensions as possible—social, ecological, functional, perceptual—and formulate adequate design objectives; 3) propose design solutions respectful of place character and based on qualities of accessibility, permeability, legibility and imageability; and 4) apply collaborative design methods to validate individual perspectives and proposals. These orientations generally underlie urban design research at our School and stem from countless collaborations with communities over the years to tackle timely and socially relevant design problems. They also derive from a framework of three outstanding methods and principles that have inspired the Laboratory's approach to urban design.

The first, urban morphology (or typomorphology), which has been taught at our school by Pierre Larochelle for many years, yields a powerful analytical tool to understand both the structure of human settlements and the logic of their transformation.[20] Our preliminary morphogenesis of Lac-Beauport points toward the necessity to further document the process of cultural and physical transformation of built and natural landscapes, at different scales. A second "responsive" design approach proposes a framework of urban qualities to create integrated, diverse, democratic, and viable environments that respond to human needs.[21] The most important of these qualities are: 1) the permeability of built forms by way of providing a well-connected network of streets and paths for pedestrians and drivers to enjoy more choices of itineraries to access both public and private places; 2) a variety of land uses, as opposed to "mono-functional" zoning, to satisfy various needs for services, employment opportunities as well as housing types and tenures; and 3) the legibility and imageability of place, so that users can forge a clear and memorable image of an environment to help them navigate through complex and layered urban spaces. Together, these qualities forge a rich environment supportive of "urbanity" and quality of life for various groups. Finally, this Laboratory tested a third method developed by Randall Arendt centering on a conservation design approach for rural communities. A team at University of Massachusetts first developed this approach in the late 1980s as part of a research project focusing on land and landscape conservation along the Connecticut River Valley. In a context of intense pressure to transform agricultural land into exclusive housing developments, they proposed conservation design guidelines for the integration of compact subdivisions into the countryside.[22] Arendt's own approach has since evolved to focus on the idea of preservation through land trusts (rather than ecological design principles) as part of a solution to the invasive suburban-like development of rural communities. Conservation design promotes compact hamlet-like configurations as an alternative to sprawling low-density patterns.[23] Many dimensions of this pro-development approach, which values the visual qualities of landscapes, seemed pertinent in the Lac-Beauport case, especially considering how there are few design methods and guideline models which tackle rural and exurban settings compared to urban areas.

The Urban Design Laboratory is committed to a collaborative design approach in line with GIRBa's work. As an example, over an eighteen-month period from June 2002 to November 2003, the Group initiated a participative design process with 60 local participants to elaborate a master plan for Québec's postwar suburbs,[24] a design process in which exurban Lac-Beauport was not included. In 2006, the Group also orchestrated a collaborative process to design a new sustainable neighborhood on Université Laval's campus.[25] Even though the Lac-Beauport Laboratory took place a few months before GIRBa's larger processes was officially initiated, its approach toward collaborative design is nonetheless modeled after the Group's own approach, although at a much more modest scale (in terms of territory limits, number of participants and meetings, etc.). Three weeks into the semester, the first activity of the Laboratory consisted of a half-day roundtable at the local community center. The objective was to test the preliminary diagnostic and planning orientations with a group of twenty "local experts" (*spécialistes du quotidien*) including a member of the local historical society as well as officials representing different departments (planning, construction, environment, and leisure/recreation) from the town and the MRC. A few residents, some of whose families had lived in Lac-Beauport for two generations, also took part in the discussions. At the very beginning of the roundtable, as an "icebreaking" exercise, we showed photos of key places in Lac-Beauport side by side with photos of other types of suburban and exurban communities. Our goal was to stimulate a free discussion that would focus on the area's qualities, constraints, and potentials. The exercise revealed different experiences and perceptions of place, as well as a variety of visions for the community's future.

Preliminary design proposals were then used as the basis of discussion during a one-day design charrette held at the School of Architecture a few weeks later. Although the same local officials were present, no residents could attend, even though we extended a few invitations. This intensive session consisted of plenary discussions and mini-workshops during which professionals who do not usually work together on a planning problem had the chance to exchange their different perspectives, aiming at a consensus.

Results of this collaborative process showed that there was a great concern over the increasing urbanization of the lakeshore, which limited visual and physical access to the water. Therefore, we reached a consensus about finding creative ways to preserve access to lakes, rivers and others natural assets, including nature trails. Participants also mentioned the necessity to physically link "enclaved" neighborhoods and diminish walking distances to different services. Neighborhood form should therefore promote permeability while maintaining privacy and tranquility, both qualities that residents highly value. Pedestrian and bike paths should link all residential areas to green spaces, recreation activities and sites of heritage value.

Problems of accessibility were also assessed in terms of affordability. Single-family houses in Lac-Beauport are relatively expensive (many new houses were selling in the C$300,000 to C$500,000 range).[26] Choices in cost, type, and tenure are very limited. There is almost no multifamily housing in Lac-Beauport, either private or rental, and none that is government sponsored. The Laboratory's design objectives therefore aimed to offer diverse housing types and densities to fit the evolving needs of residents such as the elderly and non-traditional families.

Regarding development, local officials strongly supported the creation of a village center with shops, services, and tourist accommodations. In their mind, the "marketable" image of the center could catalyze the area's economic development. In line with this view, the Laboratory decided to develop a planning strategy based on consolidating two strategically located "poles": a civic/services node and a village center, with complementary roles within the community.

The conversion of cottages into permanent dwellings and the increasing presence of generic suburban house models is in large part responsible for what residents and local officials worry is the slow eroding of the community's sense of place. At the very least, this phenomenon adds yet another layer to the vernacular character of the existing built landscape. The quest for a view of the lake or the mountains has important perceptual and ecological impacts on the landscape since expensive new houses are built higher up the mountainsides. Many newcomers to Lac-Beauport are conscious of the impact of residential development on the environment, more so, they seem to think, than the "pioneers" living around the lake and in the suburban tracts. Other newcomers, on the other hand, seem oblivious to "place" (or at least to *their* impact on place) in their willingness to pay for a great view and the proximity to both nature and urban amenities. This mutation of the built and natural landscapes was an especially sensitive and layered issue since few elements of place-identity and collective memory apart from "nature" and "leisure" appeared as significant. Despite this almost total emphasis on nature and leisure, however, many participants of the collaborative design process—especially local officials and members of the Historic Society—mentioned a few historic staples as treasures of the collective memory to be preserved as well as assets to attract tourists. Therefore, sites with heritage value, which are coveted by developers for residential development—such as the chapel sites, Lac Neigette, and a few lakeshore hotels—required special attention. The Laboratory proposed to make these elements the focus of design schemes to underline the cultural significance of Lac-Beauport's landscapes in people's experience of place.

In terms of environment and landscape, the main goal was to discourage construction within sensitive zones, such as flood plains or slopes. One of our main intentions was to keep thick wooded conservation corridors around lakes and along rivers to ensure access and sustainability. New buildings should be implemented in such a way as to keep most of the trees and minimize erosion.

In sum, these participatory design exercises helped clarify four main design objectives. The first one is to maintain and provide more access to natural resources. The concept of accessibility is strongly associated with that of permeability, whereby the network of streets and paths (in newly developed and existing areas of intervention) should provide residents with permanent and various opportunities to physically and visually access lakes, rivers, woods or mountain trails. One obvious way to ensure this accessibility is to prohibit further construction directly along waterways, for instance. The second main objective aims to favor residential development in carefully defined zones. In order to ensure adequate siting and to minimize environmental impacts, we propose to identify suitable zones to build housing, especially in un-developed wooded areas. Third, we propose to include cultural features in the affirmation of place-identity. More specifically, we propose that the natural and built artifacts that compose Lac-Beauport's landscape be respectfully integrated in new and re-designed developments as elements of a living collective memory. And finally, the last objective is to protect as much of the natural environment's integrity as possible (in terms of ecosystems interconnectedness, for instance) in a context where development is permitted.

DESIGN PROPOSALS

The proposals presented here stem from three different strategies. The first project consisted of revitalization and infill to consolidate the two new centers (a civic node/services node and a village center). The second project tackled head-on the possibilities and constraints inherent in developing housing in a sensitive forest ecosystem. The third project proposed including the two historic chapels in a lakeside park surrounded with infill housing (see Figure 7.8).

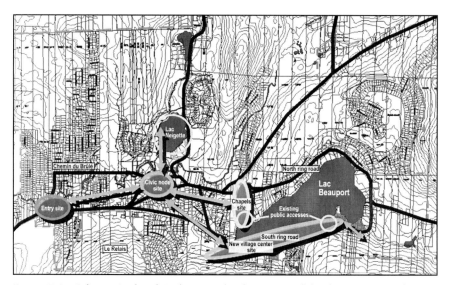

Figure 7.8 Schematic sketch indicating the three sites of the design proposals.

The Old Waterloo Settlement and Village Center

The first of two complementary "centers" is located in the Old Waterloo settlement near the Relais ski center. Our proposal would provide new affordable housing in front of and next to the school, which would be expanded to welcome a public kindergarten. This consolidated sector, which would also profit from a new public and more visible access to the Rivière Jaune, would complete a neighborhood already enjoying the proximity of services within a five-minute walking distance. We also propose to implement a new City Hall at the apex of the Y-shaped intersection formed by the entry boulevard and the ring road. This new landmark would be strategically located to welcome visitors and to project a strong image. Across the ring road, on the *Relais* ski center site, a new terminal would welcome city buses on an extended route. Combined with the new City Hall, which would also include a visitors center, these amenities could create a more legible "entrance" into the community (see Figure 7.9).

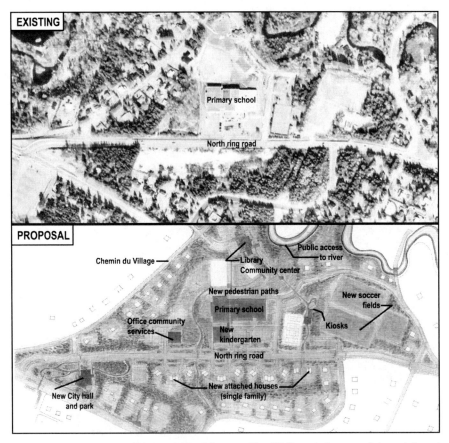

Figure 7.9 Project 1—Civic node with new City Hall: existing conditions (above) and proposed consolidation scheme.

The new village center would be located to the east, on the south ring road. It would include significant landmarks of Lac-Beauport's history such as the Manoir St-Castin lakeshore hotel and the vacated Mont St-Castin ski lodge, which would be recycled into shops and offices as a central feature of the new center. A housing development built at the base of the Mont St-Castin ski slopes (behind the restored lodge) would feature a retention basin to filter surface water. The village center would also include shops and services, an inn, a small theater as well as art galleries to show local art, among possible uses. One of the intentions is to attract different kinds of visitors through different types of accommodation and cultural amenities than those already found elsewhere in Lac-Beauport. The design team proposes to open a public access to the lake on the hotel site (there are currently only two clear public accesses to the 89 square kilometer lake) and to link the village center to existing nature paths up Mont St-Castin.

Lac Neigette: New Residential Development

To the north, the design team proposed a new residential neighborhood in a natural setting much sought after by developers. We applied ideas of conservation design whereby most of the land is left un-developed and built areas are more compactly subdivided. In a manner inspired by McHarg's[27] composite mapping, the students first identified the sensitive non-buildable areas (slopes 10 percent above grade, the limits to the Rivière Jaune and Lac Neigette flood plains, erosion areas). This mapping exercise was helped by detailed topographical surveys as well as expertise provided by local officials on the effect of development on soil erosion and the sustainability of waterways. A large conservation corridor around the lake was designed to protect it from any lakeside development. A ring road was introduced to ensure public access to the water as well as to enable higher density multifamily housing (two and three stories). In areas best suited for construction, street patterns followed the existing topography to minimize the risks of erosion. Cul-de-sacs linked by pedestrian paths are used to limit the small pockets of development that would be permitted. The choice of housing types is also dictated by topography. Highly desirable single-family houses are located in the valley to the west. Elderly and multigenerational housing composed of single- and two-family units is located nearer the civic center to minimize walking distance. Implementation criteria include orientation, visual impact on the landscape, and the necessity to minimize the overall ecological impact. Other architectural guidelines (roof line, building height, street-building ratio, etc.) are proposed to control the preservation of place character (see Figure 7.10).

The Chapel Sites

The last proposal consisted of the enhancement of the two historic chapel sites located across the main lake's North ring road from one

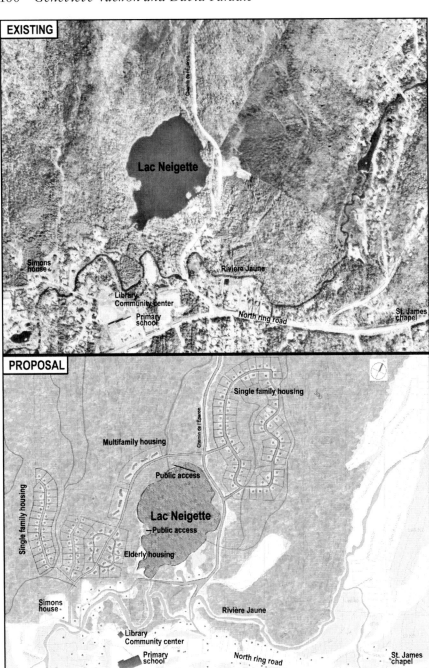

Figure 7.10 Project 2—Lac-Neigette: existing conditions (above) and proposed development.

another. Even though St. Dunstan and St. James are still in use, they are regularly transformed into concert and community halls. They are the most visible, significant, and best-known heritage buildings in Lac-Beauport. St. Dunstan is perched on a promontory near a stream running through an adjacent maple grove to form a most idyllic place. Provided the archdiocese gives permission, the St. Dunstan site could provide a rare public access to the lake. The project proposes to create a public park to commemorate the historical significance of both chapels. A main path would link St. Dunstan to St. James and the lake. Infill housing implemented across from the St. Dunstan grove would complement a nearby low density residential area. This project initiates an interesting reflection about the preservation of historic sites to maintain the collective memory while protecting access to sensitive natural assets for both residents and visitors.

CONCLUSION

The case study reported in this chapter illustrates a few of the issues and challenges underlying planning and design in exurban communities. One of them concerns the ambivalence residents and local officials feel about sustainability and nature conservation in the face of residential and economic development. Another has to do with the idea of place-identity as a manifestation of meaning and attachment but also as a statement about the image of a culturally and economically thriving community.

Our design proposals, though they were not implemented by the municipality, were nonetheless instrumental in highlighting the complex layering of officials' and residents' experience and perception of what Lac-Beauport represents as a "*paradis vert*" of exurban lifestyle, and what their community should become as part of Québec's "metropolization" process. In fact, conflicting views and expectations regarding place, landscape, and nature were expressed through struggles between officials and environmentalists regarding development, between newcomers and pioneers regarding the ecological impacts of settlement, and between locals and tourists regarding the access to nature. In spite of such differences, the Lac-Beauport community values its autonomy in attracting and managing development, a privilege that was lost by the twelve amalgamated cities that now form the new Québec City. How to reconcile this privilege with a heightened consciousness of the environmental impacts of development as well as with the many expectations of different groups of residents is now a timely challenge for the community.

This complex development context, coupled with the difficulty in tackling an under-researched type of settlement, was also quite a challenge for the students of urban design, especially at a time when professionals have to be extremely sensitive to Québec's ageing population and eventual demographic decline, which points to consolidation, revitalization, infill, and recycling strategies instead of more outward growth. The lack of design methods to study exurban landscapes as well as exemplary case projects and guidelines to nourish a theoretical framework posed another challenge. Observations, interviews and a collaborative design approach somehow filled the gaps to develop design objectives and proposals for a unique type of settlement that fits no definition. The need for further research on the morphology, image, uses and representations of exurban landscapes, as well as on the ideologies of "nature" in exurbia, remains pressing.

In the last few years, GIRBa has closely examined Québec's exurban territories through a transdisciplinary program of research combining empirical research (through qualitative methods as well as morphological and statistical analyses) with design research (through urban and architectural projects) and action-research (through collaborative design processes). In the context of Québec City's metropolitan community's strategic planning process (the plan is still pending and includes Lac-Beauport), it appears essential to better understand the reasons underlying exurbia's attraction, especially in view of the city's expected demographic decline and its efforts to promote smart growth principles. One of our continuing goals for research is to compare exurbanites' uses and representations of their environments with the representations and ideals of developers, planners, and architects who help shape these environments. Finally, we remain committed to continuing our efforts in design research to find practical, sustainable solutions that will better integrate built environments with natural landscapes and processes.

NOTES

1. The authors wish to acknowledge Claude-Bernard Lauture's special contribution to this reflection since he was responsible for writing the studio report during summer 2003. They also thank Tania Martin, Ph.D. (Berkeley), visiting professor, for her generous help with the English revisions.
2. Andrée Fortin, Carole Despres and Geneviève Vachon, *La banlieue s'étale* (Québec: Nota Bene, 2011).
3. Peter Rowe, *Making a Middle Landscape* (Cambridge, MA: The MIT Press, 1991).
4. Moura Quayle, "Campagne Urbaine-Métropole Rurale", in *Les Temps du Paysage*, eds. P. Poullaouec-Gonidec, S. Paquette, and G. Domon (Montréal: Presses de l'Université de Montréal, 2003); Larry S. Bourne, "Reinventing the Suburbs: Old Myths and New Realities," *Progress in Planning* 46.3 (1996): 164.
5. Alexandra Daris, "Mobilité et Vie Sociale: Entre le Quartier et L'ailleurs", in *La Banlieue Revisitée*, eds. A. Fortin, C. Després, and G. Vachon (Québec: Nota Bene, 2002); Nicole Brais and Nik Luka, "De la Ville à la Banlieue, de la Banlieue à la Ville: Des Représentations Spatiales en Evolution", in *La*

Banlieue Revisitée, eds. A. Fortin, C. Després, and G. Vachon (Québec: Nota Bene, 2002).
6. Alain Touraine, *Pourrons-nous Vivre Ensemble? Égaux et Différents* (Paris: Fayard, 1997): 40. Touraine questions whether "living together" is still possible in cities where the effects of globalization taint our daily experiences and often prompt us to project ourselves either as contentedly disengaged citizens of the world or stern protectors of privacy against the invasion of a global culture.
7. Watershed planning refers to the idea of an area's "natural communities," both upstream and downstream, to remain linked together by a common set of natural water-drainage courses. Usually applied to a region, watershed planning is also linked to habitat-conservation efforts. Peter Calthorpe and William Fulton. *The Regional City: Planning for the End of Sprawl* (Washington, DC: Island Press, 2001).
8. The reflections presented in this chapter owe much to the ongoing work of GIRBa members which include Carole Després, Andrée Fortin, Florent Joerin, and Gianpiero Moretti, as well as G. Vachon, D. Paradis, and many graduate students. For more information on GIRBa's members and work, consult *www.girba.crad.ulaval.ca*.
9. Andrée Fortin, Carole Després, and Geneviève Vachon, eds., *La Banlieue Revisitée* (Québec: Nota Bene, 2002).
10. GIRBa's latest book includes a morphological portrait of Québec's exurban landscapes transformation, as well as in-depth analyses of residential choices from perspectives such as mobility, social trajectories, relationship to nature. Andrée Fortin, Carole Després, and Geneviève Vachon, eds., *La Banlieue s'Etale* (Québec: Nota Bene, 2011).
11. Jean-Didier Urbain, *Paradis Verts: Désirs de Campagne et Passions Résidentielles* (Paris: Payot, 2002).
12. André Duval, *Mon Lac se Raconte. . .* (Municipalité de St-Dunstan du Lac-Beauport, 1983).
13. In the late 1990s, however, City Hall had to ban the feeding of wild ducks for fear of growing bio-pollution to the main lake (newspaper *Le Soleil*, August 21 and 22, 1999).
14. Three Québec City newspaper articles carried the story: *Le Soleil*, January 4, 2002; *Le Soleil*, February 7, 2002; *Le Soleil*, March 12, 2002.
15. The Relais peaks at roughly 300 meters above sea level.
16. All data extracted from: Municipalité Régionale de Comté (MRC) La Jacques-Cartier. Schéma d'Aménagement Révisé, Second projet (Version de Consultation), August 2001; Communauté Métropolitaine de Québec (CMQ). Portrait statistique, Lac-Beauport (http://www.cmquebec.qc.ca/), July 2009; CMQ. Portrait statistique, Ville de Québec (http://www.cmquebec.qc.ca/), August 2009.
17. The interviews were conducted by students of the *Urban Form and Cultural Practices* seminar (directed by Carole Després, professor, GIRBa). The sample of a dozen semi-directed interviews was complemented by a mental mapping exercise.
18. Duval, *Mon Lac se Raconte. . .*, 49. According to Marie-Paule Robitaille, curator of North American Natives artifacts at Québec's Musée de la Civilisation, no academics or archeologists have found, to this day, physical evidence of Native American settlements (permanent or seasonal) north of Wendake, the main Huron settlement in the Québec region, which dates from the seventeenth century and is now part of Loretteville (one of the city's boroughs).
19. Duval, 37.
20. Geneviève Vachon, Nik Luka, and Daniel Lacroix, "Complexity and Contradiction in the Ageing Early Postwar Suburbs of Québec City", in *Suburban*

Form, eds. K. Stanilov and B. Scheer (London: Routledge, 2004); Albert Lévy, "Urban Morphology and the Problem of the Modern Urban Fabric: Some Questions for Research," *Urban Morphology* 3.3 (1999): 79–85; Anne Vernez Moudon, "The Changing Morphology of Suburban Neighborhood," in *Typological Process and Design Theory*, ed. Attilio Petruccioli (Cambridge, MA: AKPIA, 1998); Carole Després and Pierre Larochelle, "Modernity and Tradition in the Making of the Terrace Flats in Québec City," *Environments by Design* 1.2 (1996): 141–161.

21. Ian Bentley, Alan Alcock, Paul Murrain, Sue McGlynn, and Graham Smith, *Responsive Environments* (London: The Architectural Press, 1985).
22. Robert D. Yaro et al., *Dealing With Change in the Connecticut River Valley: A Design Manual for Conservation and Development* (Amherst, MA: Massachusetts Department of Environmental Management, Center for Rural Massachusetts, University of Massachusetts at Amherst, 1988).
23. Randall Arendt, *Growing Greener: Putting Conservation into Local Plans and Ordinances* (Washington, DC: Island Press, 1999); Randall Arendt, *Conservation Design for Subdivisions: A Practical Guide to Creating Open Space Networks* (Washington, DC: Island Press, 1996).
24. Carole Després, Nicole Brais, and Sergio Avellan, "Collaborative Planning for Retrofitting Suburbs: Transdisciplinarity and Intersubjectivity in Action," eds. R. Lawrence and C. Després, "Transdisciplinarity in Theory and Practice," *Futures* 36, 4, (2004): 471–486.
25. G. Vachon, C. Després, A. Nembrini, F. Joerin, A. Fortin, and G.P. Moretti, "Collaborative Planning and Design: A Sustainable Neighborhood for a University Campus," in *Urban Sustainability through Environmental Design: Approaches to Time-People-Place Responsive Urban Spaces*, eds. K. Thwaites et al. (New York: Spon Press, 2007): 129–135.
26. Although housing prices and values in Québec's Upper Town neighborhoods are steadily rising, the average cost of a single-family house in Québec's metropolitan area in 2001 was C$89,906, and well over C$200,000 in 2010.
27. Ian McHarg, *Design with Nature* (New York: Natural History Press, 1969).

8 Time, Place, and Structure
Typo-Morphological Analysis of Three Calgary Neighborhoods

Beverly A. Sandalack and Andrei Nicolai

EDITORS' INTRODUCTION

The role of the ideology of nature in the expression of place has changed over time. Residential settlement in Calgary, Alberta, has evolved in just over a century from its early form consisting of a grid pattern of streets extending from the railway, to various types of speculative development, to planned neighborhood units, to more generic subdivisions.

Calgary's moving urban edge offers the opportunity to study a number of questions about the role of nature in the planned environment. This chapter does not document exurban places per se, but demonstrates the function of the ideology of nature in greening metropolitan spaces, and also shows how the relationship between public and private spaces brings up questions about whom the planning of space, nature, and infrastructure is supposed to benefit. The provision of infrastructure such as roads, water, and sewers is a central focus in the literature on exurbia. Urban designers Bev Sandalack and Andrei Nicolai wrote *The Calgary Project: Urban Form / Urban Life* and have been in professional practice as well as scholars at the University of Calgary. In keeping with the best practices of urban design, they demonstrate in this chapter how a community negotiates its motivating issues—including the negotiations involved in the designing and marketing of home life in nature—in part by reading the landscape.

As residential neighborhoods at Calgary's moving urban edge have approached the city's prairie edge, the iconography of nature has been intensified in symbolically driven ways. Shallow seasonal prairie sloughs are artificially deepened and pumped full of municipal water to produce "waterfront"—at the same time that access to the "true" natural prairie environment is lost. In their representation of shallow seasonal prairie sloughs as waterfront properties, recent exurban subdivisions on Calgary's fringes provide an almost exaggerated example of the complicated disjuncture between heavily marketed ideals of nature and the material reality of residential landscape amenity, a disjuncture which is a core concern of this book. In demonstrating how people literally construct nature out of their visions for home, Sandalack and Nicolai point out some of the difficult paradoxes facing attempts to plan and manage exurbs. If it is easy from a sustainability perspective to condemn the excavation, en-pondment, and municipal water supply required to make prairie puddles into lakefronts worthy of the kinds of homes people imagine "out at the lake," it is more complicated to consider a comparison with the densely settled and vegetated downtown homes in the neighborhood of Roxboro, an historic neighborhood that was created in an equally energy intensive way, by leveling Roxboro's erstwhile hill with fire hoses. Why does one nature strike us as more authentic (and in addition, also more successful as an urban space today), while the other seems only a façade of both nature and urbanism?

These questions are particularly important for new green sprawl landscapes. Partly because much new construction paradoxically uses exaggerated nature iconography at the same time it uses construction techniques that first remove existing landscape features, growing exurbs pose daunting challenges to both planners and residents in their efforts to talk about sense of place and connection to nature, concepts that appear to make exurbs desirable.

Sandalack and Nicolai explore the material consequences of prevailing attitudes towards nature and home by comparing three neighborhoods built during three different eras at the outer edge of Calgary. The comparison of three neighborhoods shows how planning and urban design have changed and how various qualities of urban form have either improved or deteriorated. Given public interest in the environment, the authors suggest that we would hope to see improvements in people's relationship with both the natural landscape and the contiguous city. However, what they find instead is that residential form is becoming simultaneously more removed from the natural landscape and less connected to the existing city. Their findings suggest that the resulting loss of public space, quality of built form, and access to natural environment has occurred not only because of the influence of private developers, but also because of declines of the processes of planning for truly public space.

Using accessible ways of parsing the experience of cultural landscapes from the discipline of urban design, Sandalack and Nicolai show how observations of the ordinary landscape can provide a vocabulary of the parts and processes of the built environment, a language that allows for better planning processes, involving more informed public participation in planning the form of communities. Sandalack and Nicolai advocate investment in planning that is motivated by a public vision of civic life; they argue that the practices and standards of urban design from particular time periods provide a valuable perspective on the values of those times, and on the choices that are made at specific historical moments. Reading from the landscape the values and choices involved in different eras' practices of ordering spatial, visual, and functional environmental relationships, residents and planners may be able to develop a more useful vocabulary for engaging in meaningful environmental decision-making processes.

Sandalack and Nicolai argue that as nature is sold as the answer to deteriorating urban space, the kind of nature that is provided not only exacerbates the problems of placelessness driving the search for nature in the first place, but also tends to turn residents' aspirations in the wrong direction: instead of engaging in public planning for their environments, residents seek private utopias in the superficial nature popularized by housing advertisements. Their careful consideration of the aspects of place design that make places meaningful and comfortable supports a convincing argument about the necessity for more attention to the process of planning the structure of places.

Taking on the generic critique that modern urbanization is characterized by monotony, banality, and loss of meaning, this demonstration of an effective set of tools for reading the landscape makes an excellent focus for articulating and deepening such critiques about the modern urban landscape, and for identifying those places and moments when decisions that otherwise appear already-determined or naturalized might be pried apart, contested, and negotiated.

Understanding these dynamics is one way to start addressing the problematic relationship between continuity and change that so strikingly marks the relationship between exurbs and the city. These analytical tools are particularly appropriate at the urban edge partly because the developing edge of the city exposes most dramatically the values, expressions, and unresolved dilemmas of the society that creates it.

—Kirsten Valentine Cadieux and Laura Taylor

■ ■ ■

Today's exurbs are tomorrow's suburbs. The edge of the city of Calgary has traced a lengthening radius: it now extends more than twelve miles (20 kilometers) from the city center and has eclipsed many areas that had once defined the edge. In this essay, we will explore how changing values and notions of landscape have been expressed in the evolving edge of Calgary. Analysis of three neighborhoods on the urban edge of the city, representing three eras of urban development, shows how the urban structure and the organization of land uses have changed along with the relationship to nature and quality of the urban landscape. While all developments involved drastic modifications to the landscape, the ways in which the designers attempted to create a connection to nature and the city have varied. All three developments were planned neighborhoods, as well as planned landscapes, and designed according to the practices and standards of the time. Although it may be expected that the more recent developments would embody the greatest relationship with the landscape and the greatest continuation with the existing city, given the apparent strength of the modern environmental movement and the growth of urbanity as a popular urban design paradigm, the opposite was found to be true. The earliest developments have the highest environmental quality (according to certain measures), and express the best relationship with the existing city, while the more recent developments are the most removed from the natural landscape and have the weakest connection to the existing city.

The three neighborhoods, although now part of the built up city, were all at one time on the edge of the city. That the neighborhoods differ so markedly is notable, and it is possible to read in them something of the prevailing values of their times. Calgary is a particularly interesting city in which this evolution of urban form and urban ideas can be read. Its incremental growth has taken place at the neighborhood scale, and not at the scale of the individual house. Societal trends are therefore somewhat more apparent.

Where local and regional identity were once largely a product of the unique interaction between local culture and traditions within environmental constraints, the visual qualities and spatial structure of places are now more likely to be influenced by increasingly global cultural and economic forces. Many places now tend to look more and more alike, and regional character is not as directly related to the particularities of environmental and cultural context. Critics of the placelessness of much contemporary urban form and the decline of the public realm are numerous, and have effectively drawn attention to the qualitative decline of cities in North America. While the loss of connection to the environmental and cultural context is partially responsible for a loss of sense of place, it is also the composition and relationship of important urban elements that produces places of quality and continuity, and the neglect of which contributes to urban decline. This is perhaps most notable in the exurbs and the suburbs.

Modern suburban landscapes, say their many critics, are characterized by monotony and banality, and a loss of meaning. This landscape may be generally comfortable and efficient but it lacks character and variety. Although order is present, the modern suburb doesn't bring about any sense of place, and is characterized instead by "placelessness,"[1] or the inability to identify where you are. This may be due to our loss of connection to the natural conditions of the earth, as well as to our loss of identification with the human-made things which constitute our environment. Urban designers have had a keen interest both in understanding the processes that have been responsible for the prevalence of uniformity at the expense of local and regional distinctiveness and also in developing approaches for designing places that express distinct identities. Identity is achieved by expressing local and regional character in new designs, and by reinforcing existing elements that exemplify this identity and character. A key component of authenticity and identity is the maintenance of continuity. When a neighborhood or other development easily grafts onto the established city form, there seems to be a greater sense of continuity. When a new neighborhood differs markedly, this easy transition is lost, and a more disjointed experience emerges. City evolution encompasses both continuity and change. Evolution is expected, and a coherent whole is desired.

However, as this study of neighborhood form revealed, the urban structure of the neighborhoods typical of each of Calgary's development eras expresses the relationship to nature and organization of land uses particular to the cultural values of the time, without a strong relationship to the existing urban form and character. The distinctiveness of the urban structure of particular eras is most observable on the developing edge of the city, as this is where a first generation of streets, blocks, and buildings is laid down; it is a snapshot of the values and expressions of the society that creates it. Each way of building seeks to fulfill expectations of nature and city while satisfying the need for housing and community, and the degree of the character and the quality of the urban form produced in each era have varied widely.

Urban form analysis may be used in order to reveal the characteristics of existing urban form, and in order to make different forms directly comparable. The following illustrates a proposed methodology for analysis, and includes examples of the methodology as it was applied in research on Calgary neighborhood form.

URBAN FORM ANALYSIS—AN APPROACH

The evolution of the landscape, including the public realm, reflects the evolution of ideas and ideologies. The values cultures place on the land are reflected in changing patterns of land ownership and land development, and consequently in the spatial qualities of the public realm. How one understands and interprets those landscape forms and processes depends upon one's ideological position, and consequently influences the histories that are constructed.

The form of an object or organism (including the form of the city or a neighborhood) is a diagram of the forces that have acted upon it.[2] In the case of the built environment, form is the result of the accumulation and interaction of various decisions and acts of design and building. By analyzing the built environment, it is possible to discover the values and choices that shaped it.

The basis for an understanding of urban form is to ask questions about the relationship among the parts, and between the parts and the whole. There are three primary relationships: the relationship between form and nature, the spatial relationships of production, maintenance, transformation and use of the urban forms, and the relationships involved in the processes of formation and the spatial relationship between built forms. These are outlined in more detail.

The Relationship Between Form and Nature Is Expressed in the Environmental Context, Conditions, and Features of a Place

There is an inherent logic in the early evolution of any city or town, and this often has something to do with landscape, topography, and hydrography, in addition to social, political, and economic forces. Many towns and cities were originally sited for environmental reasons: the presence of a water body, the prospect afforded by a particular geographic location, and the shelter and security provided by topography—these were often important reasons for locating settlements in specific locations, and contributed to their unique image and character. The identity of places also used to be a direct consequence of the constraints imposed by local environmental conditions: extremes of climate, quality of soil, limitation of water supplies and the availability of certain building materials influenced the form of the settlement, as well as the vernacular styles of architecture and gardening. Places tended to be distinctive and unique as a result of individual and community adaptation to local conditions. Now, this relationship is much more a matter of choice than of necessity.

Ian McHarg changed the way that environmental planners and designers gathered and analyzed environmental data, effectively demonstrating that physical planning and the design of sites should be based on a thorough understanding of the ecology of the area together with human values.[3] Through an understanding of environmental conditions and characteristics, more locale-specific and ecologically sound planning and design would be produced, and there would be a greater likelihood of producing designs which would be more harmonious with their environmental context, and more appropriate to human values and needs. The field of landscape ecology provides a broad theoretical base to urban ecology and contributes principles upon which urban planning and design decisions can be made.[4]

In order to understand the existing environmental conditions, an inventory of what is there (topography, hydrology, geology, vegetation, and so on) and their inter-relationships, on a regional and especially a local level, should be acquired. Further, it is necessary to look at the evolution of these conditions over time. Analysis of historic town plans, air photographs, and other archival material from earliest origins and moving forward in time can reveal environmental features or conditions which have been lost or transformed over time, as well as those which have persisted. As well, the sensed forms of an environment—in particular shade, shelter, sun, wind—should be graphically documented. This analysis is helpful in providing knowledge of the particularities of the place that can be helpful in making planning and design decisions, and in leading to places that provide more human comfort. Sun and wind are two of the most important elements in determining how comfortable an outdoor place is, and simple diagrams can be very helpful in providing planners and designers a better understanding.

Environmental analysis is of value in understanding places because the relationship between urban development and natural features is an expression of the prevailing values of the time. The analysis is also useful as a tool in design. Natural features can be important in contributing to open space systems and in defining neighborhood edges, and ecological analysis is essential to development of sustainable places, and can also provide design determinants.

The Spatial Relationships of Production, Maintenance, Transformation, and Use of the Urban Forms are Expressed through the Land Uses and Functional Relationships

The built environment, particularly the public realm (that is, the collection of spaces of the city, such as the streets, parks, plazas, and squares, in which we live much of our public lives), needs a distinct identity and structure to be legible, successful, and enjoyable.[5] The analysis of a town's or city's spatial structure considers land utilization and the pattern of activities that parts of a town or city generate. It describes the location and distribution of particular uses and the functional relationships between them.

A number of theories of spatial structure have been formulated, with various ways of conceptualizing space. Kevin Lynch saw the city image as a system composed of five basic elements: paths, edges, districts, nodes, and landmarks (or monuments).[6] These urban elements provide physical and psychological orientation, and together have significance to the inhabitants forming a mental map. This organizing structure provides a framework by which neighborhood urban form, its parts, and the ways they are related, can be analyzed, and then used as a basis for design.

In order to be useful as a design tool, spatial structure should be analyzed in terms of its historical evolution so as to discover how functional elements have migrated or been transformed over time, and to show how spatial relationships have changed.

The Relationships between Built Forms Include Morphology, Typology, and Visual Relationships

Morphology and the Spatial Relationships between Built Forms

Urban morphology is an approach to studying urban form which considers the three dimensional qualities of lots, blocks, streets, buildings, and open spaces, over time, and considers their relationship with each other. This approach can encourage a more integrated approach to planning and development, and help to avoid the disparate reactions to problems otherwise perceived as unrelated. There are several schools of thought in morphological studies,[7] which although rooted in different cultural and linguistic traditions and disciplines, share some common ground and common principles:

- Urban form is defined by three fundamental physical elements: buildings and their related open spaces, lots, and streets.

- These elements can be understood at different levels of resolution. Commonly, four are recognized, corresponding to the building/lot, the street/block, the city, and the region.
- Urban form can only be understood historically because the elements of which it is comprised were formed over time.

The lot is the basic cell of the urban fabric—it links built form to the land and to open spaces. Lots may be subdivided, or consolidated, over time, a process that influences how properties are developed, and it is a powerful determinant of built form. Consideration of land and its sub-division links the building scale and the city scale, and provides a better understanding of the process of city forming and re-forming.

Typology

Typology is the study of elements that cannot be further reduced. Types are general (for example the school, church, town square) but are also culture-specific, such as how the particular shape and details of a type vary from society to society. Type signifies something more permanent and long lasting than function and is therefore the basis of design.[8] Types in built form developed according to both functional requirements and aesthetic considerations. A particular type (of building or public space) is associated both with form as well as with the way of life that it allows.

However, typology is not neutral—urban spaces should be designed and analyzed in terms of their viability as containers for public life, and in terms of certain qualities. Typology must also be considered within the context of the city. A space or street by itself as a public space is meaningless—it must be conceptualized and designed in relation to its physical and spatial context, in addition to considering the qualities and characteristics of the space and its edge conditions. The typology so defined then informs the layout, the materials and other design elements.

Visual Relationships

It is at the scale of the street and the square that the visual character of the district or city is represented in a condensed form. While the city or the neighborhood is comprehended as an overall experience, only the single street or space can be held in view by an individual, and this composition is the most important when analyzing visual quality. Space is perceived as a visual relationship between objects. The nature of that space is determined by the building masses and the multitude of relationships between them, primarily their proximity, their continuity as an edge and the degree to which they create a sense of enclosure.

Currently, the character of building and spatial form is not often determined by considering visual relationships. Land-use planning in our regulated, highly administered cities largely determines function and activities within the public domain, and building forms are determined more often now by program and function (form/content relationship) as well as by building technology, density, and the vehicular—pedestrian circulation system. However, the visual components of built form should be considered together with land-use planning, since they are so profoundly important in contributing to the user's experience.

Urban form analysis comprises several ways of documenting, analyzing, and otherwise reading and understanding a place. Of course, what is not considered here are the opinions or perceptions of the residents, or of those who are involved in building the city form. In order for this approach to be practical as a way of providing guidelines for new development or redevelopment, techniques and methods for evaluating the human experiences should also be incorporated. These methods should be developed to suit the situation, and be culture-specific.

THE EVOLVING URBAN EDGE OF CALGARY

As part of a broader and more comprehensive study of Calgary's urban structure,[9] this research mapped the evolution of Calgary's suburbs, and documented and analyzed neighborhoods from each building era. The methodology for the analysis was developed through several previous projects[10] in which techniques and methods from the disciplines of urban morphology, landscape architecture, urban design, and urban planning were synthesized. The methodology, developed as an approach to pre-design, is employed in these case studies as an approach and process through which urban form and its evolution can be illustrated and understood. The research depends on multiple sources of primary information, and considers a number of inter-related aspects of the built landscape, paying particular attention to qualitative aspects of overall city form and in particular the public realm, to historic continuity of process and form, and to the importance of understanding the relationship between the natural/topographic setting and the identity that is derived from that environmental context. The research consists of documentation and analysis at various scales:

- the city
- the neighborhood
- the block
- the street
- the lot
- the building

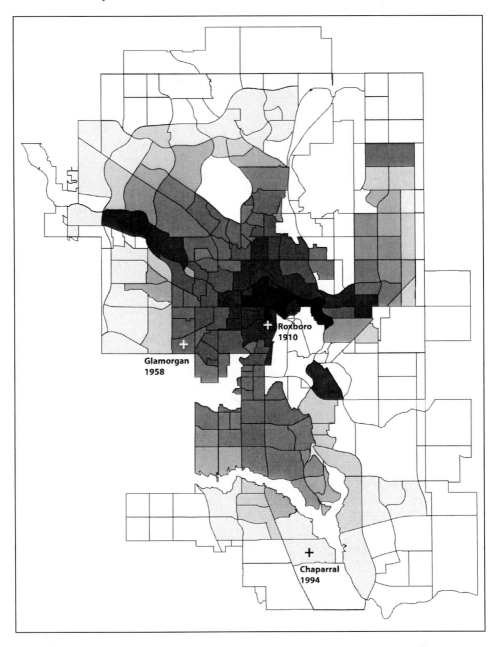

Figure 8.1 At the scale of the city, the evolution of neighborhoods was mapped.

Calgary had its beginnings in the last decades of the 19th Century as a compact core at the confluence of the Bow and Elbow Rivers. Three outlying settlements (Bowness, Forest Lawn, and Ogden) were established at some distance from the city center, but were soon annexed. With the exception of the industrial lands in the southeast, the airport in the northeast, and several large parks, most of Calgary has developed as single family residential neighborhoods, radiating from the downtown core in a series of concentric circles of development. In the drawing, neighborhoods are toned according to their age. The darkest tones represent the earliest development, and the lighter tones the most recent. The population of the city has now exceeded one million, with a footprint of approximately 330 square miles (855 square kilometers).

Examples from three distinctive development eras were selected for detailed study: Roxboro, constructed pre-World War 1, Glamorgan, constructed post-World War 2, and Lake Chaparral, constructed recently. These development eras correspond to three phases or paradigms that characterize much of North America's urban development.[11]

Roxboro

The neighborhood of Roxboro represents the first phase of urban development which lasted up to approximately 1940, and was marked by incremental changes to older forms as new technologies and socio-economic models were introduced. Architects and urbanists considered the problems of town planning and design in terms of historical precedent and context. The public realm was viewed as an important part of civic infrastructure, and many of the significant public spaces and streets of contemporary cities were established during this period. Street form and pattern usually extended and grafted onto an existing framework. The existing landscape was viewed as an environment to be tamed, subdued, and brought under control so that human habitation could take place in more comfort and security.

In the case of Calgary, which had only recently been established on the western frontier, the original plan was a grid network of streets—typical of towns and cities built on the western railway. The first commissioned city plan, prepared by Thomas Mawson in 1912, envisioned a system of streets, civic spaces, and squares, focused on the rivers, and modeled after City Beautiful ideals. The plan was clearly based on a strong vision of the public realm, and its importance, but it was never carried out due to an economic downturn and World War 1. However, some of its ideas permeated the neighborhoods that were constructed during Calgary's early boom.

Roxboro was at the southern edge of the city in the years prior to World War 1. The north edge of the neighborhood is about one mile

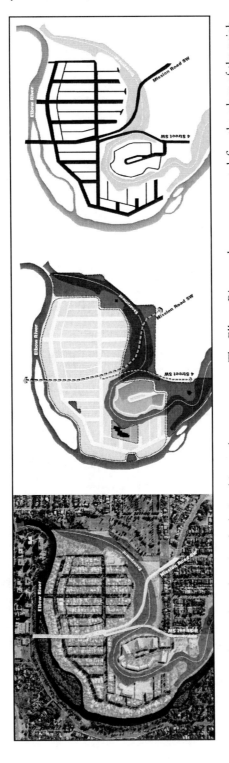

Figure 8.2 Roxboro Edges, functional relationships and street pattern. The Elbow River and an escarpment define the edges of the neighborhood. The street pattern is a modified grid, and is bisected by a major street connecting the downtown and the outlying areas. Most of the neighborhood is single-family housing; however, a high quality commercial street with most amenities is located just north of the river within easy walking distance.

Time, Place, and Structure 197

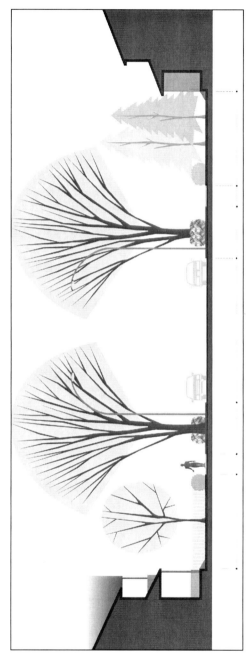

Figure 8.3 Street section: street trees (elm, interplanted with lilac) on boulevards separate the sidewalks from the streets.

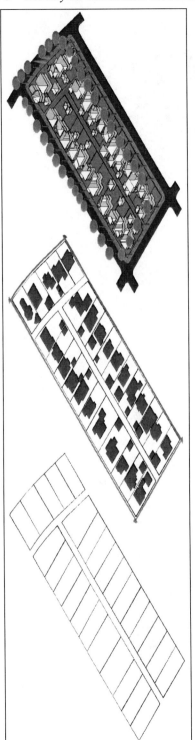

Figure 8.4 Typical Roxboro Block—property line, figure ground, and model. Buildings are placed toward the front of the lot, and garages are located in back, accessed by a lane. Houses face the street, and most include a front porch or stoop. The typical lot size is 50 by 120 feet. Lanes are included, so service access is at the rear of the lots.

Figure 8.5 Roxboro Streetscape. The combination of house type, building placement, street design, and plantings give the street a pedestrian scale, and a clear separation between public and private space.

Time, Place, and Structure 199

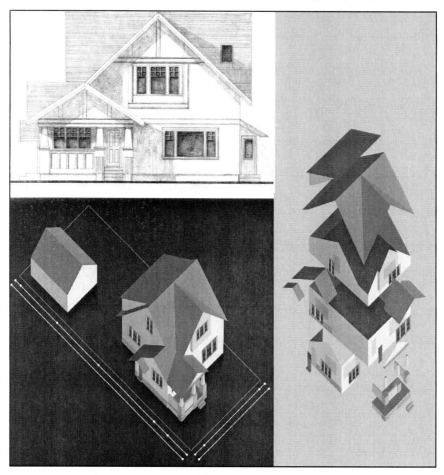

Figure 8.6 Roxboro House. Typical house types in each case study neighborhood were analyzed and graphically illustrated. The combination of house and lot shows the placement on the lot and also the presence of garages or other outbuildings. The house is broken down into its architectural components which can be useful in design of new interpretations of the type.

(1.6 kilometers) from the city center, which was defined at the time as the railway station at Centre Street and 9th Avenue SW.

The land that was to become Roxboro was owned by Freddie Lowes, who ran the largest real estate enterprise in western Canada at the time. Lowes purchased the lands from the Oblate Fathers for $100,000, and envisioned a high quality suburb of single detached houses on individual lots that would compete with the Canadian Pacific Railway's Mount Royal, the prestige address in Calgary. The neighborhood of Roxboro was created after Freddy Lowes drastically modified the landscape. Hydraulic pumps were used to direct high-pressure streams of water at Mission Hill, and after three months, 100,000 cubic yards of earth had been moved. This succeeded in raising the land closer to the river to make it flat for building purposes and bring it out of danger from the Elbow River floods. "Along the river and between the trees" was the slogan that Freddy Lowes used to promote Roxboro, a neighborhood that now has very high environmental and urban quality, but was somewhat ironically created by some of the most environmentally catastrophic moves. Lowes provided graded boulevards, street lighting and street trees, sidewalks, and curbs for Roxboro prior to the property being put on the market, and ultimately 525 lots were sold.

Landscape features of the Elbow River and its escarpment define three edges of the neighborhood, with a major street forming the western edge. The street layout is a grid, and the many through streets form a permeable structure.

Single detached dwellings constitute all of the housing, and public green space and a community hall make up the other land uses. Just to the north of the neighborhood, and easily accessible on foot, is a commercial street (4th Street SW) containing most services and amenities, and schools are located within easy walking distance.

The visual quality of the streets and the neighborhood as a whole is high. Roxboro is perceived as having a high degree of environmental quality, despite the circumstances of its creation. The river is visible and accessible from both private and public spaces, and the public street trees and private landscape contribute to a heavily treed area. There is a clear demarcation between public and private space, and the public realm of streets and public spaces is of high quality.

Glamorgan

The second phase of urban development corresponds to modernism, corporate development and the invention and institutionalization of methods for town planning, which coincided with the period of economic growth following World War 2. The paradigm of history that influenced the first phase of urban development was replaced by the paradigm of space. Space at this time was seen to have redemptive and social power, and new spatial organizations were designed to house a "better" society. Over time, town form became discontinuous (there was little attempt to graft newer developments onto the existing), building typology became less place-specific (the International style and later a context-immune and generic form predominated), the public realm declined (it was generally not required that buildings shape outdoor space), and place identity became vague and confused. Suburbs of this type developed simultaneously in countless cities and towns in North America at this time, and while they resembled each other, they had less visual relationship to their respective place.

Glamorgan was on the developing edge of the city during the 1950s through the 1960s, a distance of approximately six miles (9.6 kilometers) from the city center, and is typical of a ring of neighborhoods developed in the post-war boom that were structured after Clarence Perry's Neighborhood Unit, which was the model of the spatial and functional organization of suburban development.[12] The neighborhood units consisted of low-density development and were developed in an orderly sequence.

Much growth took place in Calgary during this period, and contributed to a broad band of similar developments ringing the early gridded core. The old grid network of permeable public streets was abandoned for curvilinear streets enclosing semi-private clusters of houses. Concurrently, town planning lost its interest in physical form; its practice became more concerned with orderly growth, social process, and the idea of space, and less with morphology, shape, and structure.[13] Planning became a technical exercise, carried out in the service of a municipal government which was pro-development and pro-growth, and street design migrated into the domain of the traffic engineer.

From air photos of pre- and post-development, it is possible to identify landscape features, and to discuss the degree of correspondence between original landscape and built form.

The pre-development landscape consisted of: an early trail at the north edge, which was later developed as Richmond Road, a major east-west thoroughfare; grid roads resulting from the Dominion Lands Survey (37th Street at the east, 50th Avenue at the south, and what was to become Sarcee Trail at the west); farmsteads and rectangular agricultural fields that had replaced the prairie landscape; and irregular occurrences of aspen bluffs as well as sloughs and seasonal ponds. 37th Street and 50th Avenue defined the city limits from 1910 to 1954. With development of the new suburb, the aspen bluffs were destroyed, and the sloughs and seasonal ponds were filled in and graded over. There is no remaining evidence of any of the natural features.

This phase of Calgary's development seems to indicate both the efforts to tame and subdue nature through aesthetic improvements, and to include neutral and inoffensive "green space" within the residential environment. Widespread use of pesticides and fertilizers enabled the entrenchment of the cult of the lawn, even in this prairie landscape. At the same time, technological innovations within the home, ranging from electric clothes dryers replacing clothes lines, to packaged foods replacing garden produce, further separated the suburban dweller from farm, country and nature.

A minimum of ten percent of the land within a planned residential development was required as public reserve for school, park and playground use, and it was typically set aside as one large parcel of land. Phase 1 of the development was organized around a centrally located school and park area. Clustered in this central area were two elementary schools,[14] a community center, and large recreation fields. This constituted the only public space in the neighborhood, with the exceptions of a piece of land at the northeast too steep for construction which was developed as green space

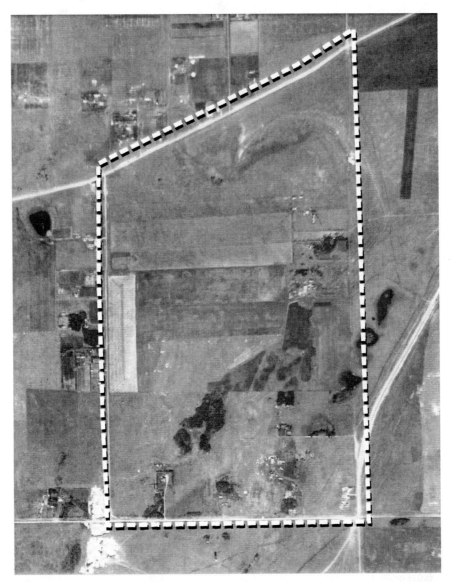

Figure 8.7 Glamorgan 1924.

with no apparent functional purpose, and another green space developed adjacent to the multifamily node. Phase 2 was clustered around another large school reserve. It has not been built on to date, and recreation fields occupy the land.

Commercial and institutional uses are typically located at the edges of planned neighborhood units. In Glamorgan a shopping center is located at the northeast corner, and consisted originally of a grocery store and strip

Figure 8.8 Glamorgan 1999.

mall containing an array of services and amenities (drug store, bank, professional offices, hairdresser, restaurants, bowling alley, drycleaners, shoe repair), with a large parking lot in front. There was a presumption that residents would have cars, and this is clearly the favored mode of circulation;

204 *Beverly A. Sandalack and Andrei Nicolai*

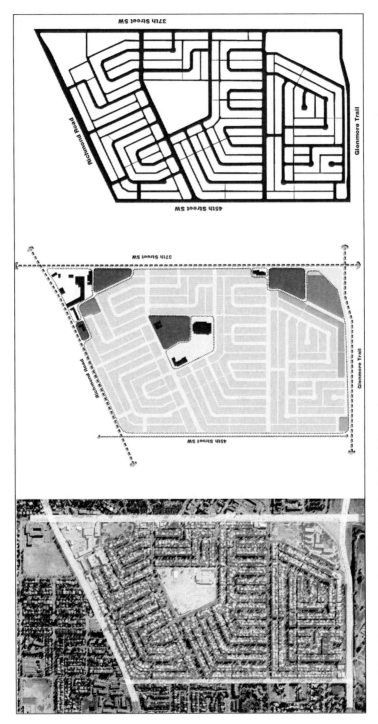

Figure 8.9 Glamorgan edges, functional relationships, and street pattern.

Time, Place, and Structure 205

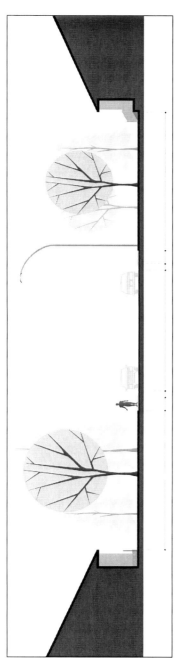

Figure 8.10 Typical Glamorgan Street Section. Provincial and municipal government subdivision regulations now dictated a minimum road width, and several changes from the earlier street typology are apparent: landscaped boulevards are not included, and sidewalks with rolled curbs line all streets.

Figure 8.11 Typical Glamorgan Block—property line, figure ground, and model.

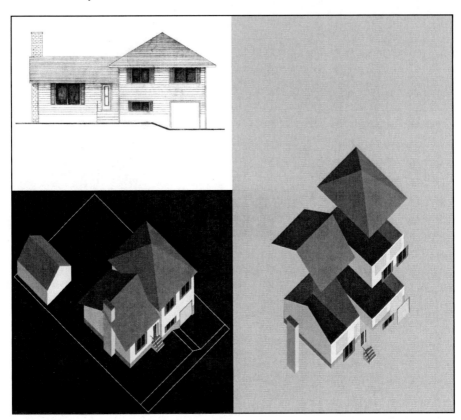

Figure 8.12 Typical Glamorgan House. The housing types included primarily bungalows and split levels, and all included large picture windows in front. 50 x 120 foot lots were the norm, and the only deviation in lot size occurred at the inner and outer corners of the crescents and cul de sacs.

Figure 8.13 Typical Glamorgan Streetscape.

it is easier to navigate the neighborhood by car than on foot. The northeast corner intersection contains three gas stations; the fourth corner is occupied by the Junior High School serving a large regional sector. An additional small strip mall with a parking lot in front is on the east edge of the neighborhood.

Arterial roads (Richmond Road and the grid roads) formed the edges of the neighborhood, and secondary grid roads (45th Street and 46th Avenue) became internal collectors. The street system is hierarchical, with few internal through streets; instead curved streets and cul de sacs were intended to break the "monotony" of the grid.[15] Phase 1 consisted almost exclusively of single detached dwellings, although a limited amount of multifamily housing was provided on the edges, and it had a much lower perceived value and status. Phase 2 included more single-family dwellings, as well as multifamily housing (duplex and townhouse), as well as a seniors' housing complex and a church.

By the end of 1955, public tree planting was included in the local improvement bylaw system in Calgary, and community associations were asked to submit petitions for tree planting, specifying the variety of tree and shrub they would prefer on each block, with residents paying a nominal amount. The developer planted one tree per lot, and in Glamorgan, there was a preponderance of birch, with some elm. The absence of the treed boulevard, as in Roxboro, created a much different street profile.

Neighborhood planning of this era reinforced the post-World War 2 emphasis on the family, with an idealization of home life finding expression in the development of backyards as family area. The front yards were primarily for show, and had a higher level of maintenance, but little use. The differentiation between public and private space was now less clear, and the amount of privately owned land per lot had increased, but the public benefit had not. Indeed, the publicly visible landscape had become much less diverse, and a private place for most to keep off the grass.

Several builders were involved in the development, with little variation in house type and price, although with some variation in quality from builder to builder. The neighborhood form is visually homogeneous, and Glamorgan is indistinguishable from other neighborhoods of its type developed at the same time, whether they are in Calgary or elsewhere in Canada.

Lake Chaparral

The third, contemporary phase of urban development corresponds to the post-modernist period of residential development at the city's edge. This is a paradigm of ambiguity: time and space are now arbitrary—they are no longer so much the result of functional requirements or of cultural constraints

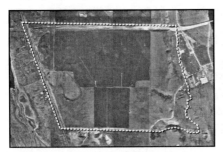 

Figure 8.14   Chaparral 1924.   *Figure 8.15*   Chaparral 1999.

as they are determined by the marketplace.[16] Almost anything is possible, and in the residential development market, the most desirable products are those that include a version of "nature." Following the modernist notion of space, and its attendant way of valuing land and landscapes, was the environmental movement, which has had a huge influence on consumer preferences of products from shampoo (herbal) to vegetables (organic) to neighborhoods (idealized rural life). Marketers stress the view of the mountains, the outdoor lifestyle, the wildlife, and the natural landscape; however, all these have disappeared with the arrival of the bulldozers and have to be recreated by the ad-makers for the benefit of the imagination of future residents. The true natural environment, the prairie—an often harsh landscape from which builders provide real shelter—does not fit with popular images of residential landscape nor provide the same opportunities for marketing.

In the case of Chaparral, the neighborhood is aggressively marketed for its "lake"—a constructed body of water twelve meters deep, filled from the municipal water supply and stocked with rainbow trout.[17]

Lake Chaparral was established in 1994 on land that was annexed by the City in 1981. It is located eleven miles (17.8 kilometers) from the city center, and as of 2004, was no longer on the city's edge, as further annexation led to additional suburban growth to the south.

Examination of air photos from 1976 and 1999 reveal little before-and-after landscape similarity. The major feature of the lake does not correspond either to topographical or hydrological attributes of the pre-urban landscape. The main features of the area are Macleod Trail at the western edge (it was an early trail, later developed as a major north-south highway); grid roads at the north and south; and a ridge of land sloping down to the Bow River valley at the east, bordering an otherwise fairly flat area. Ephemeral wetlands are present along the western edge of the neighborhood. Rectangular agricultural fields are lined by a shelterbelt of trees at the south, and small bluffs of aspen are also found. None of the landscape elements were retained as the land was redeveloped to housing.

Suburban development planning in areas such as Lake Chaparral is now carried out in large geographical units. Sector plans are prepared that

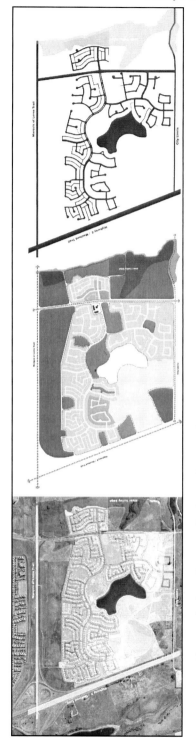

Figure 8.16 Chaparral edges, functional relationships, and street pattern. The internal street pattern is strongly hierarchical, and most internal streets are loops or cul de sacs. It is a highly impermeable street pattern.

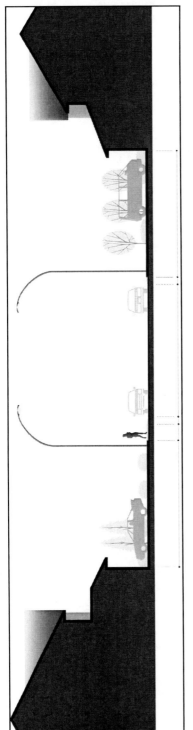

Figure 8.17 Typical Chaparral street section.

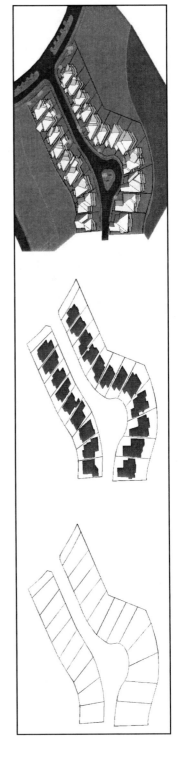

Figure 8.18 Typical Chaparral Block—property line, figure ground, and model.

Figure 8.19   Typical Chaparral House.

Figure 8.20   Typical Chaparral streetscape.

determine large residential and service areas, and development is phased, with major roads typically defining each neighborhood. The edges of Lake Chaparral are highways at the west and north edge, and a very wide buffer of land separates the housing from the northern highway. There are very poor connections with adjacent neighborhoods with only one safe road connection, should a resident be inclined to try to walk beyond Lake

Chaparral. The slope at the east defines the edge of development, and a grid road is at the south, which was upgraded to the status of a major collector road as development to the south took place.

Lake Chaparral consists entirely of large single-family houses of variations on a basic type: one or one and a half stories with attached three-car garage at the front. A limited palette of pastel colored stucco is augmented by artificial stone and brick trim. An enclave of multifamily housing is provided on the east side, but is not integrated with the rest of the neighborhood. A partial elementary school (K-4) was built in 2006. There is a small convenience commercial node near the entry at the east, in the form of a strip mall with parking in front, and another at the south.

The lake is the primary feature, and contains a constructed waterfall. It is a private amenity, and is clearly intended for use only by residents of the neighborhood. The neighborhood also includes twelve smaller playgrounds, located within walking distance of each house, and there are over two kilometers of walking paths.

Sidewalks have been eliminated from the street design in many cases, and most of the neighborhood does not include back lanes. Most of the houses incorporate three car garages that dominate the front facades. The combination of large house footprint on the lot plus the loss of plantable area to the driveway has eliminated much of the potential for street tree plantings. There is very little vegetation as yet visible on the street.

The civic function of the street is now almost completely corrupted, and its main role is that of traffic mover. There is no public face to the street; although the street is used for play by the many children in the neighborhood (75 percent of the homeowners have children under the age of twenty-five).

Development of individual neighborhoods such as Lake Chaparral is typically done by one developer who can now control the overall image, character and price point. Marketing programs aggressively advertise theme and lifestyle associated with the neighborhood, and emphasize the private lake as amenity, and the individual house as private domain.

CONCLUSIONS

Calgary's urban edge has moved from one mile from the city center in 1910, to four miles in 1958, to eleven miles in 2003, and is still moving. Analysis of three Calgary neighborhoods that represent three eras of urban development on this mobile urban edge reveals how the relationship to nature and the quality of the public realm have changed over the course of the city's evolution.

The development of Roxboro required drastic modifications to the natural landscape in order to create developable land (although an anomaly, it does represent the values and processes that typified this period), but this neighborhood exhibits the strongest connections to its natural context. The tree-lined streets in a permeable grid pattern, river parks, and visual connections to the river and escarpment contribute to a high level of environmental quality.

Glamorgan retained none of the landscape features, and although some of the perimeter and internal roads correspond to the grid roads of early landscape layers, this relationship is not clear. Glamorgan is expressive of modernist planning practices, in which urban planning lost its interest in traditional spatial relationships. Its practice became more concerned with process and the idea of land use, and less with morphology, shape, and structure. The old grid network of permeable public streets was abandoned for curvilinear streets enclosing semi-private clusters of houses. Green spaces and recreation fields, although well intentioned, contribute little to the overall urban structure, or to the integrity of the public realm. The centrally located schools, community center and recreation fields do not have the physical qualities to serve as the morphological center, due to the arrangement of the spaces, the lack of edge definition, and the lack of relationship to the neighborhood as a whole. "Open space" is too open to be perceived as meaningful space; rather, definition of space and creation of enclosure help to shape meaningful public space, especially in the prairie context, where relief from the open spaces and physical and psychological shelter are important.

Lake Chaparral represents a further departure from the original landscape. Although the environmental reserve on the east side of the neighborhood preserves the natural area, it is not related to the neighborhood form and structure. The neighborhood is focused around a new lake which is not expressive of the environmental characteristics or qualities of the prairies. It is a construction that requires significant inputs of municipal water, and great effort to maintain. The environmental quality of the streets and green spaces is low due to the present lack of street trees and to the heavily maintained small parks that are not integrated with the public streets. The visual quality is neutral and placeless, with a clear emphasis on the private realm.

Most recently in Calgary, the most desirable properties are country residential lots, where residents want the benefits of a version of a rural setting, but with all the municipal services and conveniences of the city. The "natural" environment is becoming more and more removed from the landscape—we have little idea what was the original landscape—it has been replaced with a generic development, its landscape a product of myth and invention. What were once the determining characteristics of individuality in the landscape have been paved over, straightened out, and averaged away to be replaced by a homogeneous form that has more to do with uniform engineering standards and technical criteria than with unique conditions and characteristics. Cut loose from previous defining features, identity is created through themed developments that have little relationship to their original physical context. Roxboro, despite having a radically modified topography, capitalizes on its location on the river, and emphasizes this connection through street layout, location of parks, and visual connections.

Our experience of our environment, as illustrated by Lake Chaparral, has been modified or mediated so that it is now ambiguous at best, and placeless at worst. While we can admire and enjoy landscapes that we create through our interventions, there are limits to how much we can modify the physical environment before we have created myth or ambiguity instead of the authenticity that is made possible through continuity with the landscape and connection to the land.

Paradoxically, the most "urban" neighborhood of Roxboro was developed during the earliest period of city building, although that was the period where the highest ideals of civic space were most apparent. The quality of the public realm has declined since then. Although the strong ideals of modernism and environmentalism have enthusiastically manifested themselves in subsequent open space development, the contributions that the resulting spaces have made to the public realm have been meager.

Urban planning and design must proceed from an ideological and theoretical position; otherwise, they become technical, abstract, indefensible, and ultimately irrelevant endeavors. The position that one takes as a planner or designer is based on a conviction regarding the qualities that are important to urban form and urban life. Urban analysis must be grounded in a vision of what the city should be, otherwise it remains only descriptive. In this project, each neighborhood was analyzed in terms of the degree of environmental responsiveness, permeability, legibility, variety, versatility, and expression of place, considered by the authors to be some of the most important qualities of urban form. As well, this research approach advocates continuity between a city vision, urban analysis, development strategies and design guidelines, emphasizing the public nature of all development, and the importance and potential permanence of the public realm.

The shaping of the many components of the public realm need a high degree of cooperation and coordination between the actors involved in its development and management: the planning and engineering departments; the various owners and designers involved in the development of its edges; as well as the community which uses it. Over the past several decades, city planning and the design professions have grown apart. While the various design disciplines have been pre-occupied with staking out their individual jurisdictions, the public domain has been neglected, with architects too often focusing on individual buildings and rarely considering the spaces in between, landscape architects dealing largely with site specific and market driven projects, and planners assuming roles of land use administrators, ineffectual in the promotion of public policy especially where it concerns the public civic realm in residential development. This study has shown a marked change in the creation of residential landscapes. The individual homeowner is promised an escape from the complications of everyday life through marketing that heavily uses images of the natural environment. Individual experience is directed through design to be internal to the private lot, the house, and the car, and away from civic spaces. Development values and principles have fallen by default into the domain of the private developers, who have had the strongest influence on the image and quality of the built form. As a result, the public realm has been lost as a comprehensive area of inquiry and concern, and the natural environment has become a themed and regulated private amenity.

It is the public realm, the spaces now created mostly by default, that needs to be an important focus of the planners, designers, developers, and first of all, the community. This is achieved through a renewed urban planning process grounded in the belief that the public realm should be a receptacle for civic life, and the embodiment of local and regional identity and culture. Within this framework, the role of a renewed urban planning process is to support a cultural and environmental memory, as it relates to the public realm.

There is now some evidence to suggest that we have entered into a fourth paradigm, in which urban design provides the methodology for building the city, and in which values of sustainability inform the development of a higher quality public realm. Whether those values influence the design and development of the next series of suburban neighborhoods, or if another development model perhaps replaces it, will remain to be seen.

NOTES

1. Christian Norberg-Schulz, in *Genius Loci: Towards a Phenomenology of Architecture* (New York: Rizzoli Press 1979) was one of the first to introduce the terms *genius loci* and "sense of place" into the vocabularies of contemporary environmental designers, with Edward Relph, in *Place and Placelessness* (London: Pion Ltd., 1976) among those who have discussed this issue as it relates to the North American context. Since about the 1970s, an increasing concern with the proliferation of "placeless" landscapes has developed along with an interest in constructing local and regional identity and sense of place. Michael Hough (*Out of Place: Restoring Identity to the Regional Landscape* [New Haven and London: Yale University Press, 1990]) is notable.
2. Darcy Thompson, in his great work *On Growth and Form* (1917, republished by Cambridge University Press, 1961) studied biological forms from a mathematical perspective. The growth of a city can be understood as the product of the forces acting upon it.
3. I. McHarg, *Design with Nature* (New York: Natural History Press, 1969, reissued 1992).
4. See R.T.T. Forman, *Land Mosaics: The Ecology of Landscapes and Regions* (Cambridge University Press: Cambridge, 1995); and R.T.T. Forman and M. Godron, *Landscape Ecology* (John Wiley: New York, 1986).
5. Legibility refers to the ease with which a city, or part of it, can be understood (K. Lynch, *The Image of the City* [Cambridge, MA: The MIT Press, 1960]).

The visual relationships, the way that places connect with each other, and the visual 'clues' about what a place is about, all give a user a feeling of confidence, and therefore competence, in their surroundings.

6. K. Lynch, *The Image of the City* (Cambridge, MA: The MIT Press, 1960).
7. See Ann Vernez Moudon, "Urban Morphology as an Emerging Interdisciplinary Field", *Urban Morphology* I, 1997, 3–10, for a discussion of the genealogy of urban morphological research and practice. In Canada to date, urban morphological theory has not been integrated into design and planning practice or education in a way that would allow it to inform urban design, and as Moudon points out, relative to the European situation, morphological study of the North American town or city is less related to issues of historicity than to issues of dysfunction. See C. A. Sharpe, "The Teaching of Urban Morphogenesis", *The Canadian Geographer 30*,1, 1986, 53–79, for a further discussion of morphological study in Canada.
8. Aldo Rossi (*The Architecture of the City* Cambridge, MA: The MIT Press, 1982) based his architectural studies on urban typology. He believed that the memory of the city is embodied in its architecture, whose types can either be "pathological" or "propelling" permanences. Architectural types must preserve their general forms, but also be open to new uses and new interpretations. Typological questions have always entered into the study and practice of architecture—elements of building structure, form, and organization figure into the comfort and appropriateness of any building. They are applied less often, if at all, in the study and practice of urban design, although typology applies to elements of a city as well as to buildings.
9. B.A. Sandalack and A. Nicolai, *The Calgary Project: Urban Form/Urban Life*, forthcoming.
10. The methodology for this research was first developed in a campus planning project (B.A. Sandalack, *Olds College Campus Plan*, Olds College Board of Governors, 1989) and two small town revitalization studies (A. Nicolai, "Town of Olds Period Restoration and Downtown Revitalization Study", MScArch Thesis, Centre for the Conservation of Historic Towns and Buildings, Katholieke Universiteit Leuven, 1990; and A. Nicolai, "Town of Didsbury Period Restoration and Downtown Revitalization Study", MEDes Thesis, Faculty of Environmental Design, University of Calgary, 1991; and then refined using a prairie railway town as a case study (B.A. Sandalack, "Continuity of History and Form: the Canadian Prairie Town," Ph.D. Thesis, Oxford Brookes University, 1998) and applied to several more complex projects including B.A. Sandalack and A. Nicolai, *Urban Structure—Halifax: An Urban Design Approach* (Halifax: Tuns Press, 1998) and B.A. Sandalack and A. Nicolai, *Cliff Bungalow—Mission: Townscape and Process* (The Faculty of Environmental Design, University of Calgary, 2001); and *The Cliff Bungalow* (Calgary: Mission Community Association, 2001).
11. Two authors in particular have considered the evolution of culture, planning and design, and the physical forms in general that have been produced as expressions of the values held at the time and describe them in similar terms. Edward Relph, *The Modern Urban Landscape* (London: Croom Helm, 1987) discussed three phases of urban development, and Anthony Vidler, "Architecture After History: Nostalgia and Modernity at the End of the Century," (London: RIBA Annual Discourse, 1996) described three paradigms of space, roughly corresponding to pre-World War 2 traditional development, post-World War 2 modernism, and post modernism.
12. Clarence Perry developed a Regional Plan for New York in the 1920s in which the "neighborhood unit" became the basic module. This was defined as the area served by the average elementary school, and was a form that

was intended to produce safe and healthy living environments, un-invaded by major traffic, but with ready access to thoroughfares and transit and good local shopping. In Calgary, the neighborhood unit became the model for much of the post-World War 2 residential development.
13. Where city planning was once an integrative discipline—involving planning and design skills, it lost some of those dimensions during the last half of the twentieth century. In planning education around the mid century, non-physical planning (social planning and policy planning) developed as a primary force, as more people entered the profession with backgrounds in the social sciences. Design skills in planners declined, although what emerged were several types of planners—social planners, policy planners, environmental planners, physical planners, transportation planners, and so on—all called "planners," and some kinds of necessary specializations were developed. This change has been discussed by J.M. Levy among others. J.M. Levy "What has happened to planning?" *APA Journal* (Winter 1992): 81–84.
14. The British North America Act of 1867 guarantees funding for both Protestant (public) and Catholic (separate) school boards.
15. The grid pattern dominated early North American city plans until Clarence Perry discredited its regularity, after which time street patterns, especially in suburbs, then became progressively more curvilinear and hierarchical. Recent approaches to urban design, notably New Urbanism, have rediscovered the grid pattern as a permeable, legible, and walkable framework.
16. A. Vidler, *Antoine Grumbach* (Editions du Centre Pompidou, 1996); E. Relph, *The Modern Urban Landscape* (Croom Helm, 1987).
17. Lake Bonavista was Calgary's first lake-oriented subdivision. It was constructed in the 1960s and is now a highly valued subdivision.

9 The Imagined Landscape
Language, Metaphor, and the Environmental Movement

Thomas Looker

EDITORS' INTRODUCTION

For many years, in a course entitled "The Imagined Landscape," Thomas Looker has encouraged undergraduates to become more self-conscious about their deepest assumptions concerning the nature of Nature. Examples from classic American literature demonstrate the way immensely attractive metaphors of balance, harmony, and order, "tamed" earlier, more ambivalent images of "the howling wilderness," the predatory frontier, and so forth. Today, however, a powerful dissonance has arisen between older imaginings of Nature and newer perspectives—arising out of contemporary science—that see the natural landscape as a space defined by probabilities rather than certainties, in which Nature often acts randomly and capriciously instead of harmoniously and predictably. Quite often, these "mixed metaphors" compete within the mind of the same individual, or within the programs of the same organization or institution.

How might we understand more clearly our unconscious assumptions, our deeply ingrained images, so that we may more freely re-imagine our relationship to Nature? One way is to spend time studying literature, which focuses our attention on language and metaphor, and helps us to become more comfortable with the paradoxes and ambiguities of imaginative understanding. The broader view might give us increased capacity to speak openly about ideology and nature. Every fall, Looker asks his class, "Are humans a part of nature or apart from nature?" After encountering the poetry of writers like Thoreau, students come to understand the richness of the only possible answer: "Yes."

Much has been written about the impact of literature on the psyche of society. Raymond Williams explored how English literature portrayed the city and country in the British context; Leo Marx reviewed influences of literature on current American ideals of the machine and the garden. Looker juxtaposes the writings of Henry David Thoreau and Daniel Botkin (the biologist and nature writer) critiquing the ways that ideas of nature and experience of the natural world are presented.

Looker argues that we are losing our ability to write and talk about the nuances and contradictions of our contemporary experience with nature, effectively blocking society's ability to communicate alternatives or talk about compromise—in the residential landscape, as elsewhere.

Looker writes of the imagined experience of landscape. He describes how the feeling of the sublime in nature is written in literature by Thoreau, Melville, Emerson, and others as a temporary feeling, yet this temporary feeling is what exurbia seeks to produce all the time. By fleshing out the way in which this search for the experience of nature translates into a cultural landscape of the pastoral, Looker helps explain the difficulty of embracing the idea of being in the modern and the pastoral at the same time, as if it were either/or, instead of but/and. Understanding how difficult it is to embrace the idea of but/and—the idea that our desire for all that nature represents does clash with the modern urban landscape, but that we are connected to both in any case—helps us to start to take action: we would not need exurbia if the experience of nature was widely felt in the city, or if the drain on resources of living in the country could be seen.

Thomas Looker was a writing teacher, visiting lecturer, and visiting scholar in American Studies at Amherst College for almost twenty years. He is the author of *The Sound and the Story: NPR and the Art of Radio* and creator and producer of the Peabody-award-winning public radio series, *New England Almanac: Portraits in Sound of New England Life and Landscape.* His chapter reflects on his years of teaching and writing in order to examine ways in which communication (thinking and writing) about environments reflects and influences our experience of and engagement with them. In order to appreciate its richness and potential, he suggests how we might embrace the metaphorical, provisional, and ambiguous nature of our relationship with the environment. Just because we wish for environments to be orderly and beautiful, Looker cautions, we should not suppress our knowledge that they are also filled with disturbance and uncertainty.

This chapter about environmental language provides crucial background to understanding the role of literary and intellectual history in our relationship with landscape. If the ideology of nature sustains exurbia, what sustains the ideology of nature? Arguing that the tyranny of instrumental language—the language of fundamentalists and proceduralists—deadens experience, quashes imagination, and oversimplifies life,[1] Looker explores ways in which people might stop themselves from mimicking sound bites of environmental platitudes rather than engaging in powerful and poetic language to explore and express everyday experience.

—Kirsten Valentine Cadieux and Laura Taylor

■ ■ ■

> [The laboring man] has not time to be anything but a machine. How can he remember well his ignorance—which his growth requires—who has so often to use his knowledge?
>
> —Henry Thoreau, *Walden*

> I make an extreme statement if so I may make an emphatic one.
>
> —Henry Thoreau, *Walking*

THE CRISIS OF WORDS

What might a writer (and a writing teacher), whose primary engagement is with language, metaphor, and the imagination, have to contribute to a discussion about Nature and Exurbia organized by geographers, who are concerned, I expect, with so many practical questions about how people choose to design landscapes and shape environments? Well, over the past twenty years, first as a reporter and then as a college teacher, I've listened to many people talking (and writing) about the landscapes in which they live or wished they lived. And I've noticed significant changes in the quality of that speech—its clarity, its subtlety, its ability to evoke the richness of human experience. Further, it seems to me inevitable that fuzziness of language goes hand in hand with fuzziness of thought and while there are always individual exceptions, on the whole, I find that students today reflect all too faithfully the wider society in which they've grown up: just as the level of our public discourse has atrophied—under such pressures as the compression of television news reports—so do student essays and class discussions reflect a decreased appreciation for the nuances and subtleties of language, a diminution in imaginative insight, and an increased rigidity and flatness in the ideas being expressed.

We're all familiar with the way that the short, choppy cadences of the mass media "sound bite" reduce complex ideas and events to catch-phrases ("the war on terror," "the environmental crisis") which are defined and understood only in relation to other catch-phrases ("homeland security," "saving the earth"). It seems to me that the inability of politicians (and my students) to use language with subtlety of meaning and suppleness of expression connects directly with the rise of simplistic, narrow-minded, black-and-white thinking throughout our society. The consequences of this devaluation of language strike me as dire: how can we create more wholesome places for us to inhabit on this planet if we talk about them with the nuance of tabloid newspaper headlines and the grace of telephone call-in programs?

To state my basic proposition in more positive terms: if we want to improve the way we interact with our landscapes, if we want to truly "save the earth" from misuse or overuse, I believe that we must first of all revitalize the language with which we think about, discuss, and, most importantly, *imagine* our hopes and aspirations. Of course we must continue to pursue a range of important scientific study and to develop social and political strategies; but these activities will do no good in the long term until we understand with greater depth and flexibility the complicated and often contradictory relationships that we human beings have with nature and with each other. If we continue to oversimplify our goals (and, as a consequence, to vilify with clichés and stereotypes those who disagree with us), all our attempts to preserve what we want to preserve and create what we want to create will result in continuing conflict, frustration, and failure. If we frame our arguments about how to "save the environment" using language and thoughts that are devoid of those very ambiguities that are at the heart of all human experience—and at the core of the natural world, as well—we will never be able to fully imagine, let alone tease into existence, environments that express both "humane" and "natural" values ... and that encourage the survival of both the human and non-human parts of our earth.

Here is the role I see for a writer and a teacher of writing as we try to address our environmentalist concerns: to comment on the *quality* of the language we use to think about Nature, and to provide examples from literature of writing that can help us become more self-conscious, more sensitive, and more creative in our understanding of the problems and choices we face. In what follows, I will present some readings from both contemporary and classic American writers to illustrate these possibilities. I will also suggest some new ways of listening to Henry Thoreau's unique voice that can be particularly useful in reawakening sensitivity to language.

Let me outline one broad critique of the current ecological debate to give an example of how we are affected by devalued and atrophied language. Too many environmental controversies evolve into pitched battles between people who for all their apparent differences share one important similarity: they think and speak as *fundamentalists* for whom compromise is anathema. Both sides assert positions of absolute certainty and conviction and usually base their ideas upon a host of unquestioned (and often unconscious) assumptions. On the one hand, we are told that nature is always good, civilization is almost always bad. We must save as much wilderness as we can and reduce or eliminate human influence on the natural world because if nature is left to its own devices, it will achieve some kind of ideal state. Even if nature is not exactly *perfect*, it certainly knows best; human activity only screws things up. The other side argues with a similar uncompromising self-righteousness. Tree-huggers and Bambi-lovers are sentimental fools. It is the human birthright to control and exploit nature's resources. If individuals are allowed to pursue their enlightened self-interest without interference from do-gooders and government regulations, human business and enterprise will thrive, and so will those parts of nature that we utilize. And in any event, human life is ultimately more important than a grove of trees. . . . And so forth.

Of course, most Americans are not fundamentalist ideologues. But the response to all this uncompromising certainty has not been the literate, classically liberal reasonableness that was the avowed goal of most public discourse up until thirty or forty years ago. Many of my students are shocked when I quote some of the most common ideas that I grew up with in the 1950's and 60's, such as the definition of politics as the "art of the possible."[2] Young people today seem bemused by the notion that compromise does not represent a *failure* of legislative deliberation but is the ultimate *goal* of a liberal democratic government, which assumes that no one person or interest will be entirely right all the time. "But I could be wrong, what do you think?" and "Come, let us reason together" are old pieties that have been drowned out in the cacophony of TV talking heads shouting opinions at one another. In the new environment, compromise is considered a temporary tactic at best, and, at worst, an evidence of failure. Faced with having to choose between diametrically opposed and mutually hostile sets of "agendas" (which are usually half-baked clichés or self-righteous prejudices that are passed off as "ideas"), the non-ideological public has retreated into cynicism. A pox on both their houses. Nothing anyone does can make any real difference. . . . The apathy and lack of involvement among so many citizens today I see as a direct consequence of the profound shift in our appreciation for the creative possibilities latent in an engagement with vibrant, meaningful language.

While many of the debates about the environment become unenlightening (and unenlightened) arguments between fundamentalists that most of the public either ignores or absorbs with little clarity of thought, there remains another small but important group of people whose role is to help draw up laws and enforce the regulations that control our interactions with nature. Many of these environmental managers try to pursue practical, non-ideological agenda, but their thinking and their actions are also decisively affected by the devaluation of our language. These *proceduralists* attempt to translate environmental dilemmas into sharply defined "problems" that can be "solved" by concrete actions. The language used by proceduralists derives from the imagery of machines and the industrial workplace. We can "fix" all our relationships to Nature if we can only get the various gears functioning properly.

From one perspective, fundamentalists and proceduralists couldn't be more different: the former are uncompromising in their ideologies, the latter continually try to negotiate practical solutions in specific contexts. But I see more similarities than differences between these groups, because both take what I'd call a *utilitarian* view of language and a mechanistic approach to metaphor. Nuance, subtlety, uncertainty . . . *mystery* . . . seem to play no part in the words used by fundamentalists and proceduralists when they think and write about the world around them. Both groups raise their voices to provide *answers*, to define clear possibilities, to *pin down* reality using slogans, jargon, and other verbal formulas.

What's wrong with this use of language? Why shouldn't people who know what they want try to express themselves strongly? ... Nothing—except if they are the *only* voices we hear and if their oversimplifying language effectively controls all environmental discussions.[3] While I don't deny that human beings must take steps to prevent harm to the natural world, we do no good to nature if our deliberations and our actions do harm to our *selves*—to our basic understanding of what it means to be human. And it's precisely here, in the definition of what it means to be an intelligent, thinking animal, that fundamentalist and proceduralist thinking can inflict the greatest damage ... turning us into, on the one hand, self-righteous believers who can only talk to people who share our own beliefs; and, on the other hand, molding us into narrow-minded technocrats who view the world and our behavior entirely within the confines of the reductionist, mechanical language of laws and bureaucratic memoranda (in which sentences sound like mathematical equations).[4]

I want to put more flesh on these generalities by considering some writers who have engaged with the problems posed by our relationship to nature. I'll begin with an example of a fine contemporary writer who initially speaks in a beautifully nuanced voice but then shifts his point of view to that of a fundamentalist. Eventually, I'll discuss some excerpts from Thoreau to illustrate the ways that language, artfully composed, can help us hold onto our paradoxical relationships and understandings ... to embrace the many sides of our nature and the nature that surrounds us. I should be sure my method is clear from the start: I plan to focus almost exclusively on *the words on the page* and discuss what I *hear* in the *author's voice* when I read his or her writing closely. I believe that reading is as much an art as is writing, and that when teachers insist that their students adopt a more scientific approach to "the text" (a phrase that once had a wonderfully evocative meaning but now has been reduced to jargon), they often do violence to the innate ability we all have to *listen responsively* to the words of other people. I do not expect that everyone will hear exactly what I hear when I read, nor would I want that to be the case. But if I can describe evocatively some of the ways I experience writers like Nelson and Thoreau, perhaps I will suggest possibilities that some readers may not have previously imagined. My wider purpose is to demonstrate some of the ways that a heightened sensitivity to language can help us reimagine "the environmental crisis." I will argue that our first task is to free ourselves from the tyranny of *instrumental language*, spoken by both fundamentalists and proceduralists—a language that deadens our experiences, quashes our imaginations, and oversimplifies our lives. To avoid becoming the automatons that Thoreau and many other writers have worried about, we must reacquaint ourselves with the power of *poetic language*. We must develop more fully our literate sensibilities and rediscover the rich potential of the human voice, both on the written page and in the world at large.

CERTAINTY AND UNCERTAINTY

In his fascinating book, *The Island Within*, published in 1989, a "lapsed" anthropologist turned environmentalist, Richard Nelson, describes his experiences deer hunting on an uninhabited island in the Pacific Northwest, near Kluksa Mountain.[5] He first traveled to this area of Canada in order to study the local Koyukon peoples, but after a few years of research, he decided to resettle his family and make his home in this fragment of sub-Arctic wilderness. As he describes in his first chapter, entitled "The Face in a Raindrop:"

> For me, the principles that guided Koyukon people in relation to their environment embodied a rich, important, and perhaps universal wisdom. Since Nita and I decided to make our home along this coast, I've sought to apply these principles to my own life. Now, to carry the process further, I've decided to spend as much time as possible on this island, to intensely explore and experience whatever is here, to learn without the formal intermediaries of professors or elders, to take a portion of our family's food from the land and sea, to seek guidance from both the Native American and Western scientific traditions, and to keep myself open to whatever might emerge in the process. I've undertaken a course of study, with the island as my teacher. (9)

In this passage, Nelson's voice sounds reminiscent of many other American writers who have *gone to the woods* in order to *learn what it had to teach*, as Henry Thoreau put it. Nelson is intrigued by the Koyukon; he is attracted by both their experiences of nature and their ideas. But in this part of his book, he portrays himself as a *student*—someone who is searching for answers but hasn't yet found them. The persuasive power of this passage, what allows a reader to enter into the writer's perspective, gets crystallized in a simple phrase that's propelled by a single word: "the principles . . . embodied a rich, important and *perhaps* universal wisdom. [My emphasis]" Without the "perhaps," Nelson would be asserting convictions which only readers who already agreed with him could accept.

Following this introduction, Nelson takes the reader along on several excursions through the landscape of this island and describes what he sees, sometimes in great detail. He asks many questions of himself and the place he's exploring and often compares his responses to those of his Koyukon informants. At one point, he sees two trees that have grown together through their long lives and now have been uprooted together by a recent storm. He muses on "the deeper dimensions of their partnership, whether each tree has a sense of the other" (13). Nelson turns to the Koyukon point of view for metaphors and insights and explains:

> In their world, trees are aware of whatever happens around them, and like all living things they participate in a constant interchange of power. . . . There is no emptiness in the forest, no unwatched solitude, no wilderness where a person moves outside moral judgement and law. (13)

Nelson finds this "populated" view of wild nature immensely appealing, even though he candidly admits that some of the attraction he feels proceeds not from idealism but from feelings some might called misanthropic. "[T]here *is* a kind of solitude here, one I can scarcely imagine living without: a temporary release from the scrutiny of other humans." He adds:

> Ever since childhood, I've loved to slip away into the woods or disappear in the nighttime shadows, safe from anyone else's eyes. Compared to the volatility of my own species, nature seems benign, predictable and reassuring. (14)

While in this passage I hear a subtle undercurrent of that familiar fundamentalist view that nature is good and humans are bad, Nelson manages to evoke a more complex and ambivalent point of view at this point in the book partly because he reveals an essential part of his personal history. His acknowledgement of personal quirks and dispositions adds specificity to his language, particularity to the images he evokes. He sounds less like an abstract moralizer and, again, more like a student who is acknowledging some of his own biases as he collects experiences on the island.

As the chapter continues, Nelson paints a nuanced portrait of nature on the island that includes chaotic storms and wanton destruction, along with beautiful sunsets and peaceful vistas of trees and muskeg. Throughout, he portrays his *in-between* state of mind. He can feel close to nature, he can feel distant from it. He always finds the Koyukon point of view immensely appealing, but he often acknowledges his strong ties to Western, scientific modes of understanding. At one point, while searching for signs of deer, Nelson notices a raven in circling the sky and as he watches it, he considers the Koyukon view that ravens are portents of good luck which can be used to help find elusive game animals.

> I look where the raven flies, where his beak points, for a clue . . . Should I go there or keep on in the direction I'd intended? Does the raven really care about such things, does he really know, does he move with the power Koyukon elders hold in such great regard?

And then, in a compromise that resonates with all the ambiguity and subtlety that surround western culture's attempts to figure out its relationship to "wild nature," Nelson arrives at the following conclusion:

> Accepting what the Koyukon teachers have said, I decide to take the sign of luck seriously and hunt with special care. But on one point *I will yield to my uncertainty*—I will choose my own direction. Instead of walking out toward the raven, I follow the muskeg's edge. [My emphasis.] (25)

It's a subtle, quiet moment, whose tension Nelson captures beautifully in the phrase, "I will yield to my uncertainty." The Koyukon view of nature promises a kind of community in which animals, plants, even inanimate objects are woven together in dense patterns and relationships that also can include humans. But Western culture—western science and civilization—adds to this view of coherence and interdependency a discordant element, what Thoreau calls the knowledge of our ignorance. *Yes, but . . .* We believe we have the power to understand the world around us, but our understanding grows out of our impulse continually to raise questions, to have doubts.

Nelson yearns for the certainty promised by the Koyukon view of nature, but at this critical moment, he decides to choose his own direction. Though some of Nelson's readers might not fully understand the attraction he feels for the Koyukon point of view, Nelson writes so vividly about his dilemma—the conflict between his western upbringing and an older, more traditional state of mind—that I think most of us can enter into his ambivalence and acknowledge that, in his own way, he is expressing contradictions we've all experienced, at some time or other, when encountering the natural world.

The tensions within Nelson come into particularly sharp focus during an emblematic incident towards the end of the chapter. While engaging in an early morning reverie and watching the arrival of a "still and brittle dawn," Nelson notices some herring gulls drifting "in easy, aimless circles." Then a bald eagle appears and "soars along the treetops." At first, Nelson sees "nothing ominous" about the bird . . . until a subtle shift in its behavior transforms the scene. Note how Nelson's use of precise, almost scientific language contributes to the rising tension:

> [T]he angle of its flight changes—it draws its wings inward, planes sharply down, gains a ponderous momentum, and slices straight in among the gulls.
>
> The eagle chooses a young bird with mist-colored body and mottled wings, sweeps up behind, and slowly closes to within three feet. Now separated from the other birds, they dart back and forth, up and down, the distance between them never changing, the pattern of their paired flight impeccably matched. (28)

The "deadly pursuit" continues, though it still reminds Nelson of a strange, "slow-motion ballet." As he watches, Nelson is aware that his sympathies instinctively lie with the "fleeing gull rather than the pursuing eagle" and he reflects upon this paradox:

> Both birds are predators, and a few minutes earlier it could have been the gull plucking a fish from the water or pecking a mussel to death on the beach. My reactions toward predators and prey are terrible inconsistent, especially considering that I am a predator myself. (29)

In his simple, patient language, Nelson is sketching some fundamental paradoxes. Sometimes we humans are predators, sometimes we are prey, sometimes we kill, sometimes we abhor killing—and sometimes we embrace two sets of attitudes simultaneously.

Nelson's story builds to a vivid and ambiguous climax that intensifies the experience beyond a simple tale of moral perplexity. The conclusion comes swiftly:

> Then, unexpectedly, the gull darts to one side and circles back . . . and the eagle flaps straight ahead, as if the chase had never happened. When the gull passes over me, I strain to see whether it looks terrified, desperate, exhausted, or relieved. But it seems just like any other seagull, with its expressionless beak and pale blinking eye. It only shakes itself in midair, as if its feathers were wet, then soars off into the scattering flock. The eagle sweeps to a nearby treetop, folds its wings, and stares implacably across the water. (29)

At this moment of high drama, Nelson looks into the face of nature and finds not beauty, not compassion, not horror—but only the blankness of an unfathomable *other*, "the expressionless beak and pale blinking eye" of the gull, the "implacable" stare of the eagle. These images of an impassive and uninvolved nature may strike us as a distinctly modern perception, though they occur at least as early as the mid-nineteenth century, in Edgar Allen Poe's *Descent into the Maelstrom*. (Here the narrator looks down into the center of the whirlpool's vortex, into the heart of nature, as it were, and instead of a blinding vision or a searing insight, he sees only a vague, impenetrable mistiness.)[6] But in Nelson's account, I want to emphasize his intense self-awareness of the unresolvable contradiction within which he finds himself: even as he has been trying to exercise careful, dispassionate observation, he hasn't been able to resist the impulse to project a meaning he can understand onto a natural phenomenon. He has not divorced his (objective) perception from his (subjective) imagination. But because he writes so clearly about these two points of view, readers can share both his confusion and his flickering insight. The apparent contradiction between different layers of understanding, different ways of seeing, supplies Nelson's experience, and his account of that experience (that is to say, *his language*), with its vitality and its evocative power.

Nelson has conjured up this fragile moment of perception quickly and with a few beautifully chosen words. Unfortunately, he doesn't stop here, and permit himself—and us—to dwell within the possibilities that his unresolved tension creates. Instead, he remodulates his voice and adds one more paragraph to his story, a passage which presents us with *a moral*. Denying the tensions and ambiguities he's just expressed, Nelson now selects one part of the complex experience he's described and tries to *pin down its meaning*.[7] Notice the subtle change in his tone during the following coda to his narrative; the equivocal observer is starting to abandon his in-between state of mind:

> I must be the only witness to this encounter who is so removed from the fundamental realities of life and death that he can indulge in pangs of conscience, questions of right and wrong, and the peculiar luxury of choosing sides between predator and prey. (29)

In the phrase, "so removed from the fundamental realities of life and death," I hear Nelson trying to pass judgment and impose a sense of order—and even a code of morality. *We know better* than to make choices between predator and prey, he asserts. Conscience in the woods is an *indulgence*, choosing sides is a *luxury*. Not content to remain ambivalent and complex in his relationship with nature, Nelson is now searching for certainty.

And, indeed, the succeeding chapters of *The Island Within* chronicle the author's steady pursuit of what he imagines to be a "better" relationship with the natural world: the passive, uncritical, non-modern form of humility practiced by the Koyukon. For a host of reasons that I think he never successfully articulates or evokes, Nelson views the Koyukon's approach to knowledge and understanding as not just one more interesting and perhaps valuable way to live, but as an *answer*, a solution, to the difficult "middle ground" within which Nelson has been struggling. In place of the dynamic uncertainty which characterizes his (all too brief) "studentship" in nature, Nelson starts forcing resolutions, based on his (unexamined) faith and the (mostly unexamined) inclinations of his own personality. He stops *ruminating* over the Koyukon vision of nature and, instead, asserts his determination to *embrace* it. In the following passage, towards the end of the book, Nelson speaks with the passion of the converted. To my ear, his words lack the evocative power found in his first chapter. Writing that explores and evokes has become writing that asserts and declaims:

> I seek a paradise beyond the one I've known in life. My reprieve is here, to become a part of the island—to linger in the muskeg ponds, flow in the blood of deer, blossom in the salmonberry's flowers, soak down in the sphagnum moss, cry out in the gull's face, whisper in the wind's

breath, grow upward in the hemlock's trunk, swim in the clear stream, shiver in the alder leaves. I have loved this island, willed my body and soul to become a part of it. In this way I would touch whatever is eternal and absolute. Watching a raven circle overhead, I know someday I'll soar above these shadowed forests and stare down into the throat of Kluska Mountain. I will see the image of earth fixed in the raven's gaze.

I ask no heaven but this Raven's world. (256)

While I cannot deny a certain facile lyricism in this jumble of images, I'd argue that if you do not already sympathize with the viewpoint Nelson is asserting, the *writing itself* conveys nothing but a series of fragmentary "jump cuts," lacking coherence and clear meaning. It also lacks personality. The distinctly recognizable voice of Richard Nelson (in his first chapter) has evaporated now into a cloud of generalized clichés that any number of clever wordsmiths could have put together. I find it deeply sad that the raven, which before had played such an eloquent role in Nelson's evocation of a uniquely personal encounter with nature, now has been so drained of nuance and flattened in tone that it has become a simple catch-phrase—a sound-bite, really—that Nelson uses to exhort us to a kind of New Age fundamentalism.[8] "I will yield to my uncertainty" has turned into "I ask no heaven but this Raven's world." That a writer of Nelson's evident skill and sensitivity can abandon one kind of speech for another, probably without realizing it, provides me with one more example of the atrophying of language that I see besetting our culture.

The change in Richard Nelson's prose during *The Island Within* provides us with an important lesson about how fundamentalist beliefs can undermine the power of language. When Nelson abandons "the middle" in which he embraced the tension between his western upbringing and his fascination with the Koyukon, his writing loses its creative energy. When he gives up "his uncertainty," Nelson leaves behind the self-consciousness that's available to him when he's caught between competing impulses and perceptions. In many ways, these later passages by Nelson remind me of many other voices I have encountered in all sides of the literature of environmentalism.[9] Fundamentalist attitudes by their very nature diminish our understanding of what it means to be human because, all too often, they turn a complex investigation of human non-human interaction into a simple morality play. What's out there is good, what's in here is bad (or vice versa). Nature has all the answers we need (or no answers); human knowledge and experience is forever flawed and dangerous (or the only understanding that matters).

BALANCE VS. CHAOS

It's particularly easy for environmentalists and their opponents in America to descend into uncritical fundamentalism because nature has played such an important role in the history, politics, and culture of the United States ever since the first settlers arrived in New England. The rich tapestry of inherited imagery about the natural world can become all too easily the basis for a wide range of assumptions that seem so self-evident that they're often not even acknowledged. Indeed, one of the primary motivations behind my course, *The Imagined Landscape,* was to help students become more self-conscious about their attitudes towards nature and about the hidden assumptions upon which they were based.

Though there has been a long struggle in American culture between images of conquering the wilderness and images of living peacefully with nature, a general consensus has evolved about the consummatory meaning of the natural world, its overarching importance as a source of value. The roots of these shared attitudes lie in a set of metaphors that, in essence, combine elements of Enlightenment thought (most enduringly expressed by Thomas Jefferson) with various Romantic ideals (proposed initially by Emerson and further elaborated and refined by Thoreau). From Jefferson comes the political view that the strength and originality of the new American democracy grew out of its close relationship with the land (farmers as God's chosen people).[10] Half a century later, Thoreau developed a range of images about the natural landscape that were grounded in a belief that nature could provide a source of inspiration and *re-creation* for individuals caught in the deadening web of an increasingly mechanized civilization (morning is continual opportunity to remake yourself).[11] Jeffersonian and Thoreauvian metaphors about the power of nature to inspire and to purify Americans have been reshaped into many different forms over the past hundred and fifty years, but what's most remarkable about this imagery has been its immutability. Indeed, the more complex and problematic our society has become, the more alluring these traditional assumptions about nature have seemed for many Americans. I view many parts of the environmental movement as but the latest (and, in a sense, the most desperate) attempt to find within nature a source of those perfect ideals which our imperfect human society so evidently lacks. It could be argued, for example, that the legal framework environmentalists have fought for and nurtured over the past thirty years to "protect nature" from humans represents the greatest flowering in American history of those ideals which enshrine nature not as an economic or social resource but as an essential repository of individual virtue and national morality.

What's so ironic about the current manifestation of essentially Enlightenment and Romantic longings is that they are taking shape within a cultural landscape that is otherwise dominated by modernist (and even post-modernist) language and metaphor. That is to say, much of the rhetoric of environmentalism resonates with images of nature as a positive, nurturing force which negligent human beings constantly disrupt but with which we *can* learn to coexist harmoniously, if we but exercise the proper humility. Meanwhile, much of the imagery coursing through the rest of our society is far more uncertain, contradictory, and sophisticated—both about the inherent benevolence of the natural world and about our ability to understand its complexities and modify our interdependent relationships. The implicit clash of imagery between the Romantic and the modern may be seen most clearly in the surprisingly contradictory contributions that many scientists have made to the environmental debate.

A large number of the laws passed in the early years of the environmental movement (such as those designed to protect wetlands or to curtail the expansion of "sprawl" from suburbia into exurbia) were based on scientific research which grew out of assumptions about the existence of a perfect and pristine "original state" of Nature that, *if left undisturbed*, would achieve an harmonious balance in which species would survive (or only slowly go extinct), air would remain clean, water stayed unpolluted, and (using one of the catchwords of our current culture) a "sustainable environment" would flourish. Even in those cases in which legislation allowed for a certain amount of control and management by human hands, the rhetoric surrounding the passage and implementation of the regulations was almost invariably dominated by the old images of Nature. Wetlands would be *preserved* or *restored*, open space would be *left untouched* by civilization's negative influence. And so forth.

Meanwhile, as the environmental movement has grown in strength and influence over the past thirty years, unexpected developments have been taking place in environmental science. Biologists, botanists, geologists, and others have been discovering that the old models of the natural world, based on nineteenth century paradigms, do not work properly. Theories which assumed nature to be a balanced system of relationships, operating like a complex machine, inevitably result in incorrect predictions. Increasingly sophisticated observations have shown that only by incorporating factors of chance and random disturbance into their models can scientists accurately anticipate events like the rise and fall in population of predator and prey, the complex stages of growth in a forest, or how wetlands work.[12] Cataclysmic geological and astronomical events, mass extinctions, the death of particular environments due to internal not external factors— all are now seen not as anomalies, but as central to how nature functions. In the laboratories and in the scholarly publications of environmental scientists, chaos theory and an understanding of probability have replaced the old images of balance and order.

And yet, up until very recently, whenever these same scientists spoke to the general public—during legislative hearings, at meetings of environmentalists, and the like—they altered their language, their *imagery*. Almost invariably, they dropped any references to natural world as a realm influenced by chance, sudden change, and random disturbance. Instead, they adopted the far more comforting and familiar images of Nature as a well-organized, mechanical system that had maintained a reasonable balance until humans started messing it up. A few years ago, my Amherst College colleague, sociologist Jan Dizard, began pointing out to groups of environmental scientists the discrepancy between the metaphors that underlay their laboratory work and those they used when speaking in public. He was usually greeted with both shock and recognition. The researchers had been completely unaware of the contradictions, but now confronted with the evidence of their own words, they acknowledged that they had often spoken with forked tongue.[13]

So modern society, generally, and the environmental movement in particular, finds itself in the midst of a profound imaginative contradiction, sometimes making nineteenth century assumptions about a balanced nature and at other times acknowledging modern views of uncertainty and disturbance. Biologist Daniel Botkin addresses this problem eloquently in his important book, *Discordant Harmonies*. In refreshingly straight-forward prose, Botkin argues that we all need to move away from our traditional metaphors about nature to take account of our new scientific understandings. He recognizes the difficulties that scientists and lay-people face when we abandon models based on machine-like orderliness and try to embrace images that include randomness and disruption.

> Although this nature of chance may seem less comforting than a clockwork world, it is the way that we find nature with our means of modern observation, and therefore it is the way that we must accept nature and approach the management of resources.... Once we accept the idea that we can deal with these complexities of nature, we begin to discover that the world of chance is not so bad, that it is interesting and even intriguing ... (130)

Unfortunately, Botkin's own proposed solution to the imaginative dilemma is to replace mechanical metaphors with those derived from *computers*, which, though they are essential tools for understanding chaos and probability, seem to operate in remarkably "mechanical" ways (they are based on binary thinking, after all—on-off, off-on, garbage in, garbage out). Botkin himself doesn't make clear how metaphors about nature based on a super-sophisticated machine are going to suggest any radically new relationships between humans and the natural world—and, indeed, in the final pages of his book, he falls back into the solidly mechanical image of a "space ship earth" in which engineers are reading dials and making adjustments.[14]

But Botkin is a biologist, not a poet, by which I mean that his expertise does not lie in the imaginative use of metaphors; so when he asks for new imagery about nature, it seems to me that he's essentially asking writers and artists to engage with the newest scientific insights and see where their imaginations take them.

The historian Donald Worster discusses with greater finesse the difficulties of such re-imaginings in his concise and provocative essay, "The Ecology of Order and Chaos":[15]

For centuries we have assumed that nature, despite a few appearances to the contrary, is a perfectly predictable system of linear, rational order. Give us an adequate number of facts, scientists have said, and we can describe that order in complete detail ... Now that traditional assumption may have broken down irretrievably ... For [John] Muir, the clear lesson of cosmic complexity was that humans ought to love and preserve nature just as it is. The lessons of the new ecology, in contrast, are not at all clear. Does it promote, in Ilya Prigogine and Isabelle Stenger's words, "a renewal of nature," a less hierarchical view of life, and a set of "new relations between man and nature and between man and man"? Or does it increase our alienation from the world, our withdrawal into post-modern doubt and self-consciousness? *What is there to love or preserve in a universe of chaos?* How are people supposed to behave in such a universe? [My emphasis.] (15–16)

While it's possible that a new generation of writers will help us towards the imaginative realignment that Botkin and Worster discuss,[16] I want to suggest another possibility. Perhaps what we need is not only new writers with new metaphors but fresh ways of reading the writers we already have. If those who are concerned about the environment listen again to the most complex and subtle of our literary voices, past and present, and pay more careful, self-conscious attention to the way these writers use language and create poetic meaning, we may find that we already have the imaginative resources we need to live more gracefully—perhaps, even, more "comfortably"—in a world of ambiguity and uncertainty.[17] It is in this spirit that I want to approach Henry David Thoreau.

Thoreau's art will not provide us with any quick and obvious answers to Worster's question, *What is there to love in a universe of chaos?* Indeed, in his major literary work, *Walden*, Thoreau does not present nature as a place of disorganization and disturbance.[18] But then, Thoreau was writing at a time when science expressed confidence that it could figure out the mysteries of the natural world and Thoreau's imagination engaged fully with those nineteenth-century philosophies that rested upon a firm faith in a beneficent God and a perfect Creation. Yet if the Thoreau of *Walden* seems untouched by the extreme doubts and confusions of the twentieth century, he nonetheless placed himself and his art in landscapes that contained considerable tension and contradiction. His imagination often stretched itself between conflicting facts or opposing conditions (the physical and the metaphysical, science and art, literalness and metaphor, the civilized and the wild). He would look at something as simple as the sunlight on the surface of the pond and create a vibrant web of observation and imagery that allowed him to move from motes of dust floating in the water to the spirit of God to his own squinting eyes. It seems to me that the difficulties we encounter today when we try to embrace creatively a "chaotic universe" are not all that different from the problems Thoreau wrestled with when he sat on his rowboat in the middle of Walden Pond and tried to imagine his relationship to the densely layered world around him. Thoreau's art does not record simple pastoral reveries, but instead evokes images of complexity and uncertainty not dissimilar from those presented by Botkin and Worster.

THOREAU'S ART OF NATURE

Henry Thoreau has become an iconic figure for generations of nature writers and environmentalists who proselytize one or another's forms of what has sometimes been called "Nature Religion." And when I first began reading Thoreau many years ago, I thought of him primarily as an unabashed Transcendentalist who assumed that the natural world could provide us with insights into the orderliness and purpose of a divine creation. Later, when I started teaching *The Imagined Landscape*, I assumed that I'd use *Walden* as an example of a nineteenth century Romantic view of the natural world that still resonated (often unconsciously) within my students' imaginations. I also figured that we'd spend the latter part of the course talking critically about Thoreau's metaphors and discussing the ways in which they were *not* helping us grapple with contemporary environmental problems. But a funny thing happened as I worked with my students to help them become more sensitive and self-conscious readers of Thoreau's prose. To my surprise, I found that instead of using *Walden* (and Thoreau's posthumously published essay, "Walking") as examples of old-fashioned assumptions about nature, I kept talking about the creative power of Thoreau's language, the ability of his words to suggest a complex set of responses to the landscapes he described. And when we moved to contemporary writers, many of whom might disagree with Thoreau and take altogether more "modern" views of the environment, I frequently found my students discussing how Thoreau would have added more nuance and subtlety and, ultimately, *insight* into the problem or the landscape being discussed—primarily because of his remarkable ability to *draw readers into his state of mind* so that we came to share, or almost share, his most supple and ambiguous perceptions and experiences.

Reading Thoreau, then, became one of the central and most enlightening events in *The Imagined Landscape* not because of *what* his writing told us about his view of nature, but because of *how* he expressed himself. We were reading *Walden* not for any answers it gave but because of the questions and challenges it posed to our imaginations.

I mention my pedagogical experience with Thoreau because, after twenty years, it has helped shape (perhaps decisively) a particular reading of his prose which some environmentalists might find quirky (though it's not particularly original within the literature of Thoreau criticism). This approach leads me to believe that the canonical Thoreau can help us develop new, more contemporary metaphors for nature if we spend less time focusing on the particularities of his Romantic assumptions and dig more deeply into the creative strength of his language. Put more simply, I now believe that we can read *Walden* not as an essay about nature, but as an extended investigation (an experiment, really) into the power of a poetic sensibility to shape and to illuminate our experiences in the world. In my view, it's Thoreau's *artistic imagination* and not his observation of nature, that ultimately controls and shape his passions, his intensity, his transcendental mediations. As important as the natural landscape is to Thoreau as a stimulus of inspiration, the *energy* of his creativity lies within his verbal imagination. Ultimately, *Walden* demonstrates not the power of nature, but the power of language.

I recognize that Transcendentalist philosophy would not admit to this distinction. For Emerson and others, the perfect word could become the perfect ideal in the same way as the "true" perception of *a* tree would lead an observer to "transcend" to an understanding of *the* (ideal) tree. Yet if we choose to read Thoreau solely as an exponent of a particular nineteenth century philosophy, I think we impose an unnecessary limitation on his genius as a writer. In a sense, I'm arguing that, to my ear, Thoreau's prose achieves its own "act of transcendence" by leaving behind any specifically Emersonian principles and presenting to readers a direct experience of poetry that takes place entirely—and astonishingly—within the *words on the page*. Or, at any rate, I believe that this approach to his work can be immensely valuable, especially at a time when an appreciation for the power of language seems to be so weak, even among some of our best college students.

Some examples from "Walking" and *Walden* should clarify what it is about Thoreau's poetry that I find so useful to us as we wrestle with our culture's imaginative crisis.

WILDNESS IN WORDS

"Walking"

As far as many environmentalists are concerned, Thoreau's most oft-quoted sentence certainly must be, "In Wildness is the preservation of the World."[19] Almost invariably, this phrase is snatched out of the context of the extremely complex essay, "Walking," and read literally.[20] The wildness of nature is good for us. Therefore, preserve the wilderness. Some commentators may add a bit of nuance to their reading by acknowledging that, of course, the "wilderness" through which Thoreau goes walking in this essay was never more than a few miles from the center of Concord. Clearly, Thoreau is not the same kind of "wilderness man" as John Muir, for example, who lived on his own for many years in the "true" wilderness of California's Sierra Mountains. But, to my mind, this acknowledgement still misses the most important point of

Thoreau's definition of "wildness." A few pages after he extols the Hottentots "who eagerly devour the marrow of the koodoo raw"(644), and following a section in which he praises New England farmers (and criticizes their Native American counterparts) for the success of the war they wage on the land, Thoreau abruptly shifts his concept of wildness into the realm of human *art.* "In literature it is only the wild that attracts us" he says, and then develops one of his most astonishing images, one that suggests the possibility of a union between an entirely human sphere (an *art*ificial creation) and the non-human world (a natural creation). He writes:

> A truly good book is something as natural, as unexpectedly and unaccountably fair and perfect, as a wild flower discovered on the prairies of the West or in the jungles of the East.

All right, you may say, this is merely a simile—a book is *like* a flower. The two realms remain separate. But Thoreau continues unremittingly to twirl nature and literature around each other:

> Where is the literature which would give expression to Nature? He would be a poet who could impress the winds and streams into his service, to speak for him . . . (650)

Well, okay, a nice bit of Romantic hyperbole there, even a little Wordsworthian. But wait! Thoreau will still not let go. He expands his imagery, building almost a new kind of landscape, a new kind of nature:

> He would be a poet who could impress the winds and streams into his service, to speak for him; who nailed words to their primitive senses, as farmers drive down stakes in the spring, which the frost has heaved . . .

Note the suggestion that words can have a *primitive* sense—a touch of wildness—and listen to the added specificity of that little phrase, "which the frost has heaved." Part of the power of Thoreau's art springs from the way he will move back and forth between the abstract and the concrete—throwing out an image that stretches our understanding and then following it with an utterly grounded picture. This ebb and flow of language seems to duplicate the ebb and flow of his own observations and imaginings, teasing the careful reader towards a similar kind of "vibration" between the *in here* and the *out there*. So in this sentence, Thoreau seems to be building not just a figurative landscape ("impressing" the winds and streams into his service) but simultaneously a physical one ("as farmers drive down stakes in the spring")—a landscape in which literature and nature flow seamlessly into each other. Thoreau doesn't stop here, but drives home the merging of art and nature in one of the most astonishing passages in *Walden*, a consummate example of Thoreau's passionate connection to poetry and language:

... who derived his words as often as he used them,—transplanted them to his page with earth adhering to their roots; whose words were so true and fresh and natural that they would appear to expand like the buds at the approach of spring, though they lay half-smothered between two musty leaves in a library,—ay, to bloom and bear fruit there, after their kind, annually, for the faithful reader, in sympathy with surrounding Nature. (650)

Words transplanted to the page with earth adhering to their roots ... Words that are true and fresh and natural ... These images bulge with potential fecundity and they suggest an imaginative merging between the external, non-human world—natural roots—and the internal, entirely human world of language.

In the intermingling of art and nature that Thoreau proposes, I hear expressed, with startling self-consciousness, an impulse similar to that which Richard Nelson dreams about with far less self-consciousness when he imagines being absorbed into the Koyukon world-view. As we've seen, at the beginning of his book Nelson acknowledges that he can never entirely share the experience of the Koyukon, but at the end of his adventure, he baldly asserts that his imagination *can* bridge the gap. His attempt to evoke his new state leaves me baffled and unconvinced. By contrast, when Thoreau develops a vivid and challenging imagery that conflates words and plants, and then uses this language to suggest strange connections between man's and nature's library ... his language ("ay, to bloom and bear fruit there") prods and cajoles my imagination to embrace the vision of a third kind of reality, somewhere between the entirely "natural" and the entirely "artificial." This middle landscape seems more accessible than Nelson's because though its origins lie in Thoreau's experience, the contours, content and boundaries of "the place" are defined through Thoreau's *art*—an experience which we readers are able to share.

I don't think we're ever going to see a physical landscape in which writers actually do grow words in their gardens and pluck them when they're ripe. But even as I say this, I have to admit that Thoreau's words keeping echoing in my mind, suggesting all manner of vague possibilities, and I dream of a deep forest where adjectives hang from trees and bulbous verbs grow large within the dark, fecund earth ...

TAKING THE MEASURE OF NATURE

Walden, "The Pond in Winter"

But we must be careful about equating the peaceful pastoral reveries that we've all experienced while wandering through a natural landscape with the states of mind and imagination that Thoreau evokes. For all its apparent promise of ease and beauty, the nature we find in *Walden* defines a place of considerable tension and irresolution. The apparently optimistic injunction that every morning represents an invitation to "recreate" ourselves resonates with *challenge* as much as with opportunity. Perpetual renewal implies never-achieving a final resolution—or even a resting place. Thoreau often masks ambiguity and paradox with wit or some breathtaking imagery, which so delights and fascinates us that we may not fully realize the profound contradiction within which we're being suspended. But sometimes Thoreau speaks his ambivalence directly and challenges us to follow a train of thought which, on the surface, seems to double-back on itself, even though, at a deeper level, its argument remains consistent and well-directed.

For example, consider the important passage at the conclusion of "The Pond in Winter," during which Thoreau *takes the measure* of the pond's depth. Thoreau starts off by heaping scorn on the legends about Walden's mythical depth believed by men who haven't "tak[en] the trouble to sound it." (269) Such beliefs are preposterous, he seems to say. He then describes, in precise detail, the measurements of the pond yielded up by his scientific methods. "The greatest depth was one hundred and two feet," he writes, adding with pedantic precision, "to which may be added the five feet which it has risen since, making one hundred and seven." Yet immediately after this fussiness, and with the quickness of a partridge taking flight, Thoreau's perspective shifts from science to poetry. He uses one transitional clause, which momentarily points in both directions: "This is a remarkable depth for so small an area . . ."—and note that *remarkable* is a borderland word, pertaining both to scientific observation and poetic sensibility (I remark . . . it's remarkable). After jumping the semi-colon ("yet not an inch. . ."), Thoreau has shifted his angle of observation entirely, from the dispassionate to the metaphorical . . . that is, from science to literature:

> This is a remarkable depth for so small an area; yet not an inch of it can be spared by the imagination. What if all ponds were shallow? Would it not react on the minds of men? I am thankful that this pond was made deep and pure for a symbol. While men believe in the infinite some ponds will be thought to be bottomless. (269–70)

(In passing, let me note the central importance of the *questions* Thoreau so often asks his readers—"Would it not react on the minds of men," rather than, "It would then react on the minds of men." Many sincere laborers in the vineyard of current literature about nature seem not to appreciate the power of indirection and uncertainty. Thoreau's phrasing *evokes* our agreement, or at least our understanding, of his point of view. If he asserted his conclusions more directly, he would oversimplify his ideas and weaken the power—the allure, I might say—of his argument.)

What are we to make of this final passage? How is it that Thoreau begins the measurement episode by criticizing the myth of bottomlessness, while at the end, he claims that the image of a bottomless pond provides an important stimulus to the imagination? The key here seems to be *how* observers arrive at their conclusions and at what stage in the process they connect their internal imaginings to the external facts of the pond's depth. Do we look at the *surface* of the pond and then project our picture of bottomlessness onto it? Or do we actively engage with *all* of Walden, what's visible and what's below the surface, and develop our understanding with the help of *all* the tools available to our intelligence? Scientific instruments supply us with additional information about the depth—so we should use them. The measuring line reports that the depth of the water is 102 . . . well, in fact, 107 feet. As far as science is concerned, the pond is not bottomless. But, of course, Thoreau doesn't stop there. He then applies *his imagination* to the 107 feet: "I am thankful that this pond was made deep and pure for a symbol." Thoreau has no problem embracing image of a bottomless pond *when that image is based on knowledge and not upon ignorance*. Indeed, I'd go even further and suggest that, for Thoreau, it's a crucial error to think that 107 feet defines the *entire* depth of the pond. Walden contains both a measurable *and* an un-measurable bottom. So does Thoreau embrace a view that places the reality of Walden Pond into a paradoxical middle-ground, where it exists both as a concrete scientific fact and as a limitless imaginative symbol—*and Thoreau privileges neither point of view*. Both modes of understanding are essential to his experience of the pond. Both forms of existence are "real."

LABORATORIES FOR ARTISTS

Walden, "*Spring*"

In the middle landscape that Thoreau shapes for *Walden*, poetic imagination and scientific observation intertwine in the same way that, Thoreau believes, human art should interweave itself with the natural world. Thoreau gives voice to this merging of apparent opposites most vividly and startlingly during his penultimate chapter, "Spring"—especially in the paragraphs that grow from his observation of a bank that has been cut through by the railroad (a "middle-landscape" if ever there was one).[21] In every chapter of *Walden*, Thoreau masterfully weaves back and forth between concrete fact and evocative metaphor. As in the famous section at the opening of "The Ponds" when he fishes on Walden at night and ends up casting his line into the sky as well as the water, Thoreau's consummate skill as a writer allows him to mix the concrete, the pictorial, with the abstract, the transcendental, in such a way that a reader can *almost* share the spiritual or metaphysical experience that Thoreau is evoking.[22]

In *Walden*'s early chapters, the shifts in perspective or mode of understanding occur over the course of pages (the way it does in his discussion of wildness in "Walking"). As the book progresses, Thoreau seems to change his point of view more rapidly and with less preamble. During the measuring of the pond, he speaks as a scientist and as a poet within the same paragraph, even though each voice remains in its own separate *sentences*. But when he encounters the railroad bank in "Spring," Thoreau starts oscillating more and more rapidly between science and art, observation and insight . . . and now the juxtapositions take place within the same sentence, the same phrase, the same metaphor—and even, at last, within the same word. Like a spinning disk whose opposing pictures appear to merge so long as you spin it fast enough, Thoreau's prose so intensifies that he creates astonishing images of integration that, in other contexts, would seem impossible.

Thoreau begins by describing the thawing of the railroad bank in some detail (286–87) and he conjures up an extraordinary sense of its fecundity: in this place grow both the earth's vegetable life and the poet's imaginative life. Through the concatenation of images that he perceives in the flowing sands, Thoreau suggests the vast, generative power of the natural world . . . a power not just limited to organic beings:

> Innumerable little streams overlap and interlace one with another, exhibiting a sort of hybrid product, which obeys half way the law of currents, and half way that of vegetation. As it flows it takes the forms of sappy leaves or vines, making heaps of pulpy sprays a foot or more in depth, and resembling, as you look down on them, the lacinated, lobed, and imbricated thalluses of some lichens; or you are reminded of coral, of leopard's paws or birds' feet, of brains or lungs or bowels, and excrements of all kinds. (286)

Thoreau's language brilliantly expresses his poetic observations. In a rush to convey the feeling of life springing out of the ground, the words tumble over each other, their very *sounds* suggesting the lithe bulkiness of pregnancy and possibility: "heaps of pulpy sprays," "the lacinated, lobed, and imbricated thalluses of some lichens."[23] Note once again how Thoreau imbues the analytic language of science—here, botany—with the evocative power of poetry by placing it in an unexpected context (in this case, a highly metaphoric reverie upon the thawing of a railroad bank).[24] And, in typical fashion, Thoreau follows a moment of metaphorical abstraction with a series of concrete images that accelerate into a startling incantation of animal parts and then climax in one of the most ecstatic defecations I've ever encountered on the printed page (". . .of brains or lungs or bowels, and excrements of all kinds.")

Thoreau's ability to ground his most abstruse ideas in concrete observation of the natural world offers a particularly important example for contemporary writers and environmentalists. If we consider *how* Thoreau goes about imagining his relationship to nature, we may begin to see a possible route towards developing those "new metaphors" which Botkin (explicitly) and Worster (implicitly) say that we need. For example, at the climax of this description of the bank, Thoreau encapsulates the method he's been using throughout *Walden* to create the unique middle ground from which he connects to the natural world; and he compresses the heart of this poetic vision into a deceptively simple phrase:

> When I see on the one side the inert bank,—for the sun acts on one side first,—and on the other this luxuriant foliage, the creation of an hour, I am affected as if in a peculiar sense I stood *in the laboratory of the Artist* who made the world and me,—had come to where he was still at work, sporting on this bank, and with excess of energy strewing his fresh plans about. [My emphasis.] (287)

As if . . . I stood in the laboratory of the Artist . . . Thoreau presents us with an arresting image that contains an idea which the contemporary world has forgotten, to the detriment of our culture and our relationship with the natural world. By imagining the maker of the world as an *artist* who works in a *laboratory*, Thoreau proposes a vital interdependence between poetry and science in which the two modes of understanding depend upon and enhance each other. If we are ever to develop a full understanding of the natural world, Thoreau suggests, we must enlarge our imaginations to encompass both the concrete specificity upon which scientific observations depends and the supple, creative ambiguities that inspire poetry. A union of these different perceptions is not only possible, it is necessary, according to Thoreau. And because in *Walden* Thoreau is performing as writer, his example suggests that one path to ultimate understanding leads through the power of beautifully composed language.

Where is this place where science and art (and spirit) meet? Does this "laboratory of the Artist" have a physical presence—is it somewhere we can visit? Given the overall framework in which Thoreau presents the image, I'm encouraged to supply a somewhat ambiguous (though, I hope, suggestive) answer: the laboratory of the Artist is as "real" a place as is the Walden Pond that Thoreau writes about. That is to say, though the laboratory may appear only on the page, as a literary metaphor, it is not "imaginary," a fantasy, in the way that Middle Earth or Yoknapatawpha County are fictional landscapes. The whole of *Walden* is uniquely grounded in Thoreau's direct experience of living at Walden Pond, and he is at great pains to connect all his metaphoric flights, all his transcendental ideas, to specific parts of a landscape that he has observed closely. Thoreau *sees* the laboratory of the Artist when he looks at the thawing railroad bank and the concreteness of his experience lends a peculiar weight to his evocation of the metaphor. To the extent that we can share Thoreau's momentary vision of the laboratory, we can also share, however briefly, an understanding of the possibility, the creativity, that the image inspires in Thoreau.

I'm being self-consciously slippery in my assertion that there's as much reality to Thoreau's metaphors as to his direct observations of Walden

Pond. But through my reading of Thoreau, I've become comfortable with a more porous sense of "the real" than we usually encounter in this hyper-visual age (when reality seems defined by the pictures we watch, "live," on the TV screen . . . or by the assertions we hear expressed in the glib, eviscerated language of our day). From my perspective, Thoreau's laboratory of the Artist can inspire us to imagine those new metaphors for understanding nature that Daniel Botkin urges us to find. The boldness of Thoreau's vision and the power of his art encourage us to broaden our sensibilities and embrace our contradictory experiences of the natural world—our wish that it be orderly and beautiful, our knowledge that it is filled with disturbance and uncertainty. When Richard Nelson looks at the raven and tries to both hold onto his scientific uncertainty and to believe that the bird carries a message for him . . . isn't Nelson, at that moment, standing on the threshold of Thoreau's laboratory, searching for some kind of resolution between his science and his faith, his observations and his literate imagination? I am disappointed when, in the end, Nelson retreats from the complex, intense landscape of paradox and possibility . . .turning instead to the simple, narrowing certainties that his New Age neo-primitivism asserts. But I am encouraged by observing that when Nelson is portraying his initial engagement with a challenging and ambiguous Nature, he writes prose that is supple, finely-nuanced, and filled with a wise self-consciousness that engages and persuades his readers.

Yet what about Donald Worster's more pointed question, "What is there to love in a universe of chaos?" Can a nineteenth century metaphor that unites science and art in a flash of poetry help us to coexist with a nature based on principles of disturbance and indeterminacy? Science in the twenty-first century seems dramatically different from the science Thoreau knew; surely connections are also broken to the thoughts and images which Thoreau derived from that science. Perhaps. But what if we could broaden Thoreau's metaphor of the artist's laboratory and imagine a more complicated "workspace," devoted to the propagation of all those strange anomalies (from black holes to quarks to gravitational waves) that so beguile contemporary scientists? Could we then also expand the image of a single "artist-maker" and create a metaphor flexible enough to sustain our multi-threaded contemporary longings, insights, and creativities? My own imagination falters at this point because I am talking *about* a poetic vision rather than pursuing directly a series of observations and reflections that might lead me to some imaginative break-through. But at this point, I only wish to suggest that the creative energy of Thoreau's language illustrates the possibility for a new imagining of nature and our relationship to it. The new metaphors I'm dimly perceiving will *not* succeed if they try to prescribe or to moralize . . . if they try to pin down everything with the false clarity of our devalued language. Perspectives from two other writers and a look at a final, surprisingly contemporary image from "Spring" will help amplify the difference between the voices that speak to us now from the television screen or the Internet and the kind of speech that engages more fully our *literate* imaginations.

THE BOOK OF THE EARTH

Walden, *"Spring"*

Leo Marx's classic, *The Machine in the Garden,* contains two quotations, from Robert Frost and Herman Melville, which I've long found vital to an appreciation of both Thoreau's art and our own problematic relationship to nature. The two writers address epigrammatically the essential fragility of all great insights and understandings. Frost describes the "emotional end-product of a poem" (in Marx's phrase) this way:

> It begins in delight and ends in wisdom. . .in a clarification of life—not necessarily a great clarification, such as sects or cults are founded on, but in *a momentary stay against confusion.* [My emphasis.] (Marx, 30)

And Melville, in a letter to Hawthorne in 1851, writes an amusing paragraph in which he ridicules "the all" feeling indulged in my romantics. ("Here is a fellow with a raging toothache. 'My dear boy,' Goethe says to him, '. . . you must live in the "all" and then you will be happy.' ") But then, in a fascinating PS, Melville reverses his tone and acknowledges the immense appeal of the transcendent feeling, though he then articulates the main problem with how people use it:

> N.B. This "all" feeling, though, there is some truth in. You must often have felt it, lying on the grass on a warm summer's day. Your legs seem to send out shoots into the earth. Your hair feels like leaves upon your head. This is the *all* feeling. But what plays the mischief with the truth is that *men will insist upon the universal application of a temporary feeling or opinion.* [My emphasis.] (Marx, 279)

"A momentary stay against confusion," and men insisting upon "the universal application of a temporary feeling or opinion." Frost and Melville are not discouraged by the temporariness or the impermanence of the beauty they see or the insights they have. Frost embraces fully the creative possibilities of a "momentary stay against confusion"—and notice that Frost's passage includes a dig at fundamentalism, which purports to provide "great clarifications" of life, instead of being content with the more modest claims on the truth offered by art and by poetry. It seems to me that the power of *Walden* grows out of Thoreau's ability to create intense *moments* of understanding without falling prey to the error identified by

Melville of trying to make permanent what is, by its very nature, transitory. As vividly as Thoreau may feel his insights while he fishes at night or observes the spring bank or stands within the laboratory of the Artist, he always expresses his transcendent experiences in flickerings of metaphor, tumblings of imagery . . . that invariably collapse back into the concrete solidity of earth-bound experience.

Thoreau is sometimes criticized by extreme environmentalists for not living a *truly* wild or natural life. After all, he "only" lived in his cabin for two years, and, in any case, the landscape around Concord was hardly virgin forest. I freely admit that Thoreau's relationship to the natural world is paradoxical, laden with tension and contradiction . . . but this is why I think he's such a perfect writer to help us find out what there is to love in "a universe of chaos." Time and again, Thoreau is able to suggest the possibility of creative interaction between the human and the non-human world—provided that such interaction is based on imagination and poetry . . . which also means *tentativeness* and *impermanence*. Thoreau defines himself as a *sojourner* in nature, but he also sojourns back in civilization. He inhabits a middle landscape in Concord as well as at Walden Pond, within which he thrives on continual inquiry, observation, and metaphor. His "native environment" mixes science and poetry, art and nature, thought and experience; his "ecology" is sustained by imagination and offers the possibility of insight and even transcendence . . . but such moments of understanding are never *permanent* or *easy*.

Half-way through "Spring," Thoreau presents a powerful image that, on the surface, may sound quite familiar to contemporary environmentalists . . . but I want to suggest that Thoreau's point of view contains far more complexity than is often acknowledged. Thoreau has been touring the environs of the Artist's laboratory, engaging in a continual flow of scientific observation and poetic invention. At the climax of his descriptions, Thoreau asserts that "There is nothing inorganic" and compares the "folaceous heaps" he sees on the railroad bank with the slag produced by human factories—"showing that nature is 'in full blast' within" (an extraordinary confluence of the mechanical and the natural, suggesting a whole new way of integrating the Machine into the Garden). Thoreau then expands the merging of inorganic with organic even further:

> The earth is not a mere fragment of dead history, stratum upon stratum like the leaves of a book, to be studied by geologists and antiquaries chiefly, but *living poetry*, like the leaves of a tree, which precede flower and fruit—not a fossil earth but *a living earth*; compared with whose great central life, all animal and vegetable life is merely parasitic. [My emphasis.] (289–90)

Here, at last, Thoreau heaps upon "the earth"—our natural environment—his highest accolade: it is not a fragment of dead history . . . unchangeable, rigid, fully defined and pinned down between the covers of a finished book. Rather nature is *living poetry*, a book still in the process of being written, whose forms and meanings are forever developing and changing, suggesting and regenerating.

Though Thoreau's recurrent optimism about the possibilities and potentials of his creativity may seem disconnected from the jaded and cynical age we live in, I think that the inherent flexibility and humility in Thoreau's evocations of "the poetic earth" can suggest ways of rethinking some of our current dilemmas. The power of Thoreau's prose doesn't stem from his certainty—his knowledge—but from his tensions and contradictions—his ignorance. The laboratory of the Artist bewilders as much as it enlightens. Can a poem be "as natural" as a leaf? What does it take for us to regard a book "as unexpectedly and accountably fair and perfect, as a wild flower discovered on the prairies of the West or in the jungles of the East." ("Walking," 649) And what about that mysterious process by which we pull words from our imagination and place them on the page: can it ever give us the same richly organic feeling as when we pull bulbs out from the ground, "with the soil still adhering to the roots"? If Thoreau's art allows us to intimate some of these challenging connections, then perhaps we may make similarly unlikely leaps of imagination, appropriate to our different circumstances. Infused as we are with our modern (or even post-modern) consciousness might we not be able to see *opportunities for creativity* in the disturbances and randomness of nature? Might we not transcend our uncertainties so as to experience excitement and awe, rather than fear and insecurity?

CONCLUSION: RE-IMAGINING OUR ENVIRONMENT

I would agree with those environmentalists who like to say that our fate is intimately bound up with the fate of our planet. But I do not think our salvation, our reclamation, or, indeed, our *re-creation*, depends upon the number of endangered species we save or how many acres in exurbia we fence off from direct human intervention. Before we take decisions about shaping or not shaping the landscapes around us, we need to develop a more supple understanding about *how* we arrive at these decisions. And most particularly, we must reacquaint ourselves with the capacity of language to evoke the density of our experience and to assist in the continual evolution of our thoughts. Writing *is not* a "means of communication," as it is so often taught in schools these days. The written word offers us a path towards a more sensitive understanding of ourselves and the world around us.

We might begin with some individual reflection. When we walk through our environments, are we able to engage with the complexities of how our inner consciousness perceives the physical world around us? Or do we project onto the world around us one-dimensional ideals about Nature and reject those elements of our personality and experience which add ambivalence and contradiction to our understanding? The great experimenter of transcendence, Henry Thoreau, had no problem embracing science and mystery simultaneously . . . though he also was quite happy to present the tensions in metaphors that ask more questions than they answer.

I do not know how to achieve the dynamic balance—perhaps the discordant harmony, in Botkin's phrase—between the need for us to make hard decisions and the importance of maintaining tension and irresolution in our lives. Increasingly our civilization has grown weary of uncertainty and this is surely one reason why we have seen such an explosion of fundamentalism of all kinds throughout the modern world. On the other side of the spectrum, you find people who seem to embrace ambiguity, but instead of accepting the tension that arises from the continuing attempt to move towards the truth—however momentary—they have opted for two other kinds of certainty . . . the slippery slope of relativism (everyone's experience is equally valid so there's no reason to probe too deeply) or the cold comfort of nihilism (everyone's experience is equally invalid). Relativists and the nihilists don't deny the paradoxes of being human as much as do fundamentalists. But in their protective cynicism, they externalize any contradictions they encounter and project their confusions onto other people. Those who choose to live in cities have their opinion. Those who want to live in the countryside have their opinion. They're never going to agree, so what's the point of trying to choose between them? I keep myself disengaged . . .

Relativists and nihilists oversimplify by denying the extent to which paradox lies at the heart of what it is to be human. They do not face the possibility that to fully understand our relationship to Nature, we must be more like Nelson in his first chapter and imagine ourselves to be both Koyukon and scientist, traditional and western, urban *and* exurban. Are we a part of nature or apart from Nature? *Yes*. We "are" both . . . or rather, we're in a dynamic relationship to the world around us and experience constantly shifting shades of meaning and feeling. Perhaps our goal should be to *live gracefully within our paradoxes*. Like the inside-outside space of the Japanese temple, like the idealized "middle landscape" between the city and the countryside, the moment we try to pin down the dividing line, we have surrendered the full beauty and delicacy and *meaning* of our humanity. I'm not making a plea for perpetual inaction and irresolution. I'm simply suggesting that before we make decisions, before we arrive at conclusions, we engage first of all with the full ambiguity of our condition on this planet so that when it comes time for us to act, we will do so with both humility and tentativeness . . . but *most* of all with as much literate *self-consciousness* as we can muster.

Walks in the woods, college lectures, political organizing are no longer sufficient in themselves to help us deal with our environmental crisis, or, indeed, with the wider imaginative and creative crisis of our culture. We will only start to make sensitive and sensible choices in all aspects of our lives if we develop within ourselves and between ourselves the kind of creative *language* that can allow us to savor the wealth of nuance that our sensibilities need if they, and we, are to thrive and prosper.

Certainly one important way to refine and develop our language is to read and try to write literature . . . and so experience words that evoke and do not pin down; images that challenge and enlarge our understanding; metaphors that hold our imaginations in creative tension and lead, occasionally, to momentary insights. Not only will such exercises enlarge our possibilities as we grapple with our many problems, it will also increase our ability to experience a kind of beauty and humanity which is all too often denied us in a culture that has become so cynical, so over-stimulated, and so preternaturally visual.

I've based many of my ideas and intuitions on a reading of Henry Thoreau, which I know is eccentric. I've ruminated on the imaginative power of his language, on its aesthetic appeal to our sensibilities. . . the way his words tease and inspire our own private imaginings. This approach encourages me to hear in the pages of *Walden* a much more subtle voice than others may hear—a voice that is not so much affirming a particular Transcendentalist method as it is engaging in a verbal experiment with some of those ideas. Thoreau's playfulness with language; his indulgence in metaphor to the point where his prose often seems to be led around by his images rather than his ideas; his many internal contradictions and his occasional lapses of logic—all bolster me in my view of him as an extraordinarily brilliant writer who is struggling with all the uncertainties and insecurities that humans feel when they confront nature. The way that Thoreau is able to use his art to carve out a few moments of insight and understanding gives me some comfort that we may yet be able to *imagine ourselves through* (I did not say "out of") the crises that face the contemporary world. The power of Thoreau's prose suggests to me that we may yet be able to re-imagine ourselves and our landscapes, provided we work to enlarge, combine, and make more supple the imagery of science, art, and literature. To paraphrase F. Scott Fitzgerald, our goal must be to nurture a newly-revived appreciation for language, nature, and ourselves, commensurate with our capacity for wonder.

NOTES

1. Looker has written his essay in the spirit of Thoreau's introduction to Walking. In his somewhat provocative words, "Believing passionately that contemporary culture has eviscerated the power of the written (and spoken) word, I've not tried to compose a measured statement of the problem but confess to have followed Thoreau's more polemical—and metaphorical—approach to making an argument. I know many teaching colleagues who completely disagree with me about changes I see in contemporary student writing. And, quite consciously, I've ignored many post-modern or deconstructionist views about literate speech which would argue that we've gained more than we've

lost by the break-down of 'linear writing.' I take a personal, highly-subjective position here in an attempt to provoke reflection and further discussion."
2. I blame these changes on a number of broad cultural developments, especially the rise of television and movies as the dominant mass media, the mindless over-reliance on the personal computer, and the decreased importance of book-reading. (See Sven Birkirts provocative essays collected in *The Guttenberg Elegies*.) More specifically, I also find considerable fault with the way many students are being taught to write. Increasingly, schools and teachers define writing as "a means of communication"—a "skill" or "tool" whose purpose is to "convey" ideas or opinions to other people. Students are supplied with formulas (also known as "building blocks") for the proper arrangement of sentences and the correct organization of paragraphs. Not surprisingly, most papers written in accordance with these instructions have all the flexibility and originality of an assembly line. (Among the most destructive of the "writing rules" they learn is the bizarre injunction to "Always Avoid The First Person" in an expository essay. Never say, "I think" or "I wonder" or "I understand." Is it any wonder that many students feel a strange and entirely inappropriate sense of distance from words they put down on the page?)

    Far too many students emerge from this high school pedagogy without ever having experienced writing as an intensely personal way to figure something out . . . or to explore an idea . . . or to develop a deeper understanding of their own thoughts. They don't envision language as a mediator between the subjective and the objective, the self and the other. They don't think of words as supple and mysterious symbols, whose kaleidoscopic meanings shift and change as they move back and forth between inner speech and external expression. Instead, far too many students—as, indeed, far too many people more generally—consider that "writing well" means using language to "give the right answers" in an assignment . . . which also means, most unfortunately, to *pin things down*, to eliminate ambivalence and uncertainty. Rather than using essays to *evoke* the density of experiences, or to negotiate one quandary or another with grace and critical self-awareness, many students write with the sole purpose of asserting points of view which, all too often, are as unambiguous and clear-cut as a mathematical equation. Polemics and propaganda have replaced reflection and opened-ended inquiry.
3. I believe this phrase originated with Otto Von Bismarck, though when I first heard the idea in the 1950s and 60s, it seemed part of the American political culture and was something that writers and politicians often used.
4. In part what I'm articulating here is the difference between a culture in which people get their news from a variety of sources and one in which television in general (and twenty-four-hour news coverage in particular) has become the dominant medium. When I first worked for NBC News in the 1960's reporters couldn't appear on TV unless they had had some newspaper experience; most of the television audience still read newspapers and magazines; and most people who worked in television news regarded their medium as a *supplement* to the written word. Indeed, the major broadcast talents (like Murrow, Severid, and Brinkley) openly expressed misgivings about a future in which Americans might *only* get their news from TV. It seems to me that their worst fears have now been realized, though as a former writer and producer for NPR and a long-time advocate of creative radio, I admit to a particularly jaundiced view of television.
5. I recently heard an interview on BBC radio with a student at a Christian college in Los Angeles who was asked how he could square his fundamentalist, proselytizing faith (which said, essentially, that he was right and everyone else was wrong) with the Constitutional right to freedom of religion. He said

he took this right very seriously and considered that it placed on him an obligation to spread the truth of his religion to others in society. Christians cannot be elitists, he said, we must share what we know with everyone . . . and he invoked the constitutional provision that is supposed to ensure equality among different religions as an injunction to him to convert as many people as possible to his particular faith.
6. Richard Nelson, *The Island Within* (San Francisco: North Point Press, 1989).
7. Poe describes what *surrounds* the maelstrom's bottom with prose of such intensity—both romantic hyperbole and scientific obsessiveness—that a reader can easily overlook the full horror of the ellipsis at the base of the whirlpool . . . what the narrator *doesn't* see. Leo Marx used to describe this passage as a powerful expression of "negative Romanticism," the flip side of that Romantic optimism which assumed that if we penetrated the veil of Nature, we would find the face of God. In the negative Romantic vision, the face of nature shows us only a blank stare:
>The rays of the moon seemed to search the very bottom of the profound gulf; but still I could make out nothing distinctly on account of a thick mist in which every thing there was enveloped, and over which there hung a magnificent rainbow, like that narrow and tottering bridge which Mussulmen say is the only pathway between Time and Eternity. This mist, or spray, was no doubt occasioned by the clashing of the great walls of the funnel, as they all met together at the bottom—but the yell that went up to the Heavens from out of that mist I dare not attempt to describe. (Edgar Allen Poe, *Descent into the Maelstrom*)
8. Some might prefer to say that Nelson "resolves" his tensions and uncertainties. But my point is that such an apparent resolution shatters the essential mystery of his experiences and *devalues* it—in part by using a far less richly evocative and a far more cliché-ridden language.
9. Some readers may find themselves responding positively to the string of images that Nelson winds into his smoothly-flowing prose, but I'd argue that to do so a reader must project too much of his or her *own* experiences into Nelson's words. I define a cliché as language which has no clearly recognizable individual behind it. It's a statement which anyone might say and which takes on a particular meaning only when the *reader* infuses the words with his or her own definitions and understanding. Of course, whenever we read, we are carrying on a dialogue between the external and the internal; we're mixing *another's* words (from the page) with *our own* (the thoughts in our head as we read). I don't mean to suggest that there's always a clear and obvious dividing line between prose that creates a distinctive voice and prose that doesn't speak with a clear personality because we have to fill in too many blanks for ourselves. Reading is as much an art as is writing and we are all going to hear the words on the page a little differently. Still, I'd like to suggest that one way to evaluate "successful" writing is to consider *how much* and *what kind of* work readers must do to follow a line of thought or to enter into a particular experience or feeling. . . . as well as what kind of understanding results from our efforts. And in Nelson's juxtaposition of fragments ("to linger in the muskeg ponds, flow in the blood of deer, blossom on the salmonberry's flowers") I find that I must do too much imaginative work before I can make any sense of the prose, which is to say I must project too much of my own personal understanding onto Nelson's words. As a result, I do not hear a clearly articulated sensibility arising from the page. For example, what connection does Nelson see between *flowing* in the blood of deer and *crying out* in a gull's face—so that he leaps from one to the other? The juxtaposition of the images strikes me as confused and inconsistent and the only way I can

imagine the state of mind that composed the language is to think of euphoric states I might have had when traveling in Africa years ago or when smoking marijuana or . . . but this is a long way from Nelson's sub-Arctic Island.

I don't mean to be stolidly literal in the face of prose which attempts to convey an ecstatic or transcendent state of mind. But even visionary poets like Whitman and Ginsberg ground their imaginative leaps within a strong enough verbal (and logical) framework that readers at least have a *chance* of chasing after them, of sharing their particular experience. In the flashing of Nelson's images, I'm afraid I see more MTV than *Leaves of Grass*. Indeed, the best way I can begin to figure out what he might be trying to evoke is if I think of the passage as a *film script* rather than as a passage of prose: If each comma represents a jump-cut, then perhaps the logic or meaning of the movement from picture to picture would assume a kind of coherence *when we saw it on the screen*. But in that case we would find ourselves in an entirely different medium with an entirely different set of possibilities and problems.

10. One of the practical problems of this writing style is that it gives up the possibility of persuading people who don't already share the writer's assumptions. To pick one example from a classic environmental text: in his essay "The Wilderness," Aldo Leopold argues that undeveloped land has great value for us because it allows some people to pursue "the primitive arts of wilderness travel, especially canoeing and packing." But when he tries to explain *why* such activities are valuable, Leopold retreats into language that doesn't even try to be persuasive to anyone who doesn't already share his point of view:

> I suppose some will wish to debate whether it is important to keep these primitive arts alive. I shall not debate it. Either you know it in your bones, or you are very, very old. ("Wilderness," in *Sand County Almanac*, Oxford, p. 193)

In the sixties, we used to say you were either on the bus or off the bus, and if you were off the bus, you couldn't understand what it meant to be on the bus.

11. See Jefferson, *Notes from Virginia*, and Marx's discussion in *The Machine and the Garden*, pp. 116–144.
12. See, for example, in *Walden*, the section of "Where I Lived and What I Lived For" that begins, "Every morning was a cheerful invitation to make my life of equal simplicity, and I may say innocence, with Nature herself." (pp. 84–86)
13. For a discussion of the failure of older models to predict accurately predator-prey relationships and the growth of forests, see Daniel Botkin, *Discordant Harmonies*, (New York: Oxford University Press, 1992), Chapter 3 ("Moose in the Wilderness: Stability and the Growth of Populations"), Chapter 4 ("Oaks in New Jersey: Machine-Age Forests"), and, especially, Chapter 8 ("The Forest in the Computer: New Metaphors for Nature").
14. More recently, especially ever since Daniel Botkin's *Discordant Harmonies* began being read widely, many scientists have attempted to change their public rhetoric. The problem is that old habits are extremely hard to break, especially when it comes to our deepest imagery about nature. Dizard notes that while most scientists no longer argue for a *balanced* view of nature, they still have not given up metaphors of "natural order"—and so they unwittingly have fallen into another erroneous theory from the nineteenth century, sometimes reproducing, almost exactly, old, discredited arguments in favor of the theory of Special Creation. (Conversations with Jan Dizard, Amherst College, April 2004.)
15. Botkin, *Discordant Harmonies*, 192.
16. Donald Worster, "The Ecology of Order and Chaos," *Environmental History Review* 14.102 (1990): 1-18.

17. Perhaps some already have and I simply have not yet heard their voices.
18. I sometimes think that for all the fascinating and often liberating insights it has gained, post-modern criticism has had one devastating impact on its practitioners and students: it has turned attention away from the deep magic that occurs when words on the page interact with the imagination of an individual reader. Too many students (and by now they've become faculty members, too) who have passed through courses in post-modern literary theory seem to have developed a strangely disembodied relationship to their *own* language, let alone the language they encounter on the printed page. So, for example, Thoreau may be removed from a syllabus because of *who* he was and *what* he represents . . . or if he is read, his poetry will be subsumed beneath veils of critical perspective *before* students are allowed to enjoy the alchemy of his language.
19. Thoreau presents a far more modern view of nature in some of his later writings, such as *Faith in a Seed*. Indeed, the view of nature he presents in his book, *In the Maine Woods*, differs significantly from that embodied in *Walden*, whose text he was still revising when he climbed Mount Manadnock and encountered the "cloud factories" that were inimical to human beings. In this discussion, though, I refer only to the way Thoreau wrote about the natural world in *Walden* and his posthumous essay, "Walking."
20. "Walking," in *Walden and Other Writings*, Modern Library, 1992 edition, p. 644.
21. And recall that the essay begins with Thoreau's sly assertion that he wishes to make "an extreme statement if so I may make an emphatic one." *Ibid.*, p. 627.
22. One of the most beautifully realized of Thoreau's "prose poems," the railroad bank passage deserves to be read and studied in full. See pp. 286–290.
23. I sometimes walk my students through the sentences of that famous passage and trace the sequence of concrete and abstract words and phrases. I'm convinced that the part of our subconscious minds which responds to words on the page picks up on these subtle variations, these slight shifts in clarity and vagueness; so it seems to me that in the composition and flow of his sentences themselves, Thoreau gives the reader an experience on the page that is akin to his own careful observation of the landscape and his simultaneous, imaginative response to it.
24. Webster's Third International Dictionary defines "imbricate" as: "to cause (as tiles or layers of tissue in closing a wound) to overlap." "Thallus" is "a plant body that is characteristic of the thallophytes, that does not grow from an apical point, shows no differentiation into distinct tissue systems (as vascular tissue) or members (as stems, leaves, or roots) or is composed of members resembling and performing many of the functions of but not homologous with those of the higher plants, and that may be simple or branched, may consist of filaments or plates of cells, and may vary widely in form and size from microscopic one-celled plants to complex foliated or arborescent forms (as in some of the larger marine algae)."
25. Later on in this chapter, Thoreau reverses the process and tries to analyze with scientific precision the sounds and the etymology of the words *lobe* and *leaf* . . . The result is the same, whether he is turning science into poetry or poetry into science—Thoreau uses his powers as a writer to conjure up a state of mind and imagination, a mode of perception, which penetrates the facts of the natural world and transcends those separated details into an intimation of a greater whole.

10 The Mortality of Trees in Exurbia's Pastoral Modernity

Challenging Conservation Practices to Move beyond Deferring Dialogue about the Meanings and Values of Environments

Kirsten Valentine Cadieux

EDITORS' INTRODUCTION

In this chapter, Valentine Cadieux explores the paradoxical nature of the modern pastoral, an idea drawn from Frankfurt school critical theory, architectural theory, and analysis of literature and literature of the environment. The "modern pastoral" refers to a technologically-enabled idealization of the natural environment that obscures the industrial processes that make the green sprawl landscape and its aesthetic appreciation possible. The ways that exurban residents modify landscapes to reflect their ideologies are a central theme of this book. Cadieux makes the idea of the modern pastoral accessible by taking a look at exurbanites' management of trees as an entry into understanding ideologies of nature and the ways these are made material.

The seemingly simple act of planting a tree in a yard—or just letting one grow—is a symbolic act that can be unpacked to understand more about the culture of nature. In this chapter, based on interviews with exurbanites in three sites (southern Ontario, New England, and the Canterbury region of Aotearoa New Zealand), Cadieux finds that the life and death of trees are caught up in notions of ideal home landscapes in profound ways. Exurbanites have a tree fetish—growing trees is a meaningful, even moral act, as is tree cutting—and Cadieux argues that the broader trends and conflicts of the dispersion of the modern urban landscape are particularly well represented by the idealization of trees in exurbia. After describing the meanings and symbolic functions of trees in the first half of this essay, in the second half she describes meanings and functions of *pastorals*, and explores how moving beyond a mostly aesthetic or symbolic relationship with urbanizing landscapes could help expand current conservation practices into more equitable and sustainable modes of environmental governance.

Valentine Cadieux is a researcher and lecturer in Geography and Sociology at the University of Minnesota. She is interested in the way that people respond to and shape their environments, and in how residential landscapes are used (or not used) to provide for the needs of everyday life. She is author of several essays on the meaning and production of nature and agriculture and ideological interactions with environments, and the editor, with Patrick Hurley, of a recent collection of papers on amenity migration and exurbia published as a special issue of *GeoJournal*, and with Mattias Qviström, of an issue of *Landscape Research* on combining landscape history and planning history in the study of peri-urban places and sprawl. This chapter comes out of several years of interviewing, living with, and visiting people who have chosen to live in natural settings and who have been willing to share their stories, and also builds on work with land use managers and representatives of natural resource agencies trying to figure out how people understand and make decisions about their everyday landscapes. She is a geographer in the traditions of critical and anti-colonial cultural geography, feminist political ecology, and public social science, working to better understand—and to build resilience and justice in—society-environment relationships. She uses collaborative ethnography to study societies and their environments through participating in public planning processes and facilitating dialogue about how people make sense of their social and environmental practices.

Cadieux argues that in exurbia, the pastoral symbolism of nature provides a foil for urban modernity, covering up the traits of industrial, late capitalist modernity that are difficult to think about, deal with, or acknowledge—even if they are implicit in the segregated, energy intensive form of exurbia. As discussed in the previous chapters, exurbia is a notorious site for clashes of environmental values. Attempts to collaboratively plan the amenity landscapes of green sprawl (often reforesting resource landscapes) often derail when economic or emotional values embedded in landscapes are difficult to address explicitly—and so are not discussed, but rather deferred, for example through conservation mechanisms such as restrictive zoning. Although not all conservation shares this common pitfall, a significant danger of the conservation paradigm (and particularly of private conservation) is the model it creates of privileged "green zones," exacerbating both segregation and also differential access to the range of values embedded in green infrastructure. The symbolic green landscapes where nature is protected and allowed to grow free from (certain kinds of) human intervention are justified in terms of common good, but are allocated to those who already have more access—and are made possible only through importing of lumber and other materials from other people's landscapes.

The intractability of this contradiction is at the heart of this book's analysis of green sprawl. The pastoral ideal of exurbia covers up and smooths over the parts of modern life that are aversive to think about (traffic, noise, factories, environmental destruction, smog, social inequality). Removed from view, the issues are less urgent and less immediately objectionable; with problematic issues held at bay by the veneer of a pleasant landscape, it is easier not to protest—or more possible-seeming to attempt to effect change solely by buying into green consumer lifestyles.[1] Drawing to a conclusion the conversation between the chapters in the book, this chapter discusses the exurban modern pastoral in terms of how "cultural landscape narratives push and pull people between escaping troubling environmental problems and engaging with them," and hopefully offers to the everyday experience of this tension some ways to start talking about the meanings and values involved in society-environment relationships in exurban landscapes.

—Kirsten Valentine Cadieux and Laura Taylor

■ ■ ■

I sat down to write this chapter after a weekend spent visiting a friend on his farm just northeast of Toronto. Visiting the farms and forests of friends (and strangers) is something I've taken to doing at any chance I get. A part of these visits is my ongoing effort to understand what it is that people who are escaping city—or suburban—life do with their escape. Living in the city myself, but having grown up in the urban-escape areas of the American Northeast, I enjoy these visits—as a retreat from what I perceive (for better or worse) as the geometric regularity and marketplace pace of the city, and as a return to a landscape of reforesting countryside familiar from my childhood. As I have learned to appreciate urban ecologies and landscapes, what these repeated visits have increasingly become for me is a compelling foray into the North American fascination with the naturalized pastoral landscape of exurbia.

As an idea of the good life tinged with images of scenic farms and greenery, the pastoral is a traditional way to project a desirable way of life—or, more accurately, a desirable *setting*—onto an existing lifestyle. As a way of imagining the setting to project a desirable lifestyle *achieved*, the pastoral is an environmental (and political) sleight of hand. I use this interpretation of the pastoral as a starting place for my exploration of the ideology behind the lifestyle associated with the landscape of exurbia, and the socio-environmental questions exurbanites seem to direct at their landscape. *Where is nature—or the countryside? How can I live there?* And then: *What should I do now that I'm here? How am I part of this environment? And what does being part of this environment make me?* Considering the pastoral as a potent motivator, I interrogate the landscape of exurbia, and the people who live there, to try to understand the relationship between the aspirations exurbia represents and the contradictions it embodies.

In the context of this book's discussion of exurbia and sprawl in terms of an ideology of nature, this chapter explores exurbia as a pastoral landscape, based on fifteen years of ethnographic work in a wide range of peri-urban and exurban landscapes. As a land use and as an expression of an ideology, exurbia has come to be understood as a place where people turn to a naturalized countryside in order to escape disamenities associated with urban residence. One could skeptically argue that the natural aesthetic of the exurban settlement form plays no more meaningful role than *escape* into exurban lifestyle and aesthetics as exurban havens spread with tremendous exuberance, converting vast tracts of erstwhile farmland and forest into a limbo of urbanization. While contributing to the considerable impacts of urban disinvestment, this new residential landscape neither takes on traditionally urban functions—aside from residence—nor does it retain its previous rural functional identity.

In contrast to understanding exurbia merely as escape, however, I also attempt to understand what many exurbanites have meant when they say that more intensive interaction with nature is the goal of exurbanization. The North American exurban pastoral landscape of houses set amid field and forest involves a search for nature, but also for a middle ground between the utopian haven that nature represents and the experience of globalizing urban modernity that seems to make some people feel havens necessary. This middle ground is where urban and rural models of environmental management encounter each other, and the landscape amalgam that often results involves conflicting signals of escape and engagement: many of the signs and signifiers of landscape function are about doing things with the land, such as raising livestock (perhaps horses, but also goats, chickens, etc.), managing forests or hobby farms (or just having a lot of land), or aesthetically invoking productive land uses via wagon wheels and other country trappings that have been made fashionable and commoditized. The landscape being reconfigured outside the suburban limits seems tremendously compelling for aspiring ruralites, in part because the promise of being able to *do* something with the environment—to manage a reforesting piece of land, for example, or to become involved in regional conservation efforts—stands in such contrast to the helplessness many people feel in relation to their environments and "the environment" writ large, and particularly to their perceptions of destructive environmental change.[2]

While exurbanites have been celebrated as champions of landscape conservation, the rationale for exurban conservation is often problematic. Environmentalist aesthetics may be used to provide legitimacy for conservation and to justify increased control of local environmental management regimes by exurbanites, often in competition with forestry or agricultural interests, or with tourism. However, the problem with the modern pastoral is that the projection of an imagined rural idyll onto an actual material place is not likely to resolve the problems of green sprawl. Without extraordinary socio-ecological resilience, most green sprawl destinations are likely to struggle with the plights of both rural economies (grappling with global rural restructuring in productive sectors of the economy) and urban economies (competing to be seen as prime investments and to encourage only development that will facilitate continued economic growth).[3]

Exurbanization has become a successful strategy for conserving pleasant environmental islands in an era of neoliberalizing land-use governance. Especially in metropolitan regions, exurban islands displace both resource-based land uses and land uses that are considered too urban, creating spaces where any productive dialogue between growth and conservation interests is unlikely to take place. I see this impasse as a common motive for deferring important discussions about how land uses and landscape management could better live up to people's environmental aspirations.

Deferring active management regimes that reconcile values related to resource production (and also, arguably, related to reflexive inhabitation as this has been described in the preceding chapters), exurban property accumulation and related preservation-based governance regimes often put off the development of compromise models of environmental governance in favor of preserving land unused for its aesthetic value. Despite legitimate justifications for environmental protection, exurbia's high social and environmental costs call exurban justifications into question and suggest the need for better land-use governance strategies where urban and rural land uses intersect. In addition to exacerbating sociospatial environmental justice problems, the reproduction of exurbia also poses a threat to the very environmental values expressed in the desire for something other than a metropolitan residential environment. Especially as homeowners, municipalities, and mortgage investors struggle to figure out the future of peri-urban housing political economies, it is worthwhile to understand the experiences and motives embedded in exurban landscapes.[4]

The residential call of the wild—the mythology of exurbia—beckons residents into the landscape that real estate billboards imply provides a new way of living (see Figures 10.1–10.4 below). What does exurbia offer? By many accounts, exurbia's form of urbanization is "ex" not merely because it is "out from," but also "extra": suburbia-plus, even more comfortable, with more lawn and trees, bigger houses, and the same goals of distinguished quality of life, magnified. Questions about the sustainability of exurbia have traditionally centered on whether the way of life that exurbia promises can live up to the larger-than-life goals it seems to invite of its residents—and at what cost.

How do cultural landscape narratives push and pull people between escaping troubling environmental problems and engaging with them? In this chapter, I explore the way that understanding exurban landscapes as natural undermines the possibility of *reflexively inhabiting* the landscape of exurbia. And, by extension, I examine the way that unselfconscious endorsement of the modern pastoral may inhibit reflexivity in any landscape. Interpreting the landscapes of exurbia and the narratives of its residents, I examine the role of the ideal of nature as it is manifested in the common exurban reforestation of agricultural and forestry land. What can we learn from this manifestation of a reforestation ideal? The impulse to relate to nature may entice exurbanites toward goals that turn out to be symbolic of other goals, goals that have been hidden in the complex processes involved in naturalizing a regenerated farm into an imagined old growth forest, or becoming alienated from the procurement of forest products to such a degree that the cutting of any tree is protested.

When people fight *against* goals they have identified with and taken on as their own, as in the widely supported fight against sprawl, despite the widely experienced desire to be surrounded with nature, the landscapes in question become contentious conundrums, whose contradictions make them difficult to engage. These landscapes often fail to meet the goals expressed for them on all sides. Contradictions involved in the negotiations over landscapes of sprawl and nature can reveal more explicitly some of the choices that may appear inevitable—or not available—in the processes of urbanization that construct exurbia and green sprawl landscapes. A good relationship with "nature," even in its symbolic form in green landscape ideals, often sounds a lot like an exploratory impulse toward figuring out what makes an environment a good home, and how to act to shape good home environments. But in being compressed into the shorthand of symbolic nature, this exploratory aspiration rarely achieves reflexive status, in which exurbanites and planners, for example, might consider more explicitly how their own tastes, practices, and politics shape their environmental management practices and ideals, and even their definitions of what nature itself is, or should be.

The consequences of this lack of reflexivity in the search for a home in nature can be seen in the tensions underlying one of the central conflicts in exurban land use: the widespread reforestation of exurbanized farm and forest lands.[5] This naturalizing landscape trajectory provides an opportunity to examine conflicted ideas about interacting with exurban nature: processes of urbanization and productive land uses (like the commercial agriculture and forestry that supply the material resources that support life in urban places) tend to coexist uneasily, as exurbanites justify conservation of forests and hostility toward economic forest uses for primarily aesthetic reasons. This complex conflict can be read in the ways that many peri-urban areas have been transformed over the past fifty years from active working landscapes to quietly reforesting exurbs. Exurbia very often combines the residential choice of a working landscape—homes are nestled in woodlots and ranches and reforesting fields and pastures of farms whose crops cannot compete with the profitability of residential land use—with the search for nature, for a landscape of urban escape, or at least solace and respite: nature as an insulating cushion from the humanized landscape of the city or suburb.

HOW VIEWING LANDSCAPES AS "PASTORAL" MAKES WORKING LAND AESTHETIC

Reading Trees to Interpret Landscapes

People leave records of their goals, struggles, and environmental interactions in the landscape; this aspect of landscape functions in a communicative way, and can be read and interpreted by others, albeit via their own lenses of experience. The American landscape has been slighted as a particularly incomprehensible—or even boring—text; curmudgeonly popular environmental writers such as James Kunstler and Peter Blake call it "placeless" and complain, in books such as *The Geography of Nowhere* and *God's Own Junkyard*, that suburban and peri-urban landscapes, particularly, have too much clutter, too much change and shifting investment, and too little commitment to make good "reading."[6] The meanings that individuals experience in, or attribute to, the environment are often symbolic, or metaphorical—some aspect of the environment triggers an association or stands in for something else.

The landscape as "text" provides multiple layers to read to explore what sort of stories get told about the way people live—and the way they want to live. Approaching landscapes as stories in this way also reveals different perspectives that are written into landscapes. This story-oriented metaphorical perspective also opens opportunities to decide—and change—how to tell and participate in environmental stories, rather than just be swept along by their suggestions. Understanding experience as narrative also makes the stories and metaphors of experience themselves richer, because it helps provide a framework for viewing human activities within the context of cultural and natural history—and for drawing on what other people have done facing similar opportunities and obstacles. Recognizing the ways human experience has been woven into the environment in this way may also help people see their lives as much a *part* of the natural environment as in opposition to it.

As a palpable part of lived experience and of landscapes that can be read, trees function symbolically in many ways. Trees provide a metonymy, an associative trope, of environmental relationships.[7] Although it's easy *not* to think about the everyday use of trees (or other material resources from the natural environment), people tend to *like* trees, and the green environment—they like having them around houses, especially (although, perhaps, not *too* close). Trees raise property values, raise civic loyalty, and even when they cause the need to rake, trees' color and interest are widely sought after and praised. Even where trees are not revered with Arbor Day fervor, they are valued for their fruits, nuts, and shelter from wind. Picking apples, collecting walnuts, climbing trees, and using the wood from a known tree in a building, piece of furniture, or even a fire can give satisfaction, as can appreciation of wood even from unknown or exotic trees. Trees are a central part of the cultural production of environments in many cultures, and exurbanites often represent both their cultural and environmental aspirations in arboreal terms.[8] Aside from the more obvious celebratory tree moments such as Arbor Days (or Earth Days, which many people celebrate with activities such as tree planting, always one of the easiest activities for which to elicit exurban volunteering[9]), residents of green sprawl landscapes tend to engage in a quieter and more constant forestation. This is especially true in the exurbs.

Like much of the U.S. Northeast and the eastern parts of Ontario, the farm where I spent this past weekend had been slowly reforesting around the marginal farming of the couple who grew market crops there for several decades before they retired. They held the encroachment of the forest onto their arable land at bay with crops and haying.[10] Although he continues to grow vegetables and hay on the land, in contrast to his predecessors the new owner has decided to help reforestation along, exemplifying a trend that explains at least some of the widespread reforestation in exurbia. Furthermore, having moved away from the city in midlife, my friend wants to skip the (too slow) process of natural succession that was beginning to creep over the outer pastures and to jump directly into having the sort of forest he has, by historical research, determined to have covered his hill pastures and swamps before they were cleared for agriculture. This is the forest nature *intended*, as he sees it. His farm reflects a common tendency in the exurban places I have studied to express an imagination of the future and an ideology of lifestyle in terms of trees, garden plants, and other materials of nature. This use of trees to explore the natural environment, and to shape it—or enable it to take its own shape, depending on the perspective you're reading from—illustrates a fascination and interaction with trees that flavors much exurban land use, especially in its forested manifestation. Like many exubanites, my friend has renamed the farm after a significant landscape feature; the "cedar grove" is both the symbolic and physical center of his reforestation project.

Reforestation is as defining a feature of many exurban properties as lawns, whether through intentional projects, benevolent neglect, or just capitulation to saplings. Anyone who follows controversies over expanding metropolitan areas (or who sees mainstream media) is familiar with the common use of tree imagery in sprawl and environmental management contexts. Dramatic images of trees are often used to convey significant meaning or purpose. In advertisements for everything from new suburbs to sports utility vehicles, again and again, the ideal life is represented as inhabiting nature—the place where trees are at home, but humans are only visiting (see also Chapter 6, Luka).

260 Kirsten Valentine Cadieux

Figure 10.1 Jefferson Forest's imagined aesthetic and entrance, with houses effaced by forest.

The Mortality of Trees in Exurbia's Pastoral Modernity   261

Amid the rising housing developments on the northern fringe of Toronto, "Jefferson Forest" advertises itself around its entrance on large billboards with a photograph of a family sitting in a sylvan living room with coffee table, couches, and lamps set comfortably yet incongruously in the depths of a forest—a forest much larger than any in sight and without the houses. "A very rare community nestled within forest, ravines & greenspace," it claims, tying ideas of home, family, and community to the instant achievement of a landscape that bears no mark of a residential presence, except for furniture, which appears to have sprung up magically from the carpet of leaves. This family grove is the symbolic core of the residential forest, providing, in the curtain of trees drawing around each house, the protection and solace of *home*.

All around the urban periphery, these real estate billboards advertise highly successful new residential developments, such as "Autumn Grove," with images of an imagined residential engagement with the environment. At the edge of the built-up area, where the contrast between dense urban living and the greener countryside cannot be missed, these signs tell a convincing

Figure 10.2 Jefferson Forest entrance, contrasting the advertised residential forest with the actual fringe of forest remnant and closely spaced houses. (The houses are probably spaced closely enough, with a thin enough layer soil replaced after site preparation, to preclude the regeneration of majestic pines throughout the residential site.)

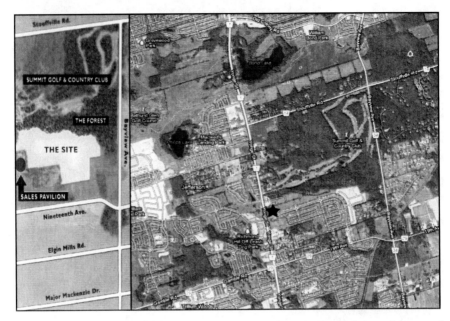

Figure 10.3 Developer's map of the Jefferson Forest site alongside aerial image of the Jefferson Forest Site (after completion of Phase 1). The Autumn Grove site is to the southwest, across Yonge Street. Google map © 2012 Google and Cnes/Spot Image, DigitalGlobe, First Base Solutions, GeoEye.

story of a wondrous natural landscape to be explored: "Explore a Magnificent New Landscape for Living!"; "Detached Homes in a Wondrous Natural Paradise"; "Up-close & Personal With Nature."[11] Depicting the geographical imagination projected over the reproduction of suburban landscapes, the billboards cannot be interpreted without the context of escape from the city, but they also present the landscape at the city's edge in terms of its potential for encounter with nature. This potential is often represented in their iconography and language by trees. Across the street from each other, Autumn Grove and Jefferson Forest are advertised almost entirely in terms of the encounter with nature and trees they offer. Jefferson Forest's descriptions of its constructed aspects are always accompanied by promises of everyday engagement with the forest, and with the ideals this forest represents: "Welcome home through a grand entrance flanked by majestic pines"; "Bordering on acres of established forest"; "Acres of Greenspace, just a short walk from every home." These signs may point to some of the cultural tropes of the modern pastoral, and the way they are reproduced through the imagery of trees, that help explain how many aspects of exurban forests (and the social relations they represent) are much more about the imaginary landscape ideals promoted to residents than about the trees themselves.

The Mortality of Trees in Exurbia's Pastoral Modernity   263

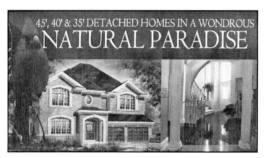

Figure 10.4 Autumn Grove, across Yonge Street from Jefferson Forest, represents itself using a similarly imaginary forest: although the deciduous species composition suggests something closer to the regenerating farmland this landscape might have become if left neither farmed nor developed for housing, the context seen in the "Up-close & Personal With Nature" sign shows that this existing landscape is not the forested one pictured.

The Mythology of Nature Symbolism: Imagining Wilderness and Immortality

> ... Big, craggy, thick-boled, they crowded the steep hillside like some dark enchantment, the sunlight filtering furtively through their crowns as if afraid to be caught touching the ground. Sleeping beauty trees, Rumplestiltskin trees.

... Forests are not ancient, they are new, every one of them, at every instant. Always being born; always dying; always *changing*. Ancient is for ruins and artifacts. What we have here is life.
—William Ashworth, *The Left Hand of Eden*[12]

Although they may appear in more prosaic form than the mythological dramatization of trees invoked in the above epigraph relating to struggles over forestry and forest protection during the spotted owl controversies of the 1990s, the problems of reconciling environmental imagination with experience are a recurrent theme when exurban residents discuss what they do with their yards and gardens. The journalist A.C. Spectorsky used the term "exurbanite" in 1955 to describe a phenomenon he saw as a "limited dream." Convinced that New York advertising executives were selling residential nature to themselves through their advertisements, Spectorsky illustrated his view of their unfounded expectations in the climax anecdote of his book, *The Exurbanites*. After planting hundreds of apple trees with his fleet of new machinery, but still not finding happiness or satisfaction on his new Connecticut exurban estate, a member of an applesauce-marketing team realizes that he has been duped by his own advertisement into a fruitless quest for meaning in nature—and a long commute.[13]

Long-term studies of classic exurban communities in scenic rural places support this view of the limited dream of transformative personal experience in unmediated natural settings as a false expectation, or an attempt to inhabit an imagined landscape. How is it that trees and nature so encourage—or mislead—the imagination? Talking to exurbanites who describe themselves as having slipped the clutches of urban modernity has given me an opportunity to consider what trees encourage in the exurban imagination—and how this imagination of trees leaves a gap between sought-after landscape experiences and the experiences exurbanites tend to have. In interviews with hundreds of residents of the exurban fringes of Toronto (Canada: 36 interviews, plus two long-term sites), New England (U.S.: 20 interviews, plus three long-term sites), and Christchurch (Aotearoa New Zealand: 46 interviews), I asked specifically about the details of what they do with their residential landscapes. Near Toronto, for example, well over half had planted trees extensively on their properties, almost all had planted or cultivated their trees in some way, and many bought properties specifically because they had already been planted with trees. This pattern was similar in New England, where several participants in the cases I researched had engaged in maintaining forests not only on their own properties but also on adjacent ones (through planting, conservation easements, and outright purchase of trees so that they would remain standing), and in Christchurch, where much afforestation was associated with native species restoration.[14]

People consistently described moving beyond the built up areas because it is more "natural" out there,[15] and described forestation as one of the central identifiers of the exurban landscape, fulfilling both the physical and imaginative expectations of connection with nature. In these areas I've studied, the forest defines the archetypal landscape difference between "exurbs" and "suburbs." Even where housing densities may be comparable, the exurban landscape aesthetic includes forest encouraged to grow up between the houses to provide the setting for the imagined good life. These exurbs fan out beyond the suburban landscape, where houses sit in a matrix of lawns, to a more closed landscape, in which houses and their lawns are islands in a matrix of trees. This new succession of the residential forest has to do with the symbolic as much as the physical characteristics of trees.

Large and solid, trees embody a tangible and tractable nature that is, at the same time, gratifyingly beyond obvious human control. Trees suggest the potential presence of nature anywhere—dig a hole with dirt in it, and you can have a tree. Trees live for a long time; they screen views over which people have no other control. Trees create a space around them, even when they are trimmed and constrained, that reads palpably as a place of nature. And enough trees together read as a significantly natural place; trees invoke a genius of a place that feels as if it transcends human making. This transcendent quality of nature is used, however, to build a relationship with the natural environment that reflects pleasing aspects of environmental experience and evades aversive aspects. Such uses construct an image of nature that reflects wishful fantasy.

The goals inscribed into this forested residential landscape of exurbia, the meanings woven between houses into the matrix of trees and gardens, show an ideal of nature informing the imagination from which the exurban landscape has been brought forth. Shaping a place as much of the mind as anywhere else, exurbanites use trees to make a place where expectations of connection to nature are translated into the landscape. Despite its potential for engagement in local and more abstract environmental issues, the ideal of redemptive reforesting nature into which residents can plunge their efforts—and from which they might extract modest realizations of utopian dreams—tends toward a view of nature that is stable, static, and although exploratory, also escapist.

It is common for residents to launch into a description of their landscape management with the assurance that they would not cut trees down, unless under dire duress. Many residents don't wish to balk the values they associate with exurban living that are represented by the stewardship of trees—or the sense that non-interference with trees means good environmental behavior. Around Christchurch, where landscapes have been spectacularly deforested for agricultural uses, exurbanites often equated the regeneration of native or imported forests as redemptive acts making up for the errors of colonization and ecological imperialism. In the Toronto exurbs, residents discussed their efforts at saving, preserving, and not cutting down even sick trees or trees that were in the way of other goals for their properties, such as gardens, play spaces, or house additions. This restraint vis-à-vis embodied nature testifies strikingly to the symbolic power these trees possess, especially as the forests in which these houses were situated were young, and in many cases, had been reforesting only since the houses had been built on agricultural lands well into the post-war era.

Through my research in exurbia, I have seen how ordinary interactions with trees can help shed light on the challenges involved in understanding the environment and interacting with it. Trees provide a clear example of expectations of the natural environment—and the failure of those expectations—in all their utopian glory and also in all their mortal failure. Despite the fact that residents don't want to cut the trees down, many feel somewhat chagrined about the way the trees have turned out: the house is too dark; the lawn and garden aren't growing under the trees; the driveway is becoming impassable. Even residents with reforestation projects, however, despite being impressed by the rapidity of growth and change they observed, often cling securely to an image of the final product—the future climax forest toward which they're working—rather than thinking of their forest in terms of process. This climax forest reflects the kind of grand immortal ideal of nature I've described. Inspired by the rugged landscape of Ontario's cottage country, many residents have planted conifers to replace their southern Ontario deciduous forests—and especially to replace its weedy and short-lived species, which don't fit the forest image these residents would like. These struggles to negotiate how the yard amidst nature should look and function emphasize the difficulties faced by residents in figuring out what they would like to do with their environments. In these yards, which epitomize the exurban dream, trees are both the most volumetrically and aesthetically abundant representative of nature and also a salient moral force for residents. Exploring what the trees mean to exurban residents helps in understanding how interactions with trees represent, in some ways, broader environmental aspirations.

Trees not only grow and add green to everything, sucking up carbon dioxide, relieving suburban ennui, and materializing imaginations of the ideal life (and the past), but they also die, not only if humans cut them down to use them or pollute them to death, but just on their own, as an attribute of being a living thing. Trees both exhibit and symbolize the larger dynamic nature of which they are a part. By showing the effects of climate, pollution, soil, insect, and water regimes, trees bring attention to socio-environmental dynamics. And when people become involved in managing or championing trees, trees provide a set of tangible icons and issues through which to investigate socio-environmental dynamics. Trees provide something to champion, and a chance to reflect on how and why such championing takes place.

Stories of why and how people revere trees, but also find them difficult, show the combination of a fascination with but also an uncritical glorification of an ideal of nature—especially of a nature that doesn't change and doesn't have problems. When exurban residents describe their concerns about the environment, and also their aspirations for their own residential environments, many tend to attribute most good things to "nature" and bad things to "people." Their land-use activities and descriptions follow these characterizations. Many residents overlook the human aspects of cultural reforestation, stewardship efforts, and the human changes to their properties, attributing them instead to natural agency: what nature is meant to do. Despite this automatic respect for natural agency and active interest in native vegetation and natural regeneration, many of these same residents vigorously oppose manifestations of natural succession such as the appearance of poplars, which, as species of early succession, are "weedy" and unpleasantly short-lived. Weediness and short lifespan in a tree seem to chafe at a deeply held attitude about nature, which I have come to think of through these interviews as a sense not only of nature's independence from people, but of its enduring immortality, or at least perfection—a conception of nature that fits in very well with a sense of "wild" nature, or wilderness, a sense of nature as being greater than, older than, and, in fact, better off without, people.

The Amenitization of Nature

The potent conflict created by searching for nature in nominally working landscapes is hidden quite well in the landscape aesthetic of exurbia, in which landscape is often valued for its primarily visual and aesthetic "amenity" value above other values. This valuation contributes to reshaping exurban landscapes around their potential amenity functions, "amenitizing" the landscape. By choosing to move to farmland (or another "resource" landscape), and then by reforesting it, exurbanites demonstrate both an interest in inhabiting the working landscape and also a sense of discomfort with what the working landscape represents vis-à-vis the relationship between ideals of nature and human agency in the environment. Reforestation—and relating to trees more generally—is clearly considered to have many benefits. However, by allowing reforestation on the farms, woodlots, ranches, and other working landscapes where exurbia has largely been built, exurbanites are acting out a conflict between, on one hand, an aestheticized interest in the kind of landscapes and landscape processes that provide the material resources on which daily life is based and, on the other hand, dismissive negative evaluations of the way such resources are procured that may be largely implicit.

Imagery of the exurbanizing rural landscape draws heavily on ideals of wilderness and on the imagined social and aesthetic world of the traditional productive landscapes that exurbia is replacing—forest and farmland. Because nature is thought of as wilderness, and because traditional rural practices have been naturalized and made aesthetic, however, exurbia ends up consuming this aesthetic of forest and farmland, rather than re-producing it: conspicuous wooden furniture and wooden architecture, grand lawns, and graceful hedgerows and woodlots replace the working landscape with lush crops of houses, nestled in amongst the nature—manifested and ornamented by trees, and flavored by wilderness.

Dominant cultural narratives of nature, such as those associated with the amenity values of the modern pastoral, appear to lead to modes of interaction with the natural environment that inhibit thorough examination of goals symbolized by nature, and by trees. In discussions of exurban land use, greenspace, and especially exurban forests, nature seems to be uncritically understood in such a way that its status is almost automatically naturalized and privileged as the way things *should* be. It is in this kind of unreflexive normative functioning of trees that the modern pastoral can be seen to work as a mode of the ideology of nature: approval of aestheticized forest landscapes as "green" and "environmentally good" smooths over potentially uncomfortable dissonances or contradictions in the everyday experiences and impacts of green sprawl.[16] Exurban treatments of trees demonstrate some of the key difficulties that define the struggle with the modern pastoral. If residents of exurban forests become alienated from the processes of forest use and forest growth, and turn away from the reality of the forest and society around them, trees end up being fetishized into a magical manifestation of nature, subject neither to processes of human culture nor death. The pastoral blurs important differences between the ways that nature is perceived and discussed and the ways in which the natural environment is used and interacted with. Refocusing on these differences—and grappling explicitly with obstacles to seeing them plainly—may provide insight into the difficulties inherent in inhabiting landscapes that are understood to be natural.

The importance of understanding the process of imbuing nature with moral imperatives becomes particularly clear in the way that forests, reforestation, and the regeneration of native ecosystems are used to naturalize assumptions about the way landscapes should be—as well as to protect landscapes (and their protectors) from changes or manifestations of social relations that are uncomfortable. Trees mark landscapes as significant, and exurban consternation in the face of both urban and rural land use processes often manifests as unwillingness to kill trees: either to make way for new housing or urban infrastructure or to manage forests for timber or habitat. Exurban trees provide an example of a way in which the ideology of nature pits environmental goals (usually framed in terms of sustainability in ways that at least imply inter- and intragenerational environmental justice) against everyday behaviors (that often call such terms into question). In fact, *most* of the broader trends and conflicts concentrated in exurbia are well represented by the idealization of trees. This idealization has permeated visual culture and environmental ethics, and was evident in my research (as well as in the literature on exurbia) in the intense growth in the popularity and amount of forested exurbia and its representation as a desirable residential landscape.

The ways trees function symbolically in exurbia opens a window onto encounter with the natural environment, and may provide insight into the socio-environmental goals that exurbia represents and the behaviors it facilitates. Perspectives on exurbia are various and struggle over the interpretation of what any landscape represents or how it functions is constant. But especially in its representation in the inviting landscapes around North American cities, the natural exurban landscape of solace and escape functions as a modern pastoral landscape, presented as an ideal toward which residents (or at least consumers of housing) are supposed to orient their aspirations. Understanding the aspirations embedded in exurban landscapes may help explain why impulses toward exploring and engaging with the myriad problems of shaping residential landscapes so often end, even unwittingly, in escape. This perspective on environmental goals and behaviors—and obstacles to these goals and behaviors—might assist in finding new and alternative ways to situate the search for nature within present environmental realities, and to turn interest in exploring the potentials of the environment toward smaller, everyday, practical goals having to do with making residential landscapes inhabitable.

Forests As Desirable Habitat

Goals for residential settings are partly defined by ideals of habitation, the everyday process of living in landscapes and of learning to shape the environment and everyday life to suit each other.[17] Aspirations for habitation include specific goals for what might be done with home and property, but also include less material aspirations, goals that may be symbolic or aesthetic, or even just vague. In the context of sustainability-inspired improvement efforts, however, there's an important difference between improving the ways people inhabit environments and aspiring to inhabit better environments. If a quest for a better life is achieved by setting up an utopian haven somewhere out *there*, and escaping the everyday world *here*—by buying into the conviction that life out in the idyllic countryside might be inherently (and effortlessly) better—a lot of the energy that might have gone into making already inhabited environments better gets shifted into finding out that new environments have many of the same problems as old ones, usually brought by the very process of escaping. This is a central paradox of exurbia: how can the cyclic reproduction of the problems of sprawl be addressed?

Many people have studied the problems caused by escaping home environments that don't live up to an ideal, and over the past fifty years a central suite of problems has come to be recognized—under names such as "urban flight," the "last settler syndrome," and "rurbanization."[18] As one hopefully ideal landscape after another is transformed and then escaped, cities, and then suburbs (and then outer suburbs), lose their desirability as a place to live. Scholarship on the "creative destruction" of exurbanization (and gentrification) suggests both that ideal landscapes, no matter how scenic, are never as ideal as they might be, and also that idealizing landscapes has significant socio-environmental costs. In the places escaped from and also those escaped to, residents may feel themselves alienated from their environmental ideals, at least partly because nature ideals often fail to include other humans. The whole process of idealizing nature appears to make it considerably more difficult to engage with actually lived-in material environments; it is difficult to read the history of exurban land use planning and not wonder about better ways to structure urbanization to help people get a grip on better aligning their ideal and material environments. Despite consensus about the desirability of containing the impact of green sprawl—or at least reducing the prevalence of exurbanites achieving private gain at public expense via environmentalist rhetoric without demonstrated benefit—methods for such planning have been scarce. Repeated defeats of various planning measures suggest the usefulness of building political will and socio-environmental literacy to put pressure on self-deceptive, escapist, and segregating manifestations of green sprawl.

Exurbia does represent, in part, a simple escape from cities to the most available "natural" landscapes. However, much more interestingly, it also represents a more complex escape from uneasy relationship with the larger environment. The impulse that drives so many exurbanites to a landscape that represents (or aesthetically demonstrates) the way resources are procured may be escapist, although it may also contain *potentially* more critical and reflexive aspirations for socio-environmental engagement. However well intentioned this aspiration, though, American and Canadian attempts to inhabit exurbia tend to falter, and instead degenerate to a commodified form of escape that resolutely sees the forests of humanized landscapes as *not* demonstrating human intervention. Intensive consultation with exurban residents around Toronto, in New England, and in Christchurch has led me to argue that some of the most troubling stumbling blocks obstructing the aspirations of exurbanites involve conflicts between cultural understandings of nature and of resource production.[19]

Exurbia provides an opportunity to learn quite a lot from the way nature and the countryside are idealized and used to define and privilege particular goals. It also demonstrates the difficulties that are involved in reconciling assumptions about urban places—places where many people live—with expectations of interacting with the environments that are natural, or rural. The desire to move away from a more humanized landscape quite often includes an urge to find or shape some version of nature. The urge to explore, to interact with and shape the environment, is often linked, in this formulation, to the impulse to escape to an ideal better place. In my experience researching the history of Euro-American urbanization and interviewing residents who have moved out of cities or suburbs into exurbia, however, I have come to believe that a great number of aspiring exurbanites wish not—or not only—to escape the problems of their previous everyday environments, but to find in a residential setting the potential for engaging (intellectually, politically, and materially) with some of the issues they see as important to a practice of reflexive habitation. In order to understand this engagement and escape, I find the idea of the modern pastoral helpful to look more closely at what exurbanites mean by "natural" and also at what it might mean to engage with something like nature, or to inhabit it.

MODERN PASTORAL

> I went to the woods because I wished to live deliberately, to front only the essential facts of life, and see if I could not learn what it had to teach, and not, when I came to die, discover that I had not lived.
> —Henry David Thoreau, *Walden*

> The story of Romulus and Remus being suckled by a wolf is not a meaningless fable. The founders of every state which has risen to eminence, have drawn their nourishment and vigor from a similar wild source. It is because the children of the empire were not suckled by the wolf that they were conquered and displaced by the children of the northern forests who were.
> —Henry David Thoreau, *Walking*[20]

The questions of what motivates the search for nature has been extensively considered, in literary and cultural studies especially, in the context of the "pastoral" as a theme concerning the conflicts between nature—often represented by an idealized image of landscape—and the artifice of human design. This analysis of the pastoral also addresses attempts to reconcile these conflicts between nature and artifice, and the term has traditionally been applied to what Leo Marx designated "middle landscapes," environments between the built up city and the remote hinterland—or environments such as nature-themed suburbs, or the exurbia that replaces timberland and farmland. This pastoral is a compromise between the human mark on the landscape and a hope that the landscape will do its own thing in the spaces left between the obviously human parts, not unlike the gap between imagined and real landscapes.[21]

In their examination of the effects of an ideology of the pastoral on the process of residential subdivision, rural sociologist John Fairweather and landscape architect Simon Swaffield discuss the Arcadian manifestation of the pastoral, confirming the tendency of pastoral landscapes to appear where urban realities (encountered daily through commuting) are intentionally effaced by largely illusory rural landscapes in which agriculture is used mainly to keep up an appearance of the rural.[22] In his classic treatise on the wishful imagination of rural living, Raymond Williams traces this Arcadian pastoral back through millennia of literature to classical Greece, exposing the irony in the pastoral. The pastoral landscapes popularized in painting and literature by the European elite were made possible by just the sort of labor this elite avoided, exploited, and tried to keep out of view. Further, the Greek region of Arcadia, archetype of pastoral landscape for this European elite, is a rough mountainous one, not at all the lush respite the poets had made it. Discussing these ironies in an analysis of the phenomenon of the fundamentally exurban "country house," William McClung calls the pastoral Arcadia a dreamscape, "a state of wish fulfillment that stands in dramatic contrast to the experience of everyday life."[23]

Literary scholar Robin Magowan elaborates the pastoral both as a contrast to everyday life and as a promise of wish fulfillment—but also as explicitly illusory. Magowan describes the development of the pastoral in literature as a "form seeking to project within certain arbitrary limits a vision of the good life."[24] The viewer creates this pastoral view by setting those particular limits that will depict the good life, and the landscape allows viewers the opportunity to "dream of what through it we can become."[25] In the way that trees and the landscape are made to bear the burden of fulfilling goals and desires, exurbanites project a pastoral vision of the environment: they may be in love with the dream of what, through ideal environments, they can become. When they forget that *they* have imagined this pastoral vision into being, however, it becomes all too possible to mistake the imagined landscapes of escape for the real, and to elude not only what it appears possible to flee from, but also the possibility of engaging with the landscape fled *to*.

Understanding exurban imagined landscapes as a manifestation of the pastoral reveals similarities across a range of different idealized escape landscapes. Aestheticized working landscape and wilderness provide destinations that invite escape—flight from conditions of modernity that might be better addressed if faced, rather than escaped. The disjuncture in the simultaneous intense appreciation for nature and substantial denial of its material appropriation, for example, is one of the contradictions that a critical understanding of the pastoral vision of exurbia appears to potentially help reconcile. If the stresses of the lifestyle of high-density modern urbanity can be seen as representing what makes residential islands in the midst of nature seem necessary, the modern pastoral is part of what makes plausible the utopian vision of exurbia as those islands. And the "modern" here, as used in the description of the exurbanites I've researched, is a big tent. What social scientists analyze in terms of the exaggerated inequalities and precariousness of late capitalism—the hyper-modernity experienced as everyday life has become more integrated in state and market operations—many exurbanites referred to as "the rat race": a system many critiqued, either because of its effects on them, or because they were self-conscious that they benefitted from this system (so that they could afford to live in exurbia) while most do not.[26]

The move out to exurbia superficially escapes the landscapes identified with modernity: exurbia is riddled with representations of the good life that make reference to the good old days with tokens of resistance—to some of the socio-environmental changes involved in urbanization, for example, which can serve as a stand-in representing an aesthetic understanding of modernity. This resistance is itself amenitized, ironically displaying class distinction by invoking folksy aesthetics that appear to value working class rurality. Neo-traditional "village" architecture and street furniture echo the *old* quality usually read into forests when their wildness is being emphasized. Forested green spaces and trees are preserved as symbols of community values, and forestation is encouraged to conserve the placidity of the unworked landscape (such as that described by Raymond Williams in *The City and the Country*), but these images of nostalgia do not explicitly elaborate, and in fact usually obscure, what is being resisted by the escape into nature.

What is different between "now" and "then," "here" and "there"? What is better in exurbia than in the place escaped? In the same way that trees represent immortality and the vital, authentic life of nature invoked so often via Thoreau quotations such as those I use to begin this section, in many ways the *city*, as such, has come to represent the changes of modernity, as well as the speed and intensity of modern life. Against the overwhelming rush of this modernity, the amenitized hinterlands of green sprawl beckon welcomingly as a place where North Americans seem to think pastoral sanity might be recovered, but people can't (or don't) escape modernity just by moving to the woods.

The Problems of a Pastoral Modernity

In her treatise on the relationship between architecture and modernity, Hilde Heynen explores the twentieth century's legacy of attempts to come to terms with modernity through construction of residential environments.[27] Describing the profound upheaval that has come with the transition to a modern way of life, Heynen contrasts, on one hand, the effort to naturalize these transformations, turning them into a story of smooth experience and untroubled progress, with, on the other hand, the attempt to emphasize—and embrace—the shocking disruption that modern technology and culture represent. Heynen chronicles the search for resolution of the contradictions of modernity, privileging in her analysis modes of resolution that do not force reconciliation of the paradoxes, disjunctures, and unassimilated difference that she sees as intrinsic to modern experience. In this way, Heynen addresses the difficulties and contradictions of a technological world that both inspires escape and also makes this escape possible. Taking her epigraph from critical theorist Theodor Adorno, Heynen emphasizes the necessity of living with contradiction, but also recognizes the urge toward the sort of escapist reconciliation that the traditional pastoral seeks to make between artifice and nature: "Beauty today can have no other measure except the depth to which a work resolves contradictions. A work must cut through the contradictions and overcome them, not by covering them up, but by pursuing them."[28]

For Heynen, a relationship with modernity requires elements of both reconciliation and disruption: a degree of reconciliation makes it easier to inhabit the discontinuity and paradox of modern enterprise and everyday life, but Heynen cautions vigorously against succumbing too much to one side or the other, to a facile tale of uncontested progress (or facile rejection of modernity) on one hand, or a revolutionary vision of constant flux on the other. Heynen describes *forced* reconciliation as the "pastoral" vision of modernity—a vision that has opted for illusory closure by covering up contradiction—and she describes abandonment of any possibility of reconciliation—intentional plunges into chaos—as "counter-pastoral." Futurists and radical modernists, with their embrace of speed and the sleekness of the machine age, stripped of all the clutter of human scale and attachment, might be counted as counter-pastoral; the exurban attempt to blend a lifestyle fully serviced by the consumer economy of capitalist modernity with the untouched pristineness of an imagined enchanted forest might be called pastoral.

Through her stretch of the word "pastoral" to cover even the *impulse* toward this sort of wishful smoothing out of the undesirable but necessary details of everyday life, Heynen's concept helps sort through the representations of desirability and undesirability in the confusing and often postmodern landscapes of exurbia to clarify what is being sought and what is being escaped, or from what affliction solace is promised via the ideology of nature. As it is made manifest in the pleasant and insulating landscapes of exurbia, the pastoral often presents an image of the good life too seamless for easy reading. In building a fortress against the unbearable aspects of modern daily experience, expectations of a different way of life are built directly into the very tableaux of yards and houses. Yet warding off the confusion and disruption of modernity with wishful pastoral visions of modernity (visions such as those of exurban wilderness and also of the tower insulated in the park) may make it more difficult to understand the details of everyday experiences that *are* modern, for better or worse. Blurring the here and now with a pastoral compromise of a get-away landscape makes it more difficult to figure out alternatives to the parts of modernity that are so troubling. These pastoral gestures also make it more difficult at the level of daily experience to learn how to inhabit—and to renovate and restructure to make more equitably inhabitable—both the urban landscapes people might choose to escape and also the exurban landscapes so many people are choosing as a get-away.

If we define exurbia as a pastoral modernity, how does this help pursue the contradictions that Adorno suggests must be worked through to overcome? In identifying and working toward the goals that can be read into the search for a pastoral modernity, can the beauty that Adorno sought to measure be found through a willingness to acknowledge contradictions, and not to cover them up? And if pastoral modernity can be understood in terms of an unwillingness to face the mortality of trees, can such an acknowledgement be understood as not only living with death, but also being less inured to the inequality inherent in exurbia—and instead committing to develop conditions that contribute to keeping open and tractable the contradictions of exurbia as a version of what Naomi Klein calls exclusive "global green zones"?[29] Writers such as Thoreau, who have looked to nature to help resolve the contradictions of modernity, have been appropriated as cultural icons of a get-away, a return to nature, but in the light of Heynen's analysis, this can be seen as a classic pastoral simplification.

Get-away, especially to nature, is often represented in terms of Thoreau's deliberate living—he went to the woods to live deliberately. However, especially in his most quoted texts, Thoreau explicitly rejected get-away. In addition to his contributions to theories of forest succession, Thoreau's enduring example as a naturalist lies in his steadfast gaze not only at the details of the natural environment, but at details of *home*—not of a get-away nature (or "wildness distant from ourselves"), but of the residential and agricultural landscape where he lived, a natural landscape thoroughly inhabited and shaped by humans, for better and for worse, a landscape that landscape photographer Frank Gohlke describes as "worn out yet inexhaustible."[30] As explained in previous chapters, Thoreau both celebrated and lamented the disruptions of modernity (such as the train), rejecting ever further-flung search for wilderness, and standing staunchly behind a philosophy of inhabiting his local environment.

In *The Machine in the Garden*, Leo Marx prefigured Heynen's ideas of pastoral and counter-pastoral with his distinction between a superficial sentimental pastoral and a more complex, engaged pastoral. Islands of pastoral narrative provide simplifications that perform a function in the face of overwhelming complexity. Sojourn provides insulation from complex details of a situation, and may allow access to a creative generativity that might be otherwise overwhelmed. Thoreau's deliberate living at Walden, however, as is argued elsewhere in this collection, transcends the sentimental because of its reflexive deliberation, because the limits and contradictions of his vision of the good life are always kept in view (in Chapters 4, 5, and 9, Blum, Donahue, and Looker elaborate

Thoreau's engagement with these limits). How do we harness these poetics of reflexive inhabitation to reform existing home landscapes?

WHERE THE WILD TREES ARE: REAL AND IMAGINED LANDSCAPES

> Landscapes are culture before they are nature; constructs of the imagination projected onto wood and water and rock. But it should also be acknowledged that once a certain idea of landscape, a myth, a vision, establishes itself in an actual place, it has a peculiar way of muddling categories, of making metaphors more real than their referents; of becoming, in fact, part of the scenery.
> —Simon Schama, *Landscape and Memory*[31]

The Vanity of Distant Wildness

In his journal entry of August 30, 1856 (almost nine years after he had left the cabin at Walden Pond), Thoreau wrote, "It is in vain to dream of a wildness distant from ourselves." These words resonate for me, being from New England, where exurbanization has transformed working landscapes in ways that I now see as gentrifying manifestations of the modern pastoral. Like many others who share concern about socio-environmental issues, I return periodically to Thoreau, not only for his palatable metaphors and his insights into the ecology of forest succession, the desire for wildness, and humans' effects on nature, but mostly for his balanced consideration of the necessities of human habitation, on the one hand, with the recognition of the desire for a romantic and deep connection to the natural environment, on the other.[32] Thoreau's comments on the vanity of distant wildness remind me that the essence of endeavors to inhabit environments is practiced through reflexivity not in an extraordinary way, but in everyday effort and imagination. Fulfilling goals and desires creates the environments of everyday life. Paying attention to how this happens may help us more effectively renegotiate the ways environments are shaped and reproduced.

Invoking a useful contrast similar to the contrast positing wishful amenitization of nature against the vanity of distant wilderness, Canadian designer Bruce Mau, in a well-known manifesto about the principles of design in housing, discusses the relationship with nature landscapes in terms of a call for a get-*to* instead of a get-*away*.[33] This call acknowledges tension in the relationship between escape and engagement, and the desire to replace uncomfortable environments with ones that encourage feelings of being at home. Like Thoreau, Mau suggests that being at home requires work—work whose necessity may be obscured or even effaced by the possibility of instead getting away, even if that get-away is only symbolic. Their

The Mortality of Trees in Exurbia's Pastoral Modernity 277

comments on the vanity of distant wildness echo much academic work that seeks to convince exurbanites to step back to see the cumulative effects of *all* exurbanization, not just each exurban get-away, and to put in the effort and imagination to fulfill exurban aspirations in ways that don't reproduce the urban problems exurbia appears to escape.

Thoreau points to a tension manifested strikingly in the everyday landscape of exurbia: the dream of a get-away one could live in, a comfortable wildness distant from humanity. But as Thoreau continued, "There is none such. It is in our brains and bowels, the primitive vigor of Nature in us, that inspires that dream."[34] The record of the desire for a natural get-away inscribed in the landscape of exurbia may provide a means of exploring the dream nature inspires—a dream filled with the tension between "get-to" and "get-away"—and of understanding how the search for nature leads to a paradoxical situation: seeing humans as alienated from nature, exurbanites seek nature as an escape from the everyday life that appears to alienate them from nature. Yet escape is sought not as a temporary respite, but as a setting for that very everyday life brought to the places where nature is sought: a sticky catch-22.[35]

The attempt to inhabit environments creates complex and problematic conflicts between hopeful visions of the environment and the practices of everyday life—and the social and economic systems of which those practices are part. People experience personal conflicts as well, conflict between escaping problems—opting for a get-away—and engaging with them, making a get-to of where we are, or at least making the places we wish to get to places we might engage with, not just anywhere we can get away. This struggle between escaping and engaging is echoed in the interviews I have conducted with exurbanites, although often as a metaphor, a wistful reflection on the immortality of trees, and the shame about our need to cut them down. I find in this idea of the immortality of trees an irresistible metaphor of relationship with the exurban environment—an ambiguous and paradoxical relationship that seems to simultaneously place value on the possibility of negotiating more diverse cultural landscapes that combine urban and rural functions (the classic "middle landscape") and at the same time overlook or dismiss this possibility, usually by categorically designating landscapes either "green" or "sprawl." In adulation of trees and environment as "nature," I interpret not only reasons that people make environments worse while wishing to make them better, but also potential keys to the rousing of political will for more vigorously negotiated inhabitation.

Teasing apart the search for nature may tell us more about what, as a metaphor, it signifies—what it is to which people want to "get to," even as they are getting away. This chapter represents my effort to understand what the "get-to" stands for, and what exurbia says about the goals and desires projected onto it—and in doing so, to understand the conflicts and contradictions in the struggle to negotiate habitation of environments.

The Illusion of Preservation

> "[Trees] grow old and die, well I guess they do, I never thought of it that way. I guess I figured they hung around forever unless we went at them, but I guess they go around that way."
>
> —Charles Farquharson in Ontario Ministry of Natural Resources, *Whose Lookin' After My Forest... And Howse it Goin*

> Deep-seated philosophical objections to harvesting are likely the greatest barrier to changing owners' approach to forest management. As long as the global consequences of consumption are ignored, widespread protectionism is heralded, and logging is abhorred (especially in one's backyard), efforts to reduce wood consumption, or to encourage sound management in areas of low ecological impact will be fruitless. Such efforts will only succeed if they are coupled with a fundamental change in attitude that reconciles the ideology of preservation with the reality that using wood means cutting trees—somewhere.
>
> —Mary Berlik et al., *The Illusion of Preservation*[36]

Ideological conflicts over nature preservation in the exurban forest demonstrate the problems of the imagined immortality of trees. A recent report arguing for the local production of natural resources in exurban Massachusetts discusses possibilities for and obstacles to sustainable forestry in terms of what its authors call the "illusion of preservation."[37] Addressing the forested landscapes that have been most appealing for exurban residential building and preservation in the last several decades, Harvard Forest researchers Mary Berlik, David Kittredge, and David Foster conclude in their study that the most crucial change necessary for a shift to more locally responsible and sustainable forest practices is ideological.

Discussing the preservation ethic that has come with the modern embrace of the forest, they explain (in the epigraph above) that preservation in affluent residential areas is illusory if it is subsidized by exploitation *elsewhere*. Successful—and often laudable—environmentalist campaigns championing and using images of grand old forests have perhaps soured images of *forestry*. Many people object, often with good reason, to either industrial forestry or agriculture in their backyards. But as productive land uses are relegated to the places where they do not bother people who can afford sensitive environmental aesthetics, they become more and more objectionable—removed (by distance, and politics) from the observation of those who could effectively protest.

Discussing how the illusion of preservation has become part of the solace that makes nature attractive, Berlik, Kittredge, and Foster call for the exploration of alternatives to current modes of environmental inhabitation. They are concerned that the current zeitgeist of exurban living, particularly, is characterized by obstacles to more reflexive inhabitation—obstacles that present themselves, for example, as the conviction that the larger sources of and solutions to environmental problems lie *out there*, and that the solutions to pressing environmental problems can be found magically in cultivating fetishes such as immortal-seeming "nature." "Mainstream environmentalist ideology must embrace multiple uses of the forest including harvesting," they insist, "and local citizens must consider the use of resources in their own backyard while maintaining a keen awareness of the global environment."[38] The difficulty seems to be in the translation between awareness of global environmental problems and representations of a version of nature that fixes environmental problems for us as long as we engage it with "environmentally friendly" intentions.

Commodifying Nature

The symbolic image of trees as timelessly powerful, and of nature as in need of human preservation because it will be the *source* of human preservation, has been tremendously important to movements that rely on a sense of empathy with nature and a desire to champion it. By making nature not only grand, desirable, and untouchable, but also consumable *in the abstract*—through a sense of moral ownership gained from the support of environmental and preservation organizations—actors that promote protection of wilderness (e.g. the Sierra Club) or organic food (e.g. Whole Foods Markets) have gained remarkable success. But this success has also reinforced the commodification and marketing of nature.[39] Political movements to protect the environment through legislation reducing industrial pollution, designating wilderness, and introducing measures of environmental accountability to agriculture have successfully drawn on the awe-inspiring and arresting largeness of photographs in the tradition of Ansel Adams to bring the grandeur of nature to an overwhelmingly urban and suburban public. Especially given their distance from most peoples' everyday experience, these images have been remarkably successful in marshaling support for everything from conservation funds to organic standards.[40] The powerful appeal of these images of pristine, ancient, uninhabited nature has its downsides, however. Environmentalism writ large may have been helped by capturing and mobilizing the immortality of trees, but this view of nature has also led not only to the illusion of preservation, but also to the commodification of nature. Both this illusion and commodification are used to greenwash the impacts of green sprawl.

Advertisements using nature to sell the tools of everyday life—houses and cars, particularly—have helped to familiarize the commodification of images of nature, making complex ecologies, political economies, spatial practices, and metaphorical frames superficial and abstract. Presenting nature that can be conquered and consumed through purchase, ads offer a manifestation of a pastoral view of nature as unattainable and awe-inspiring wilderness. If you don't look too closely, if you accept the images in the ads, you can imagine yourself engaging with nature; an exurban car or home promises to allow you to have your (roadless and uninhabited) forest wilderness, and live in (and drive to) it, too. In this way, like Jefferson Forest's sylvan living room, the idealized image of nature denies both the processes and conflicts in environmental interactions, making nature a *fetish*, a magic promise of effect without effort.

The nature fetish promises engagement with nature, while hiding the ways in which goods such as SUVs—and the infrastructure that supports them—get in the way of achieving the sustained and substantive interactions with the nature that exurban aspirations appear to imagine. The fetish of nature can *replace* such interaction, standing in for the negotiations necessary for navigating human roles in the environment. Environmental harmony can, in this commoditized version of nature, be bought with organic snack food or a Sierra Club subscription (both reliable sources of sacred tree imagery) and the moral credits achieved can be imagined to offset less rewarding environmental habits. This give and take can be part of the way ideology and practice are balanced; it can be part of the reflexive process of dialogue between experience and intention that enables us to revise our ideologies and practices as we try them out in the world. But the danger of symbolizing things like trees or nature to the point where they become merely fetishes is that they obscure the necessity of the negotiated and social *process* of this dialogue. In the gap between material environments and imagined ones, such fetishes create illusory problem closure: a smooth, grassy lawn and specimen forest where exurbanites might see instead the layers of conflict, care, and past experience written in the landscapes which surround us. These layers embody the socio-environmental context that need to be more explicitly negotiated in managing environments.

Karl Marx discussed the problem of commodity fetishism in the opening chapter of *Capital*. Using as an example the idea of magical power attributed to objects as fetishism in the "mist-enveloped regions of the religious world," Marx explored the way in which fetishization of commodities blinds people to the social and creative processes that give value to these commodities.[41] Instead, fetishes confer a naturalized power, an ability, if the fetishes in question can be acquired, to bestow what is desired without work or action. In addition, like Thoreau, Marx understood that by placing humans outside of nature, the ideology of nature alienates humans from their place within nature.[42] Such an ideology not only obscures the processes that give value to commodities (which are ultimately made out of social and ecological relations, materials, and labor), but also makes it more difficult for people to see these processes as the ways in which humans constantly do engage with nature. Instead of seeing transformation of natural materials into the stuff of everyday life as a central way in which humans inhabit nature, we often see these mediating processes as alienating aspects of modernity. This alienated view disengages people from the very ways they engage and govern environments, and feeds into the vicious circle of landscape commodification and the degradation of disinvestment and gentrification cycles involved in escaping to get-away places.[43]

From Useful to Beautiful: Aesthetic Paralysis in the Naturalized Rural Landscape

In addition to problems of environmental degradation and social conflict and inequality, new exurban residents face the dismay of finding that their escape from the modern urban landscape does not necessarily provide the nature they had imagined, as well as disillusionment in not knowing what to do, or how to engage with their new rural setting. Most exurban residents who talked with me praised the natural qualities of the landscape to which they had moved in contrast to the urban or suburban landscape they had moved from (this was particularly true in the New England and Ontario cases—in Ontario, all but one family had moved to the exurbs from the city of Toronto). However, many expressed a certain disappointment with their experience there. Some were distressed with changes in the landscape—evidence that many other people had chosen to move there, too. Others experienced a version of the disillusionment described by Spectorsky. Having hoped for the idyll of the ideal life closer to nature, and in many cases having allowed their expectations to lead into the realm of fantasy, they were dismayed that knowledge of what to do with the landscape did not come along with their properties.[44] Several expressed frustration at the difficulty of making their dreams real. They have found that escape to the sojourn of nature is not enough; meaningful experience entails not only being in the right place, but *doing* something there.

The solace of a rural way of life and a connection to nature seem particularly important to residents for whom the ideas "rural" and "nature" remain abstract and elusive. When talking about their expectations of exurban life, these residents often sound perplexed and somewhat bitter at the difference between their expectations and experiences of life "in the country." Residents like these aspire to the ruralesque exurban life because they have inscribed in it and its natural settings many things they idealize but don't know how to achieve. Likewise, they want to *do* something amidst the aesthetic of the productive landscape, but don't know what. For many of these residents, the processes of the rural landscape are as opaque as their unconscious preferences for idealized and aestheticized versions of rural land uses. Implicitly, they prefer dilapidated barns to smelly, active ones and conservation areas to working forests.

But the accumulation of conserved land in exurbia does not necessarily live up to widely supported justifications for conserving land. When I visit hobby farms and exurban forests, I am always looking for evidence that access to so much greenspace might be helping people there to learn to live more "sustainably," or to understand more about the conditions they might be able to create that would better align their goals of habitation with the functions of the ecosystems they (and everyone else) depend on from the greenspace. In short, I am looking for evidence that exurban conservation does something other than accumulate landscape around those who can afford it. However, what I tend to see is that over-romanticization of trees alienates people from straightforward acknowledgement of existing interactions they do have with trees and with the rest of the environment in which they participate. When agency in the environment is effaced and capacity to experience and understand everyday environments is undermined, it becomes more difficult to engage successfully with environmental aspirations, goals, and public planning processes. Barriers arise between present and past environmental experiences that could help people learn about inhabitation—and between the present and possible alternative futures toward which exurban aspirations could be directed.

LIVING WITH THE PASTORAL: DEATH IN THE VERY MIDST OF DELIGHT

A dramatic shift in the perception of human agency vis-à-vis trees occurred around the end of the nineteenth century, around the same time that current North American settlement forms were coalescing into their present forms. Analyzing changes in the image of the land during the rise of metropolitan Ontario, historian Allan Smith describes the process of forest clearing and settlement as vanquishing "the sense that the land was covered by a terrifying and impenetrable overgrowth," and replacing this overwhelming image of the forest with a notion of a "subdued and pleasant collection of quite manageable trees."[45] Although the common response to the "bush" at this point had been to see it as much more terrifying than sublime, Smith highlights the shift in the 1880s from a forbidding toward a more aesthetic forest with what he calls the frontiersman's ultimate question: "In this brave new world . . . where the odour of the woods is a tonic, and the air brings healing and balm, how can death exist?"

What had been impenetrable and forbidding forest was becoming an appealing forest of recreation and transient wonder. In a contemporary effort to convince the American and Canadian publics that forests were not all about pragmatism, Gifford Pinchot, a major champion of thoughtful conservation of forests, invoked this imaginative shift that came with the diminishing of the forest primeval. In his effort to promote the management agency that would become the U.S. National Forest Service, Pinchot asked the public to overcome both resourcist and overawed visions of the forest, and to consider cultural and palpable relationship with the forest instead: "The forest is as beautiful as it is useful. The old fairy tales which spoke of it as a terrible place are wrong. No one can really know the forest without feeling the gentle influence of one of the kindliest and strongest parts of nature. From every point of view it is one of the most helpful friends of man."[46] This idea of a gentle influence and

The Mortality of Trees in Exurbia's Pastoral Modernity 283

kindly friendliness clearly resonates with the large numbers of people who have decided to take up residence in the forest. Current understandings of forests have swung around to confirm Pinchot's assertion that the forest is as beautiful as it is useful. But what has happened to the understanding of a forest as helpful, or as useful as it is beautiful?

As Nik Luka asserts in Chapter 6, at least part of the gap between the forest of everyday experience and that of aspiration has been produced through the construction of landscapes of leisure, through the separation of working life from residential life, and especially through temporary sojourns in the recreational forest. If, as Luka suggests, the exurban landscape aspires to the extremely pleasant, if not temporally durable, cottage life—if everyday life aspires to an Arcadian ideal—how can exurbanites also, concurrently, reflexively inhabit a life-world produced out of the very materials to which they turn to try to buffer themselves from the impacts of urban modernity? Interpretation of Nicolas Poussin's *Shepherds in Arcadia*, a classic seventeenth-century painting in the pastoral tradition, points to the relation that bridges the working world of production and the fantasy relied upon in the exurban landscape.

In the painting, shepherds consider a tombstone in an Edenic landscape, bearing the inscription "et in Arcadia, ego": "I too, am in Arcadia." Art historians such as Erwin Panofsky and Giovanni Pietro Bellori have argued over who speaks the words inscribed on its central tombstone: the shepherds might be celebrating their heavenly abode, but it's more likely death, reminding them of its omnipresence. The historians agree that the shepherds are absorbed in the contemplation of mortality, and, as Panofsky points out, that this is a shift not unlike that to follow (from useful to beautiful) in the popular imagination of forests—"Poussin no longer shows a dramatic encounter with Death."[47] Describing the painting as Poussin's critique that the pastoral ideal is illusory, Bellori interprets the painting in a way that for me evokes forest succession ecology, as a reminder that "the grave is to be found even in Arcady and that death occurs in the very midst of delight."[48] As exurbanites and exurban land managers attempt to come to terms with the turn toward conservation and the rapprochement of urban and rural land uses and land use values—including the valuation of forests as both beautiful and useful—these moments of change in the terms of the imagination of death in nature may provide insights into how to re-incorporate the useful into aesthetics of the everyday landscape. This concluding section explores what transformations of exurban landscape ideals might be possible if changeable everyday environment were better understood through acknowledgement of the way exurbia functions as a modern pastoral.

If Trees Die, What Happens to Idealized Landscapes?

At the Cedar Grove farm this weekend, I found a back-to-the-land classic on constructing residences using recycled building materials, old tires, and earth. The book explained how the motivation for this brand of ecologically sound building rests on an emulation of trees. Disparaging the unsustainable lifestyles of earthlings, *Earthship Volume I* introduces its ideology of habitation with a parable from the point of view of aliens who visit Earth and describe the living habits of trees, animals, and humans. With nothing good to say about the humans—concentrating mainly on their tendency to soil their own nests—the aliens sing rapt praises to trees, who not only stay rooted to their home places, but also appear to embody the balanced system dynamics seen as desirable from a mid-1970s-environmental-crisis perspective. Despite their obvious differences from humans, trees, for these authors (as for so many of the aspiring exurban residents whose stories I have heard), symbolize, and perhaps even model, an ideal way of life.[49]

This idealization of trees is a problem in the changing and often perplexing modern urban landscape. In the face of change or the unknown, our impulse, as Tom Looker elaborates in Chapter 9, is too often to clamp down on complexity and ambiguity and to force resolution. In attempts to reconcile nature's promise with the shortcomings of everyday environments, openness to contradiction and complexity assists our ability to resist idealizing nature as "pure" and humanless. This openness also assists in avoiding the pitfalls of vilifying ourselves and our neighbors (as the Earthship authors do) in such a way that we see efforts to more reflexively and equitably inhabit our environments as always inadequate—and hence not worth prioritizing.

Some goals for a nice place to live may be more or less summed up by the billboard promise of a Magnificent Landscape for Living. However, if the more interesting and inspiring goals of habitation are represented as unachievable (a natural landscape *without us in it*), it is not surprising if people respond to the daunting task of understanding the human place in the environment by wanting to escape to nature. Demystifying the way that nature is made into a fetish may open more space for people to work toward their goals *through* the complexity and ambiguity that are inherent in socio-environmental relations. Perhaps if there was more opportunity within both everyday and official environmental planning for critical dialogue around how to make the benefits of natural landscapes more equitably accessible, the search for a modern pastoral would not so often lead to deferment and self-deceptive accumulation of greenspace, but could include a kind of *critical fetishism*, in which the goals associated with nature became objects of analysis themselves, used to help keep open questions about *whose nature* against the pastoral tendency toward closure. In addition, this engagement with everyday goals associated with nature, and with trees, might help people transcend the alienation and disempowerment that exaggerated awe has encouraged. The moral ambiguity of the exurbs hinges on this awe and disempowerment, which helps keep people firmly out of nature, and preserves an ideology of nature as worthy of aesthetic worship (by the elect few) but not of everyday interaction.

Wanting to inhabit nature, but thinking of humans as outside of it, we become blind to our agency in the creation of the landscapes we experience; we forget that it is *people*, as much as nature, who have written the story there: we've made this landscape we call natural. If we pretend that what we find in the exurban forest is an entirely natural environment, we deny to ourselves that the exurban landscape embodies our contingent history and the goals and desires we embed there. Reading residential landscapes provides opportunities for us to read back to ourselves the aspirations we have written into our home environments, but it can also intensify the desire to cover up with greenery problems such as poverty and pollution. The landscape of exurbia does tell us about the social and environmental problems people are trying to hide there—or to hide from. Understanding the motivations for retreating to nature or for trying to make landscapes more natural can help clarify what is disturbing about the un-natural parts of modern society, and can help sort out why "nature" seems like a good answer to problems that seem too complex to face.

Exurban demands on nature demonstrate the intensity of wishes to evade responsibility and instead to turn toward the beneficent nature our culture has constructed to escape the risk and failure we face (especially when it comes to questions of environmental management). When the desire for a comfortable home environment becomes contingent on the providence of symbolic trees in the home grove, people become more inclined to view forests as an eternal paradise. This image of perfection obscures the changing, messy forest on which we actually depend, where, as surprising as we might find it, the trees lose limbs and die, both on their own and through human activity. The Ontario Ministry of Natural Resources film on the forestry industry quoted above presents the prospect of everyday life without the death of trees in a humorous series of scenes in which the main character imagines the forestry-dependent props of everyday life disappearing out from under him, one by one: his newspaper, rocking chair, and house.[50]

Recognizing the aesthetic attachment to a fundamentally static nature and the widespread unwillingness to accept the mortality of trees, the U.S. National Park Service staff have pinned a paper sign onto a tree that has recently fallen near the visitor entrance to the Muir Woods, explaining that death happens naturally in a forest. In this token space of wilderness, an explanation of succession can be read as a monument to the persistence in the public of the ideal of trees as immortal, but also as a step in our public culture toward coming to terms with their mortality.[51] This idea of succession may be a palatable way to open up the dialogue around death in nature in a culture so afraid of its own concern for the environment, and afraid of where interest in wilderness and ecology might lead, if not to pleasant escape. The pleasant escape of exurbia, as complicated as it is, presents a more straightforward set of responsibilities than poverty and social justice, industrial contamination and exploitation, the prospect of catastrophic climate change, and the complex mesh of cultural and social relations upon which our modern society is based. As Andrew Blum and Tom Looker have argued in previous chapters, however, interest in leading a natural life is heavily tempered by significant investment in leading modern lives. Exurbia, premised as it is on Blum's "Airworld" of telecommunications, travel, and global economics, *is* a modern pastoral, and its version of the good life has as much to do with data lines and consumption as it does with natural or sylvan ideals.

More rigorous engagement with the related ecological processes and material throughput of residential environments may encourage at least some people toward reflexive inhabitation in those environments, to forge a meaningful connection to the forests in which trees grow and die, and to negotiate productive processes and social relations with which we are better able to ethically reconcile ourselves. This engagement needs to be critical—at home in the uncertainty of modernity, able to inhabit ambiguity and to acknowledge and work through the pastoral impulse—an engagement that does not need to turn to a sentimentally pastoral vision for comfort and closure. In the face of what art historian Lucy Lippard calls the "lure of the local"—the desire to retreat to the enclosure of the aesthetic locality or to nature's solace—we must consider our local landscapes in the context of the complex global and regional processes of which they are a part.[52]

But I come to the end of this chapter feeling sterner about the emotions people have for trees than I mean to. Noticing a branch broken off a white pine from last night's storm on my way home today, I felt a pang—had I spent enough time looking at that tree? Was it old? Was it going to die? Acknowledging my pang of affection for this tree, and for trees in general, I sympathize with a sense of generative connection with the natural environment, with the sojourn and stimulation provided by the patterns and motion of trees in the wind and weather, with a likely-to-be irrational attachment to what they symbolize. Considering my attachment and affection, I also recognize that what bothers me about the way we tend to relate to trees is the discrepancy I've been describing between this pleasantly thought-of relationship with the environment and the much less inspiring and conflicted way we actually behave. While preservation rhetoric has made it acceptable to consume far-off forests, air pollution has traced a widening swath into places where trees like sugar maples will no longer reproduce (an identity disaster for Toronto!),[53] and climate change threatens to rearrange the placement of our ecosystems faster than our arboreal neighbors can adapt—at the same time that we have decided to value them so highly, to seek residence amidst the landscapes they dominate.

Negotiating Pastoral Landscapes: Making Exurbia Inhabitable

Acknowledging both modernity and mortality may be an important first step in becoming comfortable in dynamic environments—and in realizing that *all* environments are dynamic. If we are reluctant to embrace change and death in images of the ideal life, we can take comfort in *succession*, that phenomenon of forest change of which naturalist Henry Thoreau remains an eloquent champion, in which death and change are integral parts of renewal and adaptation.[54] If, as Thoreau observed, trees reproduce, grow, and die, our relationship with forests changes from a simple question of to protect or to cut and becomes much more nuanced: whose forests are these? If they are public (as many forests that attract adjacent exurbanization are), what *could* we do with them? How can these goals, of which there will be many, be reconciled with each other?

In telling about his own quest to find and answer these questions in the polluted sprawling landscape of exurban Massachusetts, Frank Gohlke captures the essence of the way in which Thoreau's careful observation of succession disrupts the apathy and helplessness of the vision of trees as immortal, and pushes back against a pastoral understanding of Thoreau as a patron saint of exurban escape. Describing the way in which *learning* to be at home in an alien place invites us into an understanding of the environments we inhabit, Gohlke describes how "in [Thoreau's] manner of living and in the writing that flowed from it, he provided our first paradigm for loving, through its details, a fallen world."[55] Exploring the detritus of a complex and changing environment may help us to acknowledge the imperfection of the natural environment, and to negotiate our place in it without deferring our agency to some ideal of a modern pastoral.

> . . .because in times like these
> to have you listen at all, it's necessary
> to talk about trees.
> —Adrienne Rich, "What Kind of Times Are These"[56]

NOTES

1. See Annie Leonard's *The Story of Change* (The Story of Stuff Project, 2012) for a popular, pithy, and well-supported critique of this tendency. http://www.storyofstuff.org/movies-all/story-of-change/
2. Although I focus here mostly on trees, I have written elsewhere about the attractions, compulsions, and dilemmas of the promise of agency in semi-agricultural peri-urban landscapes: "Engagement with the Land: Redemption of the Rural Residence Fantasy?" in *Rural Change and Sustainability: Agriculture, the Environment and Communities*, eds. S. Essex, A. Gilg and R. Yarwood (Oxfordshire: CABI, 2005), 215–229; "Agrarian Problems and Promises in Exurban Sprawl," Yale Program in Agrarian Studies Seminar Paper, (New Haven: Yale Institution for Social and Policy Study, February 16, 2007), http://www.yale.edu/agrarianstudies/colloqpapers/18conspicuousproduction.pdf; "Political Ecology of Exurban 'Lifestyle' Landscapes at Christchurch's Contested Urban Fence," *Urban Forestry and Urban Greening* 7.3 (2008): 183–194; "Other Women's Gardens: Radical Homemaking

and Public Performance of the Politics of Feeding," in *Doing Nutrition Differently: Critical Approaches to Dietary Intervention* (Aldershot: Ashgate, In Press).
3. Although conservation barons like Ted Turner and John Malone receive considerable attention for their attempts to negotiate productive rural land uses and ecosystem restoration, they have been successful in their efforts largely because of their extraordinary access to land and resources (and consequent relative independence from political struggles over land-use controls). In contrast, many exurban land-use conflicts are characterized by struggles between people using land for livelihood and those using land for pleasure, with the latter often using conservation science justifications to build the legitimacy of their land uses (which come with their own downplayed economic gains) above rural land uses they portray as *merely* economically motivated; see Patrick Hurley and Peter Walker, "Whose Vision? Conspiracy Theory and Land-Use Planning in Nevada County, California," *Environment and Planning A* 36 (2004): 1529–1547; Michael Woods, "The Local Politics of the Global Countryside: Boosterism, Aspirational Ruralism and the Contested Reconstitution of Queenstown, New Zealand," *GeoJournal* 76, 4 (2009): 365–381; L. Anders Sandberg and Gerda R. Wekerle, "Reaping Nature's Dividends: The Neoliberalization and Gentrification of Nature on the Oak Ridges Moraine," *Journal of Environmental Policy and Planning* 12 (2010): 41–57.
4. J. Baird Callicott and Michael P. Nelson, eds., *The Great New Wilderness Debate* (Athens: University of Georgia Press, 1998); Kirsten Valentine Cadieux and Patrick T. Hurley, "Amenity Migration, Exurbia, and Emerging Rural Landscapes: Global Natural Amenity as Place and as Process," *GeoJournal* 76, 4 (2009): 297–302; Peirce Lewis, "The Galactic Metropolis," in *Beyond the Urban Fringe*, eds. R. H. Platt and G. Macinko (Minneapolis: University of Minnesota Press, 1983), 23–49; Donald Meinig, "Symbolic Landscapes," in *Interpretation of Ordinary Landscapes*, ed. D. Meinig (New York: Oxford Press, 1979), 164–188; Leo Marx, "The American Ideology of Space," in *Denatured Visions: Landscape and Culture in the Twentieth Century*, eds. Stuart Wrede and William Howard Adams (New York: Museum of Modern Art: Distributed by H.N. Abrams, 1991): 62–78; David Brooks, *On Paradise Drive* (New York: Simon & Schuster, 2004); John Darwin Dorst, *The Written Suburb: An American Site, an Ethnographic Dilemma* (Philadelphia: University of Pennsylvania Press, 1989).
5. The political economy of this conflict is outlined in Raymond Williams' rendition of "Ideas of Nature" in *Problems in Materialism and Culture* (London: Verso, 1980): 67–85, and in David Harvey's *Justice, Nature, and the Geography of Difference* (Blackwell: Malden, 1996), particularly chapters 6–8. The political ecology of these landscape trajectories may be considered via Paul Robbins, *Lawn People: How Grasses, Weeds, and Chemicals Make Us Who We Are* (Philadelphia: Temple University Press, 2007); S. Paquette and G. Domon, "Trends in Rural Landscape Development and Sociodemographic Recomposition in Southern Quebec," *Landscape and Urban Planning* 55 (2001): 215–238. Considerable recent research on the tensions within "multifunctional" landscapes points to the problematic contrast between justifications for protecting "working landscape" and subsequent uses: a significant amount of exurban conservation is justified via the promise of working landscapes for supplying proximal markets and the salutary effects on landscape management practices of environmentally–interested residents; however, in many places, the majority of justifications for working landscape easements—usually providing property tax benefits, at public cost—are premised on aesthetic value (Sharon Baskind–Wing, *Reclaiming*

Rural Character: Conservation, Conflict, and the Nostalgic Landscapes of Orcas Island, Washington (Ph.D. diss, Rutgers University, 2009); Susannah R. McCandless and D. Brighton, "Part of Home, Space to Roam: Landowner Participation and Stewardship at the Edge of the Northern Forest" (Chicago: Association of American Geographers Annual Meeting, March, 2006); Jeffrey A. Klein and Steven A. Wolf, "Toward Multifunctional Landscapes: Cross–Sectional Analysis of Management Priorities in New York's Northern Forest," *Rural Sociology* 72, 3 (2007): 391–417)). My fascination with exurban justifications developed in large part through my experience of exactly this tension, highlighted by the contrasts between living in the rapidly expanding city of Toronto, where everyday needs can be met within walking distance, but the landscapes of their procurement are quite far away—despite the proximity of farmland and forests to the city—and living in remote exurban New England, where despite the prevalent aesthetic of working landscape, meeting any needs requires considerable travel, and landscapes of procurement are at least as distant.

6. James Kunstler, *The Geography of Nowhere: The Rise and Decline of America's Man–Made Landscape* (New York: Simon & Schuster, 1994); Peter Blake, *God's Own Junkyard* (New York: Holt, Rinehart and Winston, 1964).

7. As part of a relationship with nature, more specific relationship with trees could also be considered a synecdoche, in which a part stands for the whole, or even in a symbolically holographic manner, in which every small part of our ordinary interactions with trees displays, if we look, the larger picture of the way we imagine nature. Robert Pogue Harrison, *Forests: The Shadow of Civilization* (Chicago: University of Chicago Press, 1992); Simon Schama, *Landscape and Memory* (New York: Vintage Books, 1995); Joel Wainwright and M. Robertson, "Territorialization, Science, and the Postcolonial State: The Case of Highway 55 in Minneapolis, USA," *Cultural Geographies* 10 (2003): 197–217; David Ley, "Between Europe and Asia: The Case of the Missing Sequoias," *Cultural Geographies* 2 (1995): 185–210.

8. Harrison, 1992; Schama, 1995; Kim Taplin, *Tongues in Trees* (Devon: Green Books, 1989); Owain Jones and Paul J. Cloke, *Tree Cultures: The Place of Trees and Trees in their Place* (New York: Berg, 2002). These sources explore arborism from a particularly Anglo perspective; for international cases in which this centrality is exhibited (especially the provisioning environment) see Shaul Cohen, *The Politics of Planting: Israeli–Palestinian Competition for Control of Land in the Jerusalem Periphery* (Chicago: University of Chicago, 1993); Emmanuel Kreike, "Hidden Fruits: A Social Ecology of Fruit Trees in Namibia and Angola, 1880s–1990s," in *Social History and African Environments*, eds. W. Beinart and J. McGregor (Oxford: James Currey, 2003) 27–42; and for references to several international examples of forest interest in several cultures (and also noting his own work in India and elsewhere), Paul Robbins, *Political Ecology*, 2nd edition (Malden: Wiley-Blackwell, 2012).

9. Personal communication with Hugh Wilson, Hinewai Reserve, Aoteroa New Zealand, May 25, 2004.

10. Clearly not all exurbia is forested. Although many archetypes of exurbia certainly include trees, exurbanization does happen in non-forested regions as well. However, reforestation (and revegetation more generally) is a common pattern in exurban land use change, and aerial photographic studies of exurbs show remarkable disappearance of houses into the forest cover over fairly short periods of time, both in the Toronto region (see, especially, John Punter, *Urbanites in the Countryside: Case Studies of the Impact of Exurban Development on the Landscape in the Toronto-Centred Region*

1954–1971 (Ph.D. diss, University of Toronto, Geography, 1974)) and New England (see Rutherford H. Platt and George Macinko, *Beyond the Urban Fringe: Land Use Issues of Nonmetropolitan America* (Minneapolis: University of Minnesota Press, 1983). Some agricultural property tax rates explicitly seek to value land in productive use, such as in Vermont, where the Current Use Program attempts to hamper conversion of open land to wooded residential use.

11. Autumn Grove and Jefferson Forest were both projects of Townwood Homes (along with Rosehaven, Aspen Ridge, and Arista Homes); I am grateful to Townwood for providing permission for these images (http://www.townwoodhomes.com/communities/autumngrove/index.htm; http://jeffersonforest.ca) (April 2005).

12. William Ashworth, *The Left Hand of Eden: Meditations on Nature and Human Nature* (Corvallis: Oregon State University Press, 1999): 106–107; 162.

13. A.C. Spectorsky, *The Exurbanites* (New York: J.B. Lippincott Co., 1955). For Spectorsky, the long commute to New York didn't necessarily have the connotation of environmental impact currently pejoratively associated with exurban living, but rather a symbolic and material distancing from the urban reality of the exurbanites' daily world: he noted that their work, social, and cultural worlds were all in the city, not the country. See also, for long term study of more distant urban to rural migration, Patrick Jobes, *Moving Nearer to Heaven: The Illusions and Disillusions of Migrants to Scenic Rural Places* (Westport, CT: Praeger, 2000).

14. Kirsten Valentine Cadieux, "Imagining Exurbia: Narratives of Land Use in the Residential Countryside" (MA thesis, University of Toronto, 2001); Cadieux, "Amenity and Productive Relationships with 'Nature' in Exurbia: Engagement and Disengagement with Urban Agriculture and the Residential Forest" (Ph.D. diss, University of Toronto, Geography, 2006); Cadieux, 2008; Cadieux, "Competing Discourses of Nature in Exurbia," *GeoJournal* 76, 4 (2011): 341–363.

15. Jeff R. Crump, "Finding a Place in the Country: Exurban and Suburban Development in Sonoma County, California," *Environment and Behavior* 35, 2 (2003): 187–202; Peter Walker and Louise Fortmann, "Whose Landscape? A Political Ecology of the 'Exurban' Sierra," *Cultural Geographies* 10, 4 (2003): 469–491; Trudi Bunting and Pierre Filion, "Dispersed City Form in Canada: A Kitchener CMA Case Study," *The Canadian Geographer* 43 (1999): 268–287.

16. Raymond Williams, 1980, 67–85 and *Keywords* (London: Verso, 1983); William Cronon, introduction to *Uncommon Ground: Toward Reinventing Nature*, ed. William Cronon (New York: W.W. Norton & Co., 1995). As Cronon eloquently captures this idea in his chapter of *Uncommon Ground* ("The Trouble with Wilderness; or, Getting Back to the Wrong Nature" 69–90): 80, "The dream of an unworked natural landscape is very much the fantasy of people who have never themselves had to work the land to make a living—urban folk for whom food comes from a supermarket or a restaurant instead of a field, and for whom the wooden houses in which they live and work apparently have no meaningful connection to the forests in which trees grow and die. . . . for the romantic ideology of wilderness leaves precisely nowhere for human beings actually to make their living from the land." Michael Pollan's discussion of "supermarket pastoral" echoes this analysis in his body of work summarized in *The Omnivore's Dilemma* (New York: Penguin, 2006).

17. Anne Whiston Spirn, *The Granite Garden: Urban Nature and Human Design* (New York: Basic Books, 1984); Michael Hough, *Cities and Natural Process* (New York: Routledge, 1995); Matthew Potteiger and Jamie Purinton, *Landscape Narratives: Design Practices for Telling Stories* (New York: John Wiley & Sons, 1998); Henri Lefebvre, *Critique of Everyday Life* (London: Verso, 1991).
18. Dolores Hayden and Jim Wark, *A Field Guide to Sprawl* (New York: W.W. Norton, 2004); Walter Firey, "Ecological Considerations in Planning for Rurban Fringes," *American Sociological Review* 11.4 (1946): 411–423; Rachel Kaplan "Human Needs for Renewable Resources and Supportive Environments," in *Land Use and Forest Resources in a Changing Environment: The Urban/Forest Interface*, ed. Gordon A. Bradley (Seattle: Washington University Press, 1984), 133–140; Lessinger, Jack, *Penturbia: Where Real Estate Will Boom after the Crash of Suburbia* (Seattle: SocioEconomics, Inc., 1991); Punter, 1974; Jobes, 2000; Marcia B. Kline, *Beyond the Land Itself: Views of Nature in Canada and the United States, Essays in History and Literature* (Cambridge: Harvard University Press, 1970).
19. Aotearoa New Zealand shares with Canada and the United States most of the same problems of exurbanization; see *Managing Change in Paradise: Sustainable Development in Peri-Urban Areas* (2001) and *Superb or Suburb? International Case Studies in Management of Icon Landscapes* (2003) (Wellington: Office of the Parliamentary Commissioner for the Environment/ Te Kaitiaki Taiao a Te Whare Paremata).
20. Henry David Thoreau, *Walden and Other Writings of Henry David Thoreau* (New York: Modern Library, 1992, originally published 1854); "Walking," *Atlantic Monthly* (1862).
21. Discussing the "folk" pastoral, Richard Judd and Christopher Beach eloquently call this middle landscape, where human marks compete with the human desire to leave the environment unmarked, "lightly humanized." Richard William Judd and Christopher S. Beach, *Natural States: The Environmental Imagination in Maine, Oregon, and the Nation* (Washington, DC: Resources for the Future, 2003).
22. Simon R. Swaffield and John R. Fairweather, "In Search of Arcadia: The Persistence of the Rural Idyll in New Zealand Rural Subdivisions," *Journal of Environmental Planning and Management* 41.1 (1998): 111–127.
23. Raymond Williams, *The Country and the City* (New York: Oxford University Press, 1973); William McClung, "The Country House Arcadia," in *The Fashioning and Functioning of the British County House*, ed. G. Jackson-Stops (London: University Press of New England, 1989). Thanks to Swaffield and Fairweather for bringing my attention to this source.
24. Robin Magowan, *Narcissus and Orpheus: Pastoral in Sand, Fromentin, Jewett, Alain-Fournier and Dinesen* (New York: Garland Publishing, Inc., 1988): 7.
25. Magowan, 1988: 16.
26. David Harvey, *The Condition of Postmodernity: An Enquiry into the Origins of Cultural Change* (Oxford: Blackwell, 1990); Edward C. Relph, *The Modern Urban Landscape* (Baltimore: Johns Hopkins University Press, 1987); see also the work of Paul Virilio and others on hyper or late modernity.
27. Hilde Heynen, *Architecture and Modernity: A Critique* (Cambridge, MA: The MIT Press, 1999).
28. Theodor Adorno, 1965, cited in Heynen's (1999) epigraph.

29. Naomi Klein, *The Shock Doctrine: The Rise of Disaster Capitalism* (Toronto: Knopf, 2008).
30. Frank Gohlke, "Living Water," in *The Sudbury River: A Celebration* (Lincoln, MA: DeCordova Museum, 1993).
31. Schama, 1995: 61.
32. K.V. Cadieux, "The Mortality of Trees: The Death of Change in Nature" (Toronto: Canadian Association of Geographers Annual Congress, 2002); K.V. Cadieux, "Antinomy and Estrangement in Forest Preservation and Production: Conservation-Based Public Policies and Productive Forest Uses across the Twentieth-Century" (St. Paul: American Society for Environmental History, 2006).
33. Bruce Mau, "anti-design manifesto" (What Makes You Wealthy) (Toronto: Interior Design Show Statement, 2004). http://dorkenwald-spitzer.com/pages/what-makes-you-wealthy-04.
34. Thoreau's journal entry from the thirtieth of August, 1856, cited in Schama, 1995; see, for a popular example of the wildness distant from ourselves: Bethany Little, "Think Small," *The New York Times*, February 16, 2007. http://www.nytimes.com/2007/02/16/realestate/greathomes/16tiny.html.
35. See also Chapters 5 (Donahue) and 6 (Luka). Scholars such as Cronon and Castree have suggested that this catch-22 may be disrupted by observing that nature is obviously where humans are, too—but exploring what enables people to overlook this point, or to behave as if we are separate (or as if nature could possibly be experienced in an unmediated way) is useful for understanding the conditions that enable acknowledgement of common assumptions about that immanence of nature, and the possibilities and challenges it signifies. Addressing the conditions that contribute to the production of fetishized "escape" versions of nature seems like an important step toward identifying ways to work toward goals commonly subsumed in the fetish of the ideology of nature, acknowledging the ways in which experience of nature is culturally mediated, and moving away from relying on "nature" always being out there, somewhere, ready to save the day.
36. Ontario Ministry of Natural Resources (OMNR), *Whose Lookin' After My Forest... And Howse it Goin* (Toronto: Queen's Printer, 1993); Mary M. Berlik, David B. Kittredge, and David R. Foster, *The Illusion of Preservation: A Global Environmental Argument for the Local Production of Natural Resources* Harvard Forest Paper No. 26 (Petersham, MA: Harvard Forest, Harvard University, 2002): 17.
37. On a related note, Brian Donahue, in Chapter 5, examines the ideology and practice of using local forests for resource production.
38. Berlik et al., 2002: 1.
39. Jennifer Price, "Looking for Nature at the Mall: A Field Guide to the Nature Company," in *Uncommon Ground: Toward Reinventing Nature*, ed. William Cronon (New York: Norton, 1995): 186–203; Michael Bell, *An Invitation to Environmental Sociology, Sociology for a New Century* (Thousand Oaks: Pine Forge Press, 1998); Cindi Katz, "Whose Nature, Whose Culture? Private Productions of Space and the 'Preservation' of Nature," in *Remaking Reality: Nature at the Millenium*, eds. Bruce Braun and Noel Castree (London: Routledge, 1998).
40. The work of Adams' predecessors, such as Carlton Watkins and Timothy O'Sullivan, was also of considerable importance in bringing American's attention to such an idea of nature. And although conservation and organic movements have had many positive social and environmental effects, the downsides of the commodification of nature include some of their less well

thought out effects, such as the removal of native or local people from lands to be "conserved." See also Eliot Porter's (rather unpeopled) photographic rendition of Thoreau: *In Wildness is the Preservation of the World* (San Francisco: Sierra Club, 1962).

41. Karl Marx, *Capital*, Chapter 1, section 4 (Marx/Engels Internet Archive [marxists.org], 1995, 1999 version of Moscow: Progress Publishers, 1887) (http://www.marxists.org/archive/marx/works/1867-c1/ch01.htm#S4) (July 2012); Matt Wray, "Fetishizing the Fetish," *Bad Subjects: Political Education for Everyday Life* 41 (1998) (http://bad.eserver.org/issues/1998/41/wray.html) (April 2005).

42. The individualist and subjective romanticism that flavors the ideology of nature also works to place humans outside of *culture*, social relations, or collective action, a point often argued in the marxian tradition, see David Harvey (1996: Chapter 6).

43. Noel Castree and Bruce Braun, "The Construction of Nature and the Nature of Construction: Analytical and Political Tools for Building Survivable Futures," in *Remaking Reality: Nature at the Millenium*, eds. Noel Castree and Bruce Braun (New York: Routledge, 1998); Neil Smith, *Uneven Development: Nature, Capital, and the Production of Space* (New York: Blackwell, 1984); Raymond A. Rogers, *Nature and the Crisis of Modernity: A Critique of Contemporary Discourse on Managing the Earth* (Montréal: Black Rose Books, 1994).

44. Cf. Jobes, 2000.

45. Allan Smith, "Farms, Forests and Cities: The Image of the Land and the Rise of the Metropolis in Ontario, 1860–1914," in *Old Ontario: Essays in Honour of J.M.S. Careless*, eds. David Keane and Colin Read (Toronto: Dundurn Press, 1990): 74–75.

46. Gifford Pinchot, *A Primer of Forestry. Part I, The Forest,* Bulletin 24. (Division of Forestry, U.S. Department of Agriculture, 1899). In addition to initiating the Forest Service, Pinchot's mission also established forestry schools in both the U.S. and Canada; his, however, was certainly not the residential forest or inhabited nature.

47. Erwin Panofsky, "Et in Arcadia Ego: Poussin and the Elegiac Tradition," in *Meaning in the Visual Arts: Papers in and on Art History* (New York: Anchor, 1955).

48. Giovanni Pietro Bellori cited in Paul Smith, *Et in Arcadia Ego.* (website: http://priory-of-sion.com/psp/id17.html) (April 2005).

49. Michael E. Reynolds, *Earthship Volume I* (Taos, N.M.: Solar Survival Architecture, 1984).

50. OMNR, 1993.

51. In the context of a forest whose visitors are expected to experience the awe of oldness, these signs explain that "the balance of life in the forest shifts;" "this is part of the forest cycle—life, death, and renewal in a dynamic forest;" "Rather than an area of death and destruction this is a wonderful place for us to now watch the rebirth of the forest." (National Park Service, Muir Woods National Monument, 2004). Brian Donahue, in Chapter 5, invokes a similarly antinomian role for Thoreau in his complementary roles as a chronicler and surveyor of trees, immortalizing them in his prose and laying out the survey lines for their harvest.

52. Lucy R. Lippard, *The Lure of the Local: Senses of Place in a Multicentered Society* (New York: New Press, 1997).

53. S.A. Watmough, T.C. Hutchinson, and E.P.S. Sager, "Changes in Tree Ring Chemistry in Sugar Maple (*Acer saccharum*) along an Urban-Rural Gradient in Southern Ontario," *Environmental Pollution* 101.3 (1998): 381–390.

54. Henry David Thoreau, *Faith in a Seed: The Dispersion of Seeds and Other Late Natural History Writings*, ed. Bradley P. Dean (Washington, DC: Island Press/Shearwater Books, 1993); David R. Foster, *Thoreau's Country: Journey through a Transformed Landscape* (Cambridge: Harvard University Press, 1999).
55. Gohlke, 1993.
56. Adrienne Rich, "What Kind of Times Are These," from *The Fact of a Doorframe: Selected Poems 1950–2001* (New York: W. W. Norton & Co., 2002).

Editors' Epilogue
An Agenda for Addressing Green Sprawl
Laura Taylor and Kirsten Valentine Cadieux

The essays in this book, *Landscape and the Ideology of Nature in Exurbia: Green Sprawl*, have, in conversation, taken a critical look at nature-seeking in residential settlement in the United States and Canada. These ten chapters have questioned decentralization in contemporary settlement patterns by focusing on the aspects of green sprawl that people appear to be trying to get *to*. Self-conscious consideration of the desire to settle in nature, these chapters suggest, has been largely missing in the debate about sprawl, and more importantly, in negotiations of "solutions" that are posited to the problems of sprawl. Using the label "exurbia" to define not only a material place where people live in the countryside while connected through their daily practices to the city, but also a conceptual and ideological space, we have looked at the meaning of nature in green sprawl and sought to highlight discussion about the continuing tendency for increasing numbers of people to seek escape from the city ever further into rural places, in spite of all the planning policies and environmental impact reports that censure such escape.

We conclude with reflections on themes in these chapters that suggest potential directions for the continued discussion about sprawl and the ideology of nature that began at the symposium. As we look back on the book, three central themes emerge in the conversation between the chapters. First, despite their varied backgrounds, the chapter authors share a dissatisfaction with common approaches to sprawl issues. Instead of reproducing usual assumptions about the characteristics and implications of sprawl, these authors bring a variety of applied perspectives to bear on what people do in the landscape; they also attempt to bring into public discourse their questions about people's engagement with the landscapes they inhabit. Although the political context is explicit only in the more planning-focused chapters, this public discourse is crucial for supporting and changing socio-environmental management regimes governing green sprawl and its alternatives.

Second, following from the emphasis on the shaping of landscapes, these chapters look at sprawl as a set of practices that materialize environmental ideologies, rather than focusing exclusively on spatial models of exurban settlement. Providing opportunities to explore contested views of landscape and the social and environmental impacts that flow from them, this approach suggests a refocusing of current research on urban dispersion and sprawl that challenges the cultural context within which individuals make decisions about their residential choices. In addition, this focus on sprawl as a process, not only a product or a place, helps develop frameworks for understanding the trajectories of people and places experiencing sprawl. Developing such frameworks, in turn, gives people better handles on how to assess and manage these histories, contemporary issues and future challenges in terms that reflect their dynamic and diverse experiences and aspirations.

Third, we note that the majority of our chapter authors ground their observations about sprawl and the ideology of nature in domains where the material and ideological issues of exurbanization are negotiated, both in public discourse (as in descriptions of public planning processes in Chapter 7 by Vachon and Paradis) and also in private management decisions (as in Luka's accounts of the production of cottage landscapes in Chapter 6). In addition to the breadth of the multiple disciplines brought together on the topic, this engagement between abstract aspects of management discussions (such as normative theories of land use planning) and materially grounded aspects (relating to the expression of landscape preferences and practices) not only allows a broader view of processes of exurbanization and sprawl, but also seems much more likely to reflect and invite public discussion of these issues.

We expand briefly upon these three points, summarizing approaches for future research and discussion that have emerged from the text. Although research programs have addressed urban dispersion across the past century, the current framework for understanding and managing change in settlement form tends to rely on rubrics of measurement and management, with a strong emphasis on addressing crises of urbanization. As we describe in the introduction, these approaches often fail to engage residents, and may exacerbate the very impacts of sprawl they seek to address by sensationalizing urbanization and valorizing the landscapes beyond. The variety of approaches taken in these chapters suggest alternatives to the outward view of urban growth boundaries, and might help stimulate discussion of land use practices across traditional divides between urban and rural and between different groups identified by affinities with particular visions of landscape use and protection.

It is taken for granted that cities have dispersed because of affordability in land price, increased mobility especially through automobile ownership, a search for privacy, a fear of crime, and a dislike of noise and pollution. Although the desire for space, especially greenspace, is recognized in the sprawl literature, underlying motivations for such desires have not been thoroughly excavated. Urban-centered metropolitan growth theory has struggled to explain non-metropolitan growth. Agriculture-restructuring arguments help explain rural decline, but as much as they complement theories of rural gentrification to illuminate the conditions that have enabled exurbanization, neither set of literatures seems to engage the imaginations or decision-making processes of exurbanites or exurban land use decision-makers. As Donahue, Vachon and Paradis, Blum, and Luka suggest, however, the planning of high amenity landscapes, by explicitly acknowledging the privileging of the natural setting in residential choice, may provide more accessible entry points for public negotiation of issues of urban dispersion. Related themes connecting other chapters contribute to the book's assertion that the ideology of nature in North American culture has had a much greater impact on decentralization than is generally acknowledged.

This central underlying assertion of this book is that exurbia is a symptom, a visible effect of the ideology of nature. To what extent does such an ideology motivate urban dispersion? And if we agree with the majority of the sprawl literature that urban dispersion has very negative ecological, social and economic impacts, what are the implications for change in these landscapes and in the processes that create them? To address these questions, we suggest not only the potential usefulness of broad public discussion of the impulses that motivate the desire to escape cities in the pastoral and natural landscapes beyond, but also the integration of scholarly insights into the uses of nature in the production of urban landscapes, particularly in programs of research focusing on gentrification, disinvestment in productive landscapes, and the production of landscapes of class and ethnic exclusion. Drawing on a rich academic literature describing the ways in which the quest for nature drives the urbanization of the landscape beyond the city (where it is imagined that landscapes *exist* beyond the reach of the city), several chapters (Cadieux, Judd, Taylor, Blum, Donahue, and Looker) discuss ways that nature has been socially constructed as a foil against the modern urban landscape and its ongoing crises.

William Cronon concluded his edited volume *Uncommon Ground* by saying that the most important contribution of his book was "to help people grapple more clearly and self-critically with the peculiarly human task of living in nature [or in our case, in sprawl] while thinking themselves outside of it."[1] In our collective work that began with the exurbia symposium and has continued through this book, we deliberately set out with Cronon's "The trouble with wilderness" thesis in hand, and so it seems only fitting to end with it. Taking his collection of ideas about the promise of the escape that underlies so much of the cultural imagination of nature, we have in our collection continued this conversation in terms of the extent to which people manifest ideas of what is natural, shaping the landscape to fit with these ideas, often remaking—and ignoring—the realities of their material landscapes. As with Cronon's wilderness, green sprawl "expresses and reproduces the very values its devotees seek to reject." Representing sprawl as a scourge on the landscape, exurbanites absolve themselves of their responsibility by identifying with symbolic nature. In the process of constructing "green sprawl," many exurbanites paradoxically construct sprawl as the embodiment of everything wrong with the contemporary city—a collective externalization of gentrification, disinvestment, and the imagination of residential havens outside of the urban fabric of civil society. This makes it much harder to see the troubling landscapes they produce. Using the lens of the ideology of nature to question this escape, the preceding chapters offer potential alternatives to the self-perpetuating logic of sprawl.

NOTES

1. William Cronon. "Toward a Conclusion", in *Uncommon Ground: Rethinking the Human Place in Nature*, William Cronon ed. (New York: W.W. Norton, 1996), 459.

Contributors

Andrew Blum is a journalist and independent scholar based in New York City. He is the author of *Tubes: A Journey to the Center of the Internet* (Ecco/Viking, 2012), a correspondent at Wired Magazine, and a contributing editor at Metropolis.

Kirsten Valentine Cadieux studies the cultural geography of land use change and the politics of everyday environmental decision making at the urban-rural interface. Her work at the University of Minnesota focuses on the relationships between land use and landscape ideologies as understood through material and representational practices. Her interests include food and feeding practices and related land uses and social justice politics, as well as the use of concepts of place, landscape, and nature, and the epistemological issues involved in the interplay between political ecology and other approaches to nature-society relations.

Brian Donahue is Associate Professor of American Environmental Studies at Brandeis University and Environmental Historian at Harvard Forest. He co-founded and directed Land's Sake, a community farm in Weston, Massachusetts. He is the author of *Reclaiming the Commons*, *The Great Meadow*, co-editor of *American Georgics: Writings on Farming, Culture, and the Land*, and co-author of *Wildlands & Woodlands: A Vision for the New England Landscape*.

Richard Judd received his Ph.D. in 1979 and before joining the University of Maine faculty in 1984 worked on the *Journal of Forest History*. His publications include *The Untilled Garden: Natural History and the Spirit of Conservation in America*; *Natural States: The Environmental Imagination in Maine, Oregon, and the Nation*; and *Common Lands, Common People: The Origins of Conservation in Northern New England*.

Thomas Looker taught at Amherst College for twenty years as Visiting Lecturer and Visiting Scholar in American Studies. He is the author of *The Sound and the Story—NPR and the Art of Radio* (Houghton, 1992) and created and produced the public radio series *New England Almanac: Portraits in Sound of New England Life and Landscape*, which won a Peabody award in 1983.

Nik Luka is cross-appointed to the School of Architecture and School of Urban Planning at McGill University, where he is also an Associate of the School of Environment. His work focuses on reurbanism: the critical diagnostic and generative work of (re)urban(ising) landscapes at the nexus of urban design and civil society.

Andrei Nicolai, until his death in 2008, was an urban designer and planner. He was an Adjunct Professor in the Faculty of Environmental Design at the University of Calgary, and Research Associate in the Urban Lab. He collaborated on several award-winning books and urban design projects with Bev Sandalack.

David Paradis is an urban planner and instructor of urban planning at Université Laval's Superior School of Regional Planning and Development in Québec City. Holding two masters of planning and urban design, he is the director of the "Sustainable communities" research project at *Vivre en ville*, a non-profit organization advocating sustainability in planning.

Beverly A. Sandalack is a Professor in the Faculty of Environmental Design at the University of Calgary and Director of the Urban Lab, an innovative and award-winning research group. She also practices urban design, landscape architecture and planning with her Calgary-based firm Sandalack + Associates Inc.

Laura Taylor is Assistant Professor and Planning Programs Coordinator in the Faculty of Environmental Studies at York University, Toronto, with research in cultural landscape studies and nature politics. Her current research interest is in exurbia, studying the processes and discourses of landscape settlement and landscape conservation at the urban-rural fringe. She is a consulting planner in the greater Toronto area, and a registered professional planner with the Ontario Professional Planners Institute.

Geneviève Vachon is an architect and professor of urban design at Université Laval's School of Architecture in Québec City. As co-director of GIRBa (http://www.girba.crad.ulaval.ca), she is interested in collaborative design, sustainable mobility, suburban regeneration, landscape design and housing. She co-edited '*La banlieue s'étale*' (Nota Bene, 2011) on Québec's exurbs.

Index

A

aboriginal, 82
   Iroquois, 38
   Koyukon, 224–226, 228–229, 237, 246
   Mississauga Anishinaabe, 38
   Native American, 102, 114, 171, 183n18, 224, 236
   settlements, 38
Abbey, Edward, 71–72
advertisements, 11, 17–18, 48–49, 138–139, 187, 212, 259, 261–262, 264, 279
   advertising industry 16, 89, 133, 264
Adorno, Theodor, 274–275
aesthetics, 53, 54, 79, 95, 135, 143, 192, 201, 247, 256, 257, 258, 269
   aestheticization, 49
   environmental, 6, 122, 130, 255, 278
   foundations of the ideology of nature, 21, 94
   ideals of natural living, 11
   landscape, 5, 16, 36, 252, 265, 267, 270, 272
   nature, 10, 13, 24, 25n6, 255, 266
   of forest landscapes, 268
   of modernity, 273
   paralysis, 281–286
Agee, James, 62
agency, 267, 282–287
agriculture/agrarianism, 18, 64, 69, 95, 104–108, 114, 117, 257, 259, 271, 278–279. *See also* production
air pollution, 6, 17–18, 108, 111, 253
   smog, and fresh air as a contrast, 68, 122, 134, 137, 281
   as a right around which environmental policy activism has been organized, 73, 231
Airworld, 78–93, 286
alienation
   from nature, 4, 61, 66, 72, 233, 257, 268, 270, 277, 280, 282, 284, 287
   from labor, 6, 10, 70–72, 78–79, 97, 108, 146, 258–269, 283
ambiguity/ambivalence/uncertainty, xix, 4, 25n7, 134, 148, 155n29, 219–234, 238, 242, 245, 248, 286
amenity migration, xxi, 143–145, 253
analysis. *See* methods
anti-idyll, 63, 87, 281. *See also* rural idyll
anti-urbanism, 6, 11, 15, 50, 54, 61
architecture, 51–52, 66, 141, 152, 190, 193, 216, 267, 273. *See also* built environments
Arendt, Randall, 18, 29n47, 56n7, 174
associations (e.g. residents, civic, cottagers, neighborhood), 65, 122, 207
attitudes, 6, 31, 36, 55, 91, 130, 148, 173, 186, 227, 229–230. *See also* perceptions
Atwood, Margaret, 134–135
Audubon, 68, 72
authenticity, 61–62, 64, 78, 188, 214
automobile culture, xv, 3, 44, 64, 70, 107, 109, 141, 296

B

baby boomers, 143–144
Barnes, Trevor, 26n15, 35
Bartram, Edward, 148

Bell, Tom (High Country News), 71–72
Benton, Thomas Hart, 62
Berlik, Mary, 278
biodiversity, 73n8, 95, 99, 112, 113, 148, 155n29
Blake, Peter, 258
block pattern, 189–210
Bobcaygeon, Ontario, 144
Bookchin, Murray, 70
Botkin, Daniel, 218, 232–234, 240, 242, 246
boundaries, urban, xvi, 5, 14, 25, 48, 103, 129, 296
Brampton, Ontario, 33, 35, 37, 39, 41, 44, 46–48, 50, 54
bridges, 31–47, 50–56, 164
built environments, 2, 7, 17, 63, 79, 125, 173, 182, 186, 189, 191. *See also* architecture
Bunce, Michael, xxi, 25n5, 135

C

Calgary, 185–215
Calthorpe, Peter, 23n1, 183n7
Campbell, Claire, 129
Canadian Pacific Railway, 138, 199
capitalist/ism, 53–54, 73, 84–86, 89, 253, 272, 274
  capital, 86, 280
  ecological capital, 117
Carson, Rachel, 64
Casey, Ed, 81, 83
Castree, Noel, 10, 26nn14–15, 27n20, 29n50, 55, 292n35
change in nature (and difficulty with), 12, 121, 148, 151, 224–234, 263, 276
charrette, 159–175
Chiras, Daniel D., 65
Christchurch, Aotearoa New Zealand, 252, 264–265, 270, 291n19, 264–265, 270
Churchville, Ontario, 31, 33–38, 40–54
citizenship
  exurban, 69
  citizens/citizenry, 166, 183n6, 222, 279
city, 1, 5, 13, 16, 54, 146, 147, 150, 186, 192, 214, 254, 273, 298
  abandonment of, 6, 70, 130, 137
  anti-city, 50, 54
  city-dwellers, 107, 126, 140
  city-regions, 3
  city's edge, xvii, 2, 3, 31, 35, 187, 189, 200, 207, 262
  connections to, 3, 12
  decentralization/dispersion, 5–11, 22, 162, 252, 295–297
  everyday life, 4, 137, 128–9
  evolution of, 188, 189, 212
  limits, 63
  migration to. *See* migration
City Beautiful, 195
Cleveland, Ohio, 126, 137
climate change, 5–6, 21, 108, 266, 285–286
CMQ (Communauté Métropolitaine de Québec), 165
collaborative design, 160, 162–163, 174–176, 182. *See also* public participation
collaborative environmental management, xvi, 4, 111, 253
collaborative ethnography, 253
commodification of nature, xviii, 22, 59–71, 279–281
communications, 34, 78, 80, 81, 83–85, 145, 219, 286
  cell phone towers, 79–80
commute/commuting, 14, 34, 68, 129, 130, 137–139, 264, 290n13. *See also* automobile culture, transportation
complexity, 4–7, 81, 109–113, 117–118, 134, 151, 228–235, 244–245, 253, 275, 284–287
compromise/dialogue, 32, 54–55, 122, 160, 219, 221, 222, 226, 256, 271, 274
Concord, Massachusetts, 89, 94, 96–97, 99–100, 102–109, 118, 235, 244
conservation, 7, 14, 18, 106, 109, 111, 115, 163, 174, 176, 179, 253, 255, 257, 264
conservation biology, 17
consumption, 4, 8, 59, 67–73, 117, 143, 286. *See also* shopping
  of forest resources, 95, 110
context
  culture/history, 9–10, 21, 37, 49, 59, 96, 122, 148, 174, 188, 195, 200, 280, 296
  landscape, 51, 54, 88, 142, 150, 271
  nature, 96, 132, 134–135, 188, 190, 193, 213, 240, 180

cottage, 36, 41, 54, 61, 121–152, 159, 162–164, 167, 168–169, 171, 173, 176, 266, 283
Cottage Life, 133, 146, 148
countryside, 2–3, 33, 53–54, 102, 104–107, 135, 161
   ideals, 2, 4, 11, 16, 31, 50, 61, 69, 113, 135, 254–255, 269–270
Cowan, Stuart, 148, 151
crabgrass frontier, 63
creative destruction, 269
creative energy (loss of), 229, 242, 275
Credit River, Ontario, 37–38, 44, 47, 50
Cronon, William, 24n5, 29–30n51, 58n28, 125, 147–148, 151–152, 290n16, 292n35, 297
cultural landscape. *See* landscape
culture/nature (humans as part of or as outside of nature), 26n15, 218, 219, 246, 280

D

decentralization
   of work, 68
   of settlement, 295, 297
demography, 9, 143, 144, 160, 182
   demographers, 11
   demographic shift, 69
density, 6, 34, 70, 130, 159, 162, 166, 174, 179, 181, 193, 201, 272
Depression (populism), 61, 107
design/design process, 7–8, 18, 22, 36, 48, 50, 62–64, 129, 141, 148, 151–152, 159–163, 165–166, 173–182, 186–193, 200–201, 212, 214–215, 276
desire, xv, xvi, xix, 7, 8, 12, 14, 17, 12, 23, 270, 272, 277
   and fetish, 280
   for a connection to nature, xviii, 1, 3, 19, 50, 54, 55, 122, 147, 159, 219, 257, 276, 279, 285, 295
   for protection of nature, 4
   for space, 296
developers, 10, 63, 70, 72, 144, 166, 176, 179, 186, 205, 212, 215
development, 161, 261–263. *See also* growth
   economic, 142, 146,148, 150, 162, 255
   eras of, 187, 189, 195
   ideals, 61–73, 130, 215

   redevelopment, 160, 193
   speculative, 185
   sustainable, 159, 160, 161, 163, 191
dialogue between growth and conservation interests, 36, 60, 79, 256, 284–285
dialogue impasse (blind spots, deferments, contradictions, conundrums, failures to meet goals, especially goals that are invisible/inaccessible/inevitable), 256–258, 265
discourse, xix, 3–4, 10–11, 13–15, 35–36, 128, 130, 170, 220, 222, 295–296
   discursive field, 10
disequilibrium model, 231
dispossession, 6, 18
disturbance ecology (vs. balance/order), 102–103, 112–113, 116–117, 231–233, 242
diversity, 50, 157n65
   housing 18
   forms of environmentalism 71
   landscape 151
Dizard, Jan, 232, 250n14
dream, 237, 272, 276, 290n16. *See also* escape
   American, 3, 63
   exurban, 16, 78, 122, 264, 266, 277
   of a house in the country, 3, 54
Duany, Andre, 23n1
Dubos, Rene, 62
Ducks Unlimited, 165
Duncan James, 26n15, 35, 49, 57n9
Duncan, Nancy, 49
dwelling, 13, 21, 55, 91, 121, 125–126, 128, 148, 151–152, 201, 228

E

Eagleton, Terry, 27n19, 29n30
ecology (health), 33, 118, 190, 285
ecology (perceptions of), 64, 68, 233, 244, 285
ecology metaphors, 90, 105, 218, 225, 230, 232–235, 238–242, 258, 277
economy, 54, 60–61, 64, 68, 109–111, 188, 255
   economic growth, 34, 66, 104, 142, 165, 173, 176, 200
   real estate/property, 97, 146, 165
ecosystems and natural/environmental process, xvi, 4, 17, 64, 69–71,

81, 95–102, 111–116, 125–129, 134, 147–152, 162, 177, 259, 266–268, 282, 286
eco-social processes, 9, 12, 35, 52, 201, 280, 296–297
education
 environmental, 107, 111, 114–115
 post-secondary, 163, 173, 182, 218, 220, 234
Ehrlich, Paul, 64
Eldorado Park, 33, 38, 41, 46
elite, 50, 59, 67, 68, 70, 272
 elitist, 60, 248–9n5
Emerson, Ralph Waldo, 97, 100, 104–105, 107–109, 115, 219, 230, 235
emotion, 14, 15, 57n9, 243, 253, 286
encounter with nature, 31, 35, 107, 218, 226, 228–229, 240, 262, 269, 271
energy
 creative, 229, 235, 242, 269
 intensive use in development, 186
 intensive use in exurbia, xv, 6, 253
 massive amounts, 146
 scarce resources, 18
 sustainable, 73
 time and, 55
engineering, 35, 38, 52, 68, 201, 213, 214, 232
 bridge 45, 51
Enlightenment, 230
environmental aspirations and goals, 6–7, 14–15, 95, 111–117, 151–152, 160, 176, 221–222, 256, 259, 266–267, 269–271, 277, 282–285
environmental crisis (and perception of), 151, 223, 246, 284
environmental experience, 7, 142, 265, 282
environmental imagination, 4, 10, 17, 23, 50, 95, 134–135, 145, 233, 235, 238, 244, 259, 264–266, 272, 283, 297
environmental impacts, 6, 13, 17–18, 34, 60, 71, 95, 105, 108
environmental management (competition/conflict/collaborative), 9, 55, 95, 104, 112–116, 231–232, 255–257, 295
environmental movement, 61–63, 65, 208, 230–232
 environmentalism, 68, 71, 229

environmental politics, 59–63, 73, 18–20, 65–66, 222, 230, 279
environmental psychology (and cognitive science), 129
equity, 13, 21. *See also* justice
escape, 4–7, 13–15, 17, 21–22, 31, 34–36, 44, 54, 59, 65, 71, 79, 81, 85–90, 121–122, 131–134, 137, 146–150, 159, 214, 254–257, 262, 270–287, 292n35, 295–298
ethnicity, 13, 297. *See also* race
Evans, Walker, 62
everyday life, 4–5, 20, 22, 48, 135, 147, 150–151, 253, 269, 272, 274, 276–277, 279–280, 283, 285
 behavior, 268
 engagement, 7, 32, 35, 262, 284
 environments, 7–8, 23, 36, 79, 271, 282–284
 experience, 7, 32, 151, 268, 274, 279, 283
 landscape, 35–36, 147, 253, 277, 283
 practice, 55
 relationship (with nature), 7, 284, 96
exclusion, 28, 50, 86, 297
exurbs, exurban, exurbia. *See also* growth
 effects of exurbia (natural/social), 6, 14, 18, 55, 70, 97, 111, 188, 255–257, 278
 exurban ideal, xvii, 10, 35, 69, 72, 78, 81, 125, 146, 254, 266–270, 281
 literature on exurbia, xvi, 12, 18, 129, 185, 268
 origins of exurbia, 16, 61, 68, 161, 265

F
Fainstein, Susan, 150, 158
Fairweather, John, 27
farming. *See* agriculture
farmland, 11, 33, 39, 96, 99, 103, 104–105, 107, 112, 255
Farrell, James T., 66
fetish(ism), 279–281
Fishman, Robert, 70
Fitzgerald, F. Scott, 247
Foster, David, 120n21, 23, 278
Fordism, 16, 25n14, 26, 155n28

Index 305

forest, 1, 64, 136, 225, 231, 257, 266, 278–280, 282–283
  California, 80
  Jefferson Forest, 261–263
  Muir Woods, 285
  Ontario, 33, 39, 41
  Quebec, 163–164
  Walden Woods/Massachusetts, 94–97, 99, 102–118
forestry (resource-based land uses), 95, 109, 113–117, 264, 278, 285
Fortmann, Louise, 20, 60
freeway, 63, 87. See also highway
fringe, 11, 18, 60, 63, 163, 185, 261, 264
Fromm, Eric, 61
Frost, Robert, 243
Frye, Northrop, 134–135
fundamentalism, 23, 219, 221–229, 243, 246, 248

G

Garden City, 7. See also Howard, Ebenezer
Garreau, Joel, 67
gentrification, 64, 146, 159, 270, 281, 297
  privilege, 6–7, 10, 78–79, 146, 253–254
geographical imagination. See environmental imagination
geography, 4, 31, 54, 70, 84–85, 145, 166
GIRBa, 160–161, 163, 168–170, 174–175, 182
Glasmorgan, Alberta, 200–207, 213–214
global, globalization/(global) rural restructuring, 6, 10, 78, 81–82, 85, 188, 255, 275, 279, 286
  North, 6, 10
  South, 6, 10
global green zones, 275
Gohlke, Frank, 275, 287
Golden Horseshoe, 124–126
Goodman, Paul, 63
governance, 80, 110, 135, 141–142, 221–222, 252, 256, 295
  municipal government, 41, 53–54, 63, 160, 165, 201
government. See governance
green space/greening, 2, 66, 175, 200–202, 213, 273
grid, 39–44, 53, 185, 195–217

growth. See also urbanization
  exurban growth/development, 6, 55, 69, 87, 107, 109, 121, 126, 144, 185–189, 193–215
  exurbanization, 6, 11, 13, 17, 104, 122, 162, 255–256, 276–277
  population, 18, 33 40, 72, 142, 169, 173, 195
  smart, xv, xvii, 3, 14, 182
  urban growth/development, xv, 3–5, 15, 146, 190, 195, 200, 205
Guthrie, Woodie, 62

H

habitat, xv, 9, 17, 56n7, 65–69, 111–113, 183n7, 268. See also ecology, inhabitation
Harvey, David, 86, 288n5
Hays, Samuel, 68
health
  human, 63, 126, 137
  ecosystem, 111–112, 148, 155n29
Heidegger, Martin, 81, 91
heritage
  cultural, 94, 106
  building, 168, 176, 181
  district, 31–34, 37–38, 41, 46–51
Heynen, Hilde, 20, 91, 273–275
highway, 32, 64, 68, 151, 161, 162, 164, 173, 208, 211. See also freeway
  expressway, 46, 70
history
  cottage country, 136–145
  environmental movement, 60–66, 230–233
  Lac-Beauport, 163–166, 171–177
  of suburbs, 59–73
  settlement in Ontario (Churchville), 32–33, 37
  trends of development/values, 187–189, 193–215
  of Walden Woods, 94–118
Hope, Jack, 72
Howard, Ebenezer, 7, 62
Hurley, Patrick, 20, 253, 288n3

I

idealism, 4, 50, 59, 69, 109, 121–122, 225, 246, 270–272, 283
ideology, xvii, 1–27n19, 28n33, 30n54, 94, 121, 155n36, 160, 185, 218,

219, 254, 259, 268, 271, 274, 280, 284, 297
imagined landscape, 218–245. *See also* environmental imagination
indigenous. *See* aboriginal
inequality, 14, 275, 281. *See also* justice
inhabitation, xvi, xix, 10–13, 21–23, 35–38, 57n9, 84, 96, 100–102, 115–118, 123, 129–130, 140–145, 195, 218–225, 246, 256–259, 264, 267–286, 293n46, 295–297
integrated, 66, 122, 177, 212, 213, 272
  design, 48
  exurban with urban, 148
  planning approach, 174, 191
  rural with urban, 18
intentions related to environments and landscapes. *See* environmental aspirations
interdisciplinary research, 9, 14, 163. *See also* methods
Iroquois, 38
Irland, Lloyd, 69
Iyer, Pico, 84

J
Jackson, J.B., 36, 156n49
Jackson, Kenneth, 61
Jakle, John, 129
Jefferson, Thomas, 61, 127, 135, 230
justice, social and environmental, 1, 4, 13–15, 22, 91, 157n65, 253, 256, 268, 285

K
Kawarthas, 127
Kirn, Walter, 78–80, 85, 87, 89, 91
Kittredge, David, 278
Kluksa Mountain, 224
Kunstler, James, 258

L
L.L. Bean Company, 67, 69
Lac-Beauport, Québec, 159–177, 179, 181–182
Lake Chaparral, Alberta, 207–214
Lake Rosseau, Ontario, 140
land-use management, 4, 13. *See also* environmental management
land-use governance, 6, 26n14, 256
  regulation 69
  civic/public land use (public realm), 195–215

landscape
  as text, 35, 48–49, 54, 258
  cultural landscape, xvi, 5–6, 8–13, 35, 51, 95, 123, 144, 219, 230, 186, 256, 277
  landscape approach/studies, 4, 9, 15, 35, 48–49, 122
  landscape metaphor. *See* ecology metaphors.
  material landscape, 5, 9, 11–12, 50, 95, 297
  meaning, 9, 23, 34, 37, 49, 51, 66, 81–82, 84–86, 92, 125, 142, 145, 152, 160, 188, 213, 252, 258–259
  middle landscape 7, 50, 161, 237, 239, 244, 246, 271
Larochelle, Pierre, 174
law, 222–3
  by-law, 207
  Canadian, 154n18
  environmental, 66, 231
Lefebvre, Henri, 20, 24n5
Le Relais, Quebec, 163, 166, 170, 172
legibility, 174, 179, 191, 214, 215n5, 217n15
Leopold, Aldo, 113
Lessinger, Jack, 69–70
literature of environmentalism, 61–62, 134, 150, 218–219, 221, 229, 236, 238, 272
Lowes, Freddie, 199–200
Lynch, Kevin, 36, 191

M
magazines, 68, 72, 126, 133, 146, 148, 248n4
  lifestyle, 1. See also *Cottage Life*
Marin County, California, 67–68, 80
marketing, 48, 50, 68, 87, 133, 185, 208, 212, 215, 264, 279
Marx, Karl, 20, 280–281
Marx, Leo, 20–21, 218, 243, 249n7, 271, 275
Mau, Bruce, 276
Mawson, Thomas, 195
Mayakovsky, Vladimir, 56
McHarg, Ian, 18, 56n7, 179, 190
McLuhan, Marshall, 133
Meadows, Donella H., 64
media, 11, 133
Melville, Herman, 219, 243–244
memory, 49, 71, 87, 117, 215, 216n8
  collective, 166, 170, 176–181

mental map, 170, 191
metaphor, 90, 218–247, 258, 276–277, 279–281
methods/methodology (*see also* chapter introductions, for overview of each chapter's approach), 9–11, 14–15, 22–23, 31–32, 59–60, 89–90, 159–162, 173–177, 182, 183n17, 185–193, 215, 216n10, 218–219, 223, 238, 239–243, 247, 252–253, 264–267
    analytic methods, 20, 174, 187, 240
metropolitan, 15, 34, 62, 67–68, 70, 121, 125–126, 142, 144–150, 164, 256, 259, 282
middle landscape, 7, 50, 237–277, 291n21
migration, 54, 168, 171
    amenity, 12, 253
    of design ideas, 191, 201
    to cottage country, 145, 148
    to exurbs, 13, 72
    to farms, 136
    to suburbs, 62, 69
Mills, C. Wright, 61
Mississauga, 38
modernity, 2–7, 15–16, 20–23, 25n4, 78–79, 82–84, 90–91, 264, 272, 274, 280, 286. *See also* urban modernity
modernization, rejection of 50
Mont Tremblant, 165
Montréal, Québec, 165
morphology, 174, 191, 193, 201
    integrated into planning, 216n7
Muir, John, 80, 233, 235, 285
Mumford, Lewis, 61
Municipalité régionale de comté (MRC), 165
Muskoka, 125, 136–141

N

narrative, 31, 59, 79, 84–85, 88, 94, 128, 133–136, 218–245, 256, 268, 268, 275
natural environment, xvi, xvii, 3, 7–8, 47, 50, 59–60, 69, 122, 166, 177, 208, 213, 215, 258, 265, 268–269, 276, 285, 287
natural resources/natural resource management. *See* production
nature
    construction of, 4, 10–11 23, 62, 72, 152, 185–187

    ideology of, 1–15, 17, 19–23, 59–60, 121, 160, 185, 219, 254, 268, 280, 284
    in exurbia, xviii
    politics of, xix, 5, 9–11, 19, 35, 59, 62, 115, 230, 279
    preservation of, 19, 94, 96, 100, 165, 278
    social constructions of, 4, 10–11
    symbols of, 1, 9, 11, 99, 239
Nelson, Richard, 224–229, 237, 247, 246
Neumann, R., 24n5, 26n17
New England, 72, 102, 110, 116, 230, 236, 252, 264, 270, 276, 281, 289
new urbanism, 3
New York, 16, 49, 51, 80, 87, 129, 133, 150, 264
normative values, 6, 13, 27n21, 268, 296

O

Old Waterloo, 178
Olmsted, Frederick Law, 13, 18
Ontario, 31, 37–38, 41, 46, 53, 121–128, 130–135, 141–142, 145–146, 150–152, 266, 281–282
outdoor apparel, 67
outdoor recreation (hunting, fishing, camping, boating, and hiking) (resort) (leisure), 12, 33, 64–65, 70–71, 132, 137–139, 162–163, 167, 172–173, 175, 283

P

paradox of urban/rural combination, 3, 276–285
Parry Sound, 131
participation, 15, 159–160, 162–163, 175–176, 186. *See* public participation
pastoral/modern pastoral, 271–287
    land use, 254–257, 269–273
    unequal land use/inequality, 255, 272–273, 278–279
    reflexive inhabitation, 256–257
    reforestation, 257, 259
    landscape as symbol/metaphor, 258–287
    Jefferson Forest, 261–263
    Autumn Grove, 261–263
    idealization of nature, 262–287

Peet, Richard and Michael Watts, 27n21
Penning-Roswell, Edmund, 129
penturbia, 69
perceptions, xvi, 3, 4, 10, 64, 83, 142, 181, 227–228, 235, 241, 255. *See also* attitudes
periphery, 256, 258, 261
peri-urban, 253, 256, 258, 287n287
photography, 62, 114, 167, 168, 190, 261, 275, 279
Pigeon Lake, Ontario, 127, 144
Pincetl, Stephanie, 65
Pinchot, Gifford, 282–283, 293n46
Pittsburgh, Pennsylvania, 126, 137
place, 14–16, 21–22, 34–37, 51–55, 59–61, 68, 78–92, 121–123, 125–126, 132, 148, 150, 162, 163, 174–177, 181, 185–193, 200, 214, 254, 258, 265, 270, 284, 287. *See also* space
  hybrid place, 79, 81–82, 84–85, 89, 91
  place identity, 162–163, 176–177, 188–215
  sense of place, 31, 34, 51–52, 60, 64, 66, 72, 81, 88, 176, 186, 188
planning
  land-use planning, 4, 5, 12, 25n14, 186–193, 270
  planners, 5
  planning process, 187, 253, 280–282
  urban planning, 2–5, 12, 31, 33, 150, 190, 193, 213–215
planning models, 82, 115, 117, 174–176, 195, 201, 253, 295
Poe, Edgar Allen, 227, 249n7
poetics, 134, 218–245, 250, 251, 272–276
political ecology, 9–10, 19–20, 26n18, 288n5
  feminist, 253
political economy, 9, 19–21, 256, 279, 288n5
politics, 18, 62, 65, 73, 222
Polk Center, Minnesota, 87–88
Pollution, water, 6, 63, 65, 111
Portland, 24n2
post-fordist cities (continued economic growth via competitive attraction of investment/development), 16, 34, 155n28
  urban economic growth, 34, 165, 255

urban investment, 7, 25–26n14
Poussin, Nicolas, 283
prairie, 185–6, 201, 208, 213
preservation-based governance regimes, 94–97, 256
production, productive processes, natural resource management, 95–96, 117, 252
provisions/provisioning, 139–140
working rural landscape, 64
property rights. *See* rights, land use and property
property values, 6, 10, 14, 97, 258
  accumulation, 10, 256
  acquisition, 21, 124
public, 4, 59
  public interest, 65, 110, 186
  public realm, 188–189, 191, 193, 195, 200, 212–215
  space, 186, 192
public participation, 4, 15, 23, 65, 175–176, 186–187, 214–215
  participatory design, 186–187, 214–215. *See also* collaborative design, collaborative environmental management
pull of nature, 18
  of landscape narratives, 254–256

Q

quality of life, 1, 2, 6, 10, 17, 60, 62–73, 162, 174, 256
Québec, 159–173, 175, 181–182

R

race, 5, 13, 27n26, 28n31, 105
  diversity, (ethno-cultural), 157n65
rational planning, 31–34
Rapoport, Amos, 129, 156n49, 157n55
rationale for exurban conservation, 13, 97, 255–256
reconciling landscape values, 3, 5, 19, 22, 79, 142, 147, 166, 181, 256, 271–272, 278, 286–287
reductionism / instrumental language, 219, 223
reflexivity/reflexive inhabitation or relationship with nature. *See* inhabitation
region
  conservation, 113, 190, 255
  experiential scale, 161
  forest stewardship, 110, 116
  government, 53–54, 148

planning, 34, 53
processes, 286
regional identity, 59, 66, 70, 86, 188, 215
Reich, Charles, 66
Relph, Edward, 81–82, 84, 129, 150
representation, 62, 130, 133, 145–146, 160, 163, 185, 274
non-representational theory, 57n9
of exurbia, 11, 19, 182, 268–269, 273
of nature, xviii, 121, 148, 279
of sprawl, 15
spatial, 166–170
residential (home) landscapes, 2–3, 6–10, 12, 14, 60, 62–66, 68–73, 109, 130, 150, 161–162, 164–166, 176–177, 185, 201, 208, 256–257, 261, 264–265, 269, 271–273, 286
resilience, socio-ecological, 253–255. *See also* sustainability
Resort Spas, 59–60, 64, 70–72, 130, 138, 147, 163, 168
Richert, Evan, 69
rights, land use and property, 10, 65, 150
Rivière Jaune, 165–166, 170, 173, 178–179
Romanticism, 21, 60–63, 69, 82, 105, 129, 132, 148, 230–231, 234–236
Rome, Adam, 61, 63
Roseland, Mark, 23n1, 27n28
Roxboro, Alberta, 195–200, 212–214
rural idyll, 63–64, 87–88, 145, 181, 255, 269, 281

S

science, 10, 17, 52, 218, 231–233, 238–41, 247
cognitive, 129
environmental, 231
natural science as prescriptive, 20
role of in environmental protection, 19, 288n3
social sciences, scientists, 11, 20 217, 253, 272
Thoreau's, 242, 244, 246, 251n29
Western, 226, 246
Scott Brown, Denise, 89
second nature. *See* culture/nature
Sennett, Richard, 147
settlements, 3–4, 12–13, 16–17, 19, 31–34, 38, 40–46, 51, 53–54, 82, 126, 128–132, 142, 146–149, 152, 161, 169, 171, 174, 178, 181, 190, 255, 282, 295–296
shopping, 23, 66–68, 70–73, 208, 254, 269, 274. *See also* consumption
Smith, Allan, 126, 141, 282
Smith, Neil, 20
space, production of, 24n5, 145. *See also* place
spatial structure, 188, 191
Spectorsky, Auguste, 16–18, 28n38, 129–133, 264, 281
sprawl, xv–xvi, 2,13–19, 21–23, 87, 104, 109, 148, 160–161, 237, 257, 269
anti-, xv, xvii, 13, 19, 256
green, xv, xvii–xviii, 1, 2, 6–7, 10, 94–95, 121–122, 252, 253, 259, 268, 270
street trees, 197, 200, 212–213
streetscape, 53, 149, 195–213, 207, 211, 262
Steinbeck, John, 62
studio. *See* charette
survey, 114, 179
Calgary, 201
Ontario, 37–39, 44, 53, 105
Thoreau as surveyor, 106, 293n51
suburbs, 33–34, 47, 49–51, 60–71, 97, 109, 126, 147–149, 161–164, 167–168, 176, 188, 254, 266, 269
vs. exurbia, 7–8, 12, 16, 129, 161, 256–259
subdivision, 1, 13, 20, 46–50, 54–55, 63, 70–1, 126, 144, 159, 173–4, 185, 205, 271
succession (including post-eqilibrium), 55, 109, 231–233, 259, 265, 267, 275–276, 283, 285, 287
Susman, Warren, 62
sustainability, xvii, 13–14, 79, 94, 165, 231, 256, 281, 284
sustainable design, 159–160, 163, 166, 173, 175, 186, 215
sustainable forestry, 99, 109, 113, 115–117, 278
Swaffield, Simon, 20, 271
symbolic landscapes, role of symbols in environmental management, 1, 9–11, 21, 66, 99–112, 151, 185, 252–285

T

tax, 6, 7, 69
  as data, 104
  property, 114, 288n5, 290n10
  urban base, 6
technology, 21, 52, 54, 78–79, 81, 83–85, 91, 145, 201, 223, 252, 273
Telluride, Colorado, 71–72
tension between city/country/exurb, xv, 8, 11, 14, 221, 54, 79, 85, 91, 95, 147, 238, 257, 276–277
Thoreau, Henry David, 5, 21–22, 84, 89–91, 94–97, 99–118, 218–221, 223–224, 230, 223–247, 271–277, 280, 286–287
Toronto, 33, 34, 37, 39, 41, 50, 54, 123–126, 133, 137–138, 142, 146, 148–149, 261, 264–265, 286
tourism, 130, 136, 146, 150, 162, 165, 225
  residential, 130, 150
  recreational, 63, 71, 82, 136–138, 146, 162–181, 255
Transcendentalism, 95, 97, 105, 107, 234–235, 239, 241, 247
transit
  -oriented communities, 3
  public, 13
  use, 18
transportation
  air travel, 12, 14, 52, 70, 78–92
  automobile travel, 3, 44–46, 64–70, 107, 109, 141, 151, 161–164, 181, 296. See also automobile culture
  mobility patterns, 161, 174–179
typology, 191–193, 200, 205, 216n8

U

uncertainty. See ambiguity
urban containment. See boundaries
  definition of, 25n5
urban
  design, xviii, 1, 36, 128, 159–184, 185–217
  disinvestment (blight/flight), 255, 269–272, 281, 296–297
  ecology, 190
  ecologies, 1, 2, 254, 279
  edge, xvi–xviii, 1–15, 31–43, 126, 159, 185–214, 261–262
  form, 2, 7, 23n1, 25n14, 146, 148, 151–152, 161, 186–217. See also grid
  fringe. See fringe
  migration. See migration, urban
  modernity, 2, 4, 15–16, 20–23, 53, 79, 91, 187, 219, 252–253, 255, 264, 273–275, 281–284, 297
urbanization, xv–xviii, 1, 3, 15, 31, 34–35, 148, 175, 187, 195, 255, 257, 270, 273, 296–297. See also growth
  modern, 35, 187, 273
  of cottage country, 128
  of landscape, 297
  planners' management of, 7
  problems related to, 5–6
  processes of, 257
Urry, John, 150

V

value and values (including value of nature and competing values, including cultural, land, and economic values), 6–7, 13–14, 19–23, 31–35, 49–55, 56n6, 61, 66–72, 85–89, 95, 99, 107–109, 115, 122, 129, 141, 146, 151, 162, 165, 168, 173–181, 184n26, 186–190, 207–208, 212, 215, 216n11, 217n17, 221, 230, 249n8, 253–254, 256, 258, 265–268, 273, 377, 280, 283, 286, 288n5, 290n10, 297. See also property values
van der Ryn, Sim, 148, 151
Venturi, Robert, 89
Vietnam War, 65
village, 33, 38, 41, 46–48, 50, 55, 71, 84, 99, 104–106, 166, 176–179

W

Walden. See Thoreau, Henry David and Forest, Walden Woods
Walker, Peter, 20, 60, 288n3
Wylie, John, 26n15, 57n9
water amenity (lakefront, lakeside), 121–152
Whyte, William Allen, 62
wicked problems, 5, 25n7
wilderness, 4–5, 32, 60–61, 68, 94–96, 134–135, 148, 151, 230, 267, 279–285

ideology/perception/symbolic association, 5, 11, 24n5, 58n28, 62, 64, 66, 79–80, 99–100, 105, 112, 118, 126, 135, 145, 147, 148, 155n33, 218, 224–225, 235, 250n10, 263–268, 274, 276, 295
Williams, Raymond, 19–20, 53–55, 218, 272–273
Wilson, Sloan, 63
World War II (development following), 16–18, 59–69, 128–148, 160–164, 172–173, 195–207, 266
Worster, Donald, 232–234, 240, 242

WPA, 62
Wright, Frank Lloyd, 62

Z
Zizek, Slavoj, 27n19, 29n30
Zoline, Joseph T., 72
zone
   green, 253, 275
   exurban, 12, 159
   land-use, 12
   of ecological impact, 28n41, 176
   residential, 177
   riparian, 99
zoning, 9, 63, 174, 253